# Reasoning in Biological Discoveries

*Reasoning in Biological Discoveries* brings together a series of essays written and cowritten by Lindley Darden that focus on one of the most heavily debated topics of scientific discovery today. Collected, and richly illustrated for the first time in this edition, Darden's essays represent a ground-breaking foray into one of the major problems – biological mechanisms – facing scientists and philosophers of science alike. Divided into three sections, the essays focus on broad themes, notably historical and philosophical issues at play in discussions of biological mechanism; and the problem of developing and refining reasoning strategies, including interfield relations and anomaly resolution. In a group of essays published here for the first time, Darden summarizes the philosophy of discovery and elaborates the role that mechanisms play in biological discovery. Throughout the book, she uses historical case studies to extract advisory reasoning strategies for discovery. Examples in genetics, molecular biology, biochemistry, immunology, neuroscience, and evolutionary biology reveal the process of discovery in action.

Lindley Darden is Professor of Philosophy and in the Committee for Philosophy and the Sciences at the University of Maryland, College Park. She is the author of *Theory Change in Science: Strategies from Mendelian Genetics*, as well as numerous articles in history and philosophy of science journals. Elected a Fellow of the American Association for the Advancement of Science in 1995, she served as President of the International Society for History, Philosophy, and Social Studies of Biology from 2001 to 2003.

# CAMBRIDGE STUDIES IN PHILOSOPHY AND BIOLOGY

*General Editor*
Michael Ruse   *Florida State University*

*Advisory Board*

Michael Donoghue   *Yale University*
Jean Gayon   *University of Paris*
Jonathan Hodge   *University of Leeds*
Jane Maienschein   *Arizona State University*
Jesús Mosterín   *Instituto de Filosofía (Spanish Research Council)*
Elliott Sober   *University of Wisconsin*

*Other Books in the Series*

Alfred I. Tauber   *The Immune Self: Theory or Metaphor?*
Elliott Sober   *From a Biological Point of View*
Robert Brandon   *Concepts and Methods in Evolutionary Biology*
Peter Godfrey-Smith   *Complexity and the Function of Mind in Nature*
William A. Rottschaefer   *The Biology and Psychology of Moral Agency*
Sahotra Sarkar   *Genetics and Reductionism*
Jean Gayon   *Darwinism's Struggle for Survival*
Jane Maienschein and Michael Ruse (eds.)   *Biology and the Foundation of Ethics*
Jack Wilson   *Biological Individuality*
Richard Creath and Jane Maienschein (eds.)   *Biology and Epistemology*
Alexander Rosenberg   *Darwinism in Philosophy, Social Science, and Policy*
Peter Beurton, Raphael Falk, and Hans-Jörg Rheinberger (eds.)   *The Concept of the Gene in Development and Evolution*
David Hull   *Science and Selection*
James G. Lennox   *Aristotle's Philosophy of Biology*
Marc Ereshefsky   *The Poverty of the Linnaean Hierarchy*
Kim Sterelny   *The Evolution of Agency and Other Essays*
William S. Cooper   *The Evolution of Reason*
Peter McLaughlin   *What Functions Explain*
Steven Hecht Orzack and Elliott Sober (eds.)   *Adaptationism and Optimality*
Bryan G. Norton   *Searching for Sustainability*
Sandra D. Mitchell   *Biological Complexity and Integrative Pluralism*
Joseph LaPorte   *Natural Kinds and Conceptual Change*
Greg Cooper   *The Science of the Struggle for Existence*
Jason Scott Robert   *Embryology, Epigenesis, and Evolution*
William F. Harms   *Information and Meaning in Evolutionary Processes*

# Reasoning in Biological Discoveries

## Essays on Mechanisms, Interfield Relations, and Anomaly Resolution

LINDLEY DARDEN

*University of Maryland, College Park*

Illustrated by
Darren Hudson Hick

CAMBRIDGE
UNIVERSITY PRESS

CAMBRIDGE UNIVERSITY PRESS
Cambridge, New York, Melbourne, Madrid, Cape Town, Singapore, São Paulo, Delhi

Cambridge University Press
32 Avenue of the Americas, New York, NY 10013-2473, USA

www.cambridge.org
Information on this title: www.cambridge.org/9780521117272

First published 2006
This digitally printed version 2009

*A catalog record for this publication is available from the British Library*

*Library of Congress Cataloging in Publication data*

Darden, Lindley.
Reasoning in biological discoveries : essays on mechanisms, interfield relations, and
anomaly resolution / Lindley Darden; illustrated by Darren Hudson Hick.
    p.   cm. – (Cambridge studies in philosophy and biology)
Includes bibliographical references (p. ) and index.
ISBN-13: 978-0-521-85887-8 (hardback)
ISBN-10: 0-521-85887-9 (hardback)
1. Biology – Philosophy.   2. Mechanism (Philosophy)   I. Title.   II. Series.
QH331.D26   2006
570.1 – dc22                                                      2006001022

ISBN 978-0-521-85887-8 hardback
ISBN 978-0-521-11727-2 paperback

*For my family: Tom, David, and Morgan*

# Contents

# Contents

# Long Contents

## Introduction

Discovery is an extended process of construction, evaluation, and revision. Reasoning strategies exemplified in biological cases, discussed in the following chapters, provide advice that may be useful in future discovery episodes. Especially fruitful is the perspective that what is to be discovered is a biological mechanism. The nature of the product guides the process of discovery.

### PART I: BIOLOGICAL MECHANISMS

### *Chapter 1: Thinking About Mechanisms,* with Peter Machamer and Carl F. Craver

The concept of mechanism is analyzed in terms of entities and activities, organized such that they are productive of regular changes. Examples show how mechanisms work in neurobiology and molecular biology. Thinking in terms of mechanisms provides a new framework for addressing many traditional philosophical issues: causality, laws, explanation, reduction, and scientific change.

### *Chapter 2: Discovering Mechanisms in Neurobiology: The Case of Spatial Memory* with Carl F. Craver

Discovery is an extended process of construction, evaluation, and revision, as illustrated by the case of the discovery of mechanisms of spatial memory. The discovery of mechanisms is constrained by phenomenal, componency, spatial, temporal, and hierarchical constraints. Experimental testing of hypothesized mechanisms requires techniques for intervening via exciting or inhibiting one stage of the mechanism and then detecting effects at a different stage, either later or up/down in a hierarchy. Incompleteness and anomalies drive refinements during discovery.

### Chapter 3: Strategies in the Interfield Discovery of the Mechanism of Protein Synthesis with Carl F. Craver

In the 1950s and 1960s, an interfield interaction between molecular biologists and biochemists integrated important discoveries about the mechanism of protein synthesis. This extended discovery episode reveals two general reasoning strategies for eliminating gaps in descriptions of the productive continuity of mechanisms: schema instantiation and forward/backward chaining. Schema instantiation involves filling roles in an overall framework for the mechanism. Forward and backward chaining eliminates gaps using knowledge about types of entities and their activities. Attention to mechanisms highlights salient features of this historical episode while providing general reasoning strategies for mechanism discovery.

### Chapter 4: Relations Among Fields: Mendelian, Cytological, and Molecular Mechanisms

Philosophers have proposed various kinds of relations between Mendelian genetics and molecular biology: reduction, replacement, explanatory extension. This chapter argues that the two fields are best characterized as investigating different, serially integrated, hereditary *mechanisms*. The mechanisms operate at different times and contain different working entities. The working entities of the mechanisms of Mendelian heredity are chromosomes, whose movements serve to segregate alleles and independently assort genes in different linkage groups. The working entities of numerous mechanisms of molecular biology are larger and smaller segments of DNA plus related molecules. Discovery of molecular DNA mechanisms filled black boxes that were noted, but unilluminated, by Mendelian genetics.

### PART II: REASONING STRATEGIES: RELATING FIELDS, RESOLVING ANOMALIES

### Chapter 5: Interfield Theories with Nancy Maull

This chapter analyzes the generation and function of *interfield theories*, theories which bridge two fields of science. Interfield theories are likely to be generated when two fields share an interest in explaining different aspects of the same phenomenon and when background knowledge already exists relating the two fields. The interfield theory functions to provide a solution to a characteristic type of theoretical problem: How are the relations between fields to be explained? In solving this problem the interfield theory may provide answers to questions which arise in one field but cannot be answered within it alone, may focus attention on domain items not previously considered important, and may predict new domain items for one or both fields. Implications of this analysis for the problems of reduction and the unity and progress of science are mentioned.

## *Chapter 6: Theory Construction in Genetics*

Progress occurred in theory construction in Mendelian genetics when the appeal to vague analogies was replaced by fruitful interfield connections to cytology. William Bateson's appeal to vague analogies to coupling and repulsing forces, vortices, and vibrations was in contrast with T. H. Morgan's use of interfield relations to chromosomes in his construction of the theory of the gene. Analogies may provide a source for new ideas in theory construction (as many philosophers have argued), but interfield relations, if available, are likely to be more fruitful.

## *Chapter 7: Relations Among Fields in the Evolutionary Synthesis*

The synthetic theory of evolution is a multifield theory. According to Dobzhansky (1937), evolutionary mechanisms at three levels are studied by different fields. Genetics and cytology study mutations and chromosomal changes in organisms; population genetics studies the impact of the environment on populational changes, such as via selection or migration; and, finally, the study of isolating mechanisms that prevent interbreeding between populations shows how new species arise. The stage of development of these fields was crucial to the role they played in the synthesis, showing why Mendelism and Darwinism, although in conflict from 1900 to 1910, could be productively related in the 1930s.

## *Chapter 8: Selection Type Theories* with Joseph A. Cain

Selection type theories solve adaptation problems. Natural selection, clonal selection for antibody production, and selective theories of higher brain function are examples. An abstract characterization of typical selection processes is generated by analyzing and extending previous work on the nature of natural selection. Once constructed, this abstraction provides a useful tool for analyzing the nature of other selection theories and may be of use in new instances of theory construction. This suggests the potential fruitfulness of research to find other theory types and construct their abstractions.

### *Chapter 9: Strategies for Anomaly Resolution: Diagnosis and Redesign*

Anomaly resolution entails both diagnosis and redesign tasks. Steps and strategies for localizing and fixing anomalies for scientific theories are outlined. The resolution of a monster anomaly in genetics resulted in the discovery of lethal gene combinations. A simulation model for genetic processes is systematically debugged as an illustration of anomaly localization in a computational philosophy of science experiment.

### *Chapter 10: Exemplars, Abstractions, and Anomalies: Representations and Theory Change in Mendelian and Molecular Genetics*

Representations of scientific theories are closely tied to reasoning strategies for theory change. A scientific theory may be represented by a set of concrete exemplary problem solutions. Alternatively, a theory may be depicted in an abstract pattern, which, when its variables are filled with constants, becomes a particular explanation. The exemplars and abstractions may be depicted diagrammatically, as they are in the cases from Mendelian genetics and molecular biology. One way that a theory grows is by adding new types of exemplars to its explanatory repertoire. Model anomalies show the need for a new exemplar; they turn out to be examples of a typical, normal pattern that had not been included in the previous stage of theory development. A special-case anomaly indicates the need for a new exemplar or abstraction, but it has a small scope of applicability. Thus, ideas discussed here are exemplars, abstractions, diagrammatic representations, and anomalies and the roles they play in the representation of explanatory theories and in the change of such theories. Examples come from Mendelian genetics and molecular biology, including a special-case anomaly for the central dogma of molecular biology, namely, the discovery of the enzyme, reverse transcriptase, that copies RNA into DNA.

### *Chapter 11: Strategies for Anomaly Resolution in the Case of Adaptive Mutation*

The phenomenon of adaptive mutation is an anomaly that has received many diverse responses. These range from radical challenges to the theory of natural selection and the central dogma of molecular biology to the claim that adaptive mutations are produced via operation of previously known mechanisms. Examination of this anomaly provides a range of responses, from viewing the anomaly as a monster, a special case, or a model for all mutations. The case suggests refinements of strategies for anomaly resolution.

This chapter brings together the discussions of Part I on mechanisms and Part II on reasoning strategies. It summarizes the view of discovery via iterative refinement and elaborates the way that the characterization and features of mechanisms aid their discovery. In Section 12.2, the MDC (Machamer, Darden, Craver) discussion of mechanisms from Chapter 1 is refined and defended against recent criticisms. Subsequent sections of the chapter summarize and expand earlier discussions of reasoning strategies for construction, evaluation, and refinement. Construction strategies include schema instantiation, modular subassembly, and forward/backward chaining. Evaluation strategies serve to assess adequacy. They detect incompleteness and aid in moving from how possibly, to how plausibly, to how actually the mechanism works. Anomaly resolution strategies guide diagnosis and repair during revision. Examples come from molecular biology, biochemistry, immunology, neuroscience, and evolutionary biology discussed in more detail in earlier chapters.

# List of Figures

## List of Figures

# List of Tables

# List of Tables

# Acknowledgments

This book would not have been possible without my wonderful collaborators. Certainly my work would have been much less fun. I've learned a lot from my colleagues and students. I thank all my coauthors for their permission to reprint our joint articles: Nancy Maull, Joe Cain, Peter Machamer, and Carl Craver. Beginning in graduate school in the Committee on the Conceptual Foundations of Science at the University of Chicago, Nancy Maull and I began comparing the scientific cases we were studying. As we began our jobs as assistant professors, we completed "Interfield Theories," reprinted in Chapter 5. Joe Cain was an undergraduate and master's degree student at the University of Maryland, College Park, when we wrote "Selection Type Theories," reprinted in Chapter 8.

My visit to the hospitable Center for Philosophy of Science at the University of Pittsburgh led to my collaboration with Peter Machamer and Carl Craver on "Thinking About Mechanisms." Peter and I had been in graduate school together at the University of Chicago and revived shared perspectives in our collaboration. It was a pleasure to work with him and enjoy his legendary hospitality in Pittsburgh. While Carl was a post-doc at the University of Maryland (for which I thank the Committee on Cognitive Studies), he drafted "Discovering Mechanisms in Neurobiology" (Chapter 2), and I drafted "Strategies in the Interfield Discovery of the Mechanisms of Protein Synthesis" (Chapter 3). Then, with much lively conversation producing new insights, we rewrote both together. Carl probably talked more about discovery as a result of my influence, and Carl saw more general ideas about mechanisms than I did as I reveled in the particular details of the protein case study. Peter contributed vociferous comments. It was a pleasure working with Carl again as we edited and wrote the introduction for the special issue on mechanisms of *Studies in History and Philosophy of Biological and Biomedical Sciences* (June 2005). Chapter 12 makes use of several excellent articles from that issue. Carl influenced my ideas in Parts I and III more than citations to his work can indicate. Most recently, his comments on Section 12.2, where I respond to some critiques of MDC, greatly improved it, even though I didn't always take his advice. I look forward to reading the new book he is writing on mechanisms in neuroscience, which I have purposely not read until I finished Part III.

My former student Nancy S. Hall has done much to aid the production of this book. She is now a Research Associate at the University of Maryland while maintaining a full-time teaching job at the University of Delaware. She perfected her abilities as an editor in her work on Stephen Brush's (2003) *The Kinetic Theory of Gases: An Anthology of Classic Papers with Historical Commentary,* edited by Nancy S. Hall (London: Imperial College Press). That experience served her well in editing this book. She assembled the review copy. She proofread every chapter. She compiled the master reference list and checked citations against references. She helped with the index. She was there with a friendly call when I needed a prod to stop rewriting Chapters 11 and 12 again! It is difficult to thank her enough for her friendship and all that she did to bring this book to completion.

Other former students have read various drafts of various chapters and given me helpful comments, sometimes on short notice. My thanks to Jason Baker, David Didion, Pam Henson, and Rob Skipper. Members of the DC History and Philosophy of Biology discussion group were supportively critical readers of drafts of Chapters 4, 11, and 12. The group changes as people come and go in the area. I thank especially Dick Burian, Ilya Farber, Greg Morgan, Jessica Pfeiffer, Eric Saidel, Ken Schaffner, and Joan Straumanis. Also, Jim Bogen read an early draft of Chapter 12 and supplied new ideas from his own developing perspective on mechanisms; conversations with him are always enlightening. Jim Woodward gently challenged me to think more about causality and the ways our perspectives may be related, something requiring more work. Bill Bechtel and I have shared similar interests throughout our careers, first on interfield relations and later on mechanisms. His support and encouragement have been important to me. Comments from him, and more recently from his wife Adele Abrahamsen, provided insights for me from the cognitive science literature on understanding mechanisms, as reflected in Chapter 12. My work on anomaly resolution has profited from ideas of Douglas Allchin and Kevin Elliott. Others who helped on specific chapters are noted in the first footnote of each chapter. A number of philosophers of science working on discovery, scientific change, and mechanisms share a Chicago background and I have profited from their work and from discussions with them. Perhaps we constitute something of a "Chicago school": Douglas Allchin,

Bill Bechtel, Peter Machamer, Bob Richardson, Stuart Glennan, Scott Kleiner, Ken Schaffner, Dudley Shapere, Stephen Toulmin, and Bill Wimsatt. I am very grateful to Michael Ruse, the editor of the Cambridge Studies in Philosophy and Biology series, who encouraged me to put together this collection and guided it through the review process. Thanks to Ben Cranston, who scanned several articles, did careful proofreading, and gave me general encouragement. I thank all of these friends for making possible my lively intellectual life.

Darren Hick used his considerable artistic skills to redraw all the figures and tables. I thank him for his work as illustrator. Matthew Barker helped with indexing.

My work over the years was supported by a U.S. National Endowment for the Humanities summer grant, an American Council of Learned Societies grant, several grants from the U.S. National Science Foundation, General Research Board awards from the Graduate School at the University of Maryland, as well as sabbatical leaves. I visited the Heuristic Programming Project at Stanford University as the guest of Bruce Buchanan, the Department of History of Science at Harvard University, the Laboratory for Artificial Intelligence Research at Ohio State University as the guest of John Josephson and B. Chandrasekaran, the Laboratory of Molecular Genetics and Informatics at Rockefeller University as the guest of Joshua Lederberg, and the Center for Philosophy of Science at the University of Pittsburgh. The time for research and intellectual stimulation engendered by these colleagues were invaluable.

Chapters 11 and 12 are original to this book. I thank the following publishers of the previously published articles in Chapters 1–10 for their kind permission to reprint them:

To the Philosophy of Science Association and the University of Chicago Press for:

Chapter 1: Machamer, Peter, Lindley Darden, and Carl F. Craver (2000), "Thinking About Mechanisms," *Philosophy of Science* 67: 1–25.

Chapter 5: Darden, Lindley, and Nancy Maull (1977), "Interfield Theories," *Philosophy of Science* 44: 43–64.

Chapter 8: Darden, Lindley, and Joseph A. Cain (1989), "Selection Type Theories," *Philosophy of Science* 56: 106–129.

To Elsevier, Ltd., for:

Chapter 3: Darden, Lindley, and Carl F. Craver (2002), "Strategies in the Interfield Discovery of the Mechanism of Protein Synthesis," *Studies in History and Philosophy of Biological and Biomedical Sciences* 33: 1–28.

Chapter 4: Darden, Lindley (2005), "Relations Among Fields: Mendelian, Cytological, and Molecular Mechanisms," in Carl F. Craver and Lindley

Darden (eds.), Special Issue: "Mechanisms in Biology," *Studies in History and Philosophy of Biological and Biomedical Sciences* 36: 357–371.

To the University of Pittsburgh Press and Universitaetsverlag Konstanz for:

Chapter 2: Craver, Carl F., and Lindley Darden (2001), "Discovering Mechanisms in Neurobiology: The Case of Spatial Memory," in Peter Machamer, R. Grush, and P. McLaughlin (eds.), *Theory and Method in the Neurosciences*. Pittsburgh, PA: University of Pittsburgh Press, pp. 112–137. © 2001 University of Pittsburgh Press.

To the University of Pittsburgh Press for

Chapter 10: Darden, Lindley (1995), "Exemplars, Abstractions, and Anomalies: Representations and Theory Change in Mendelian and Molecular Genetics," in James G. Lennox and Gereon Wolters (eds.), *Concepts, Theories, and Rationality in the Biological Sciences*. Pittsburgh, PA: University of Pittsburgh Press, pp. 137–158. © 1995 University of Pittsburgh Press.

To Springer Science and Business Media for:

Chapter 6: Darden, Lindley (1980), "Theory Construction in Genetics," in T. Nickles (ed.), *Scientific Discovery: Case Studies*. Dordrecht: Reidel, pp. 151–170.

Chapter 7: Darden, Lindley (1986), "Relations Among Fields in the Evolutionary Synthesis," in W. Bechtel (ed.), *Integrating Scientific Disciplines*. Dordrecht: Nijhoff, pp. 113–123.

To the University of Minnesota Press for:

Chapter 9: Darden, Lindley (1992), "Strategies for Anomaly Resolution: Diagnosis and Redesign," in R. Giere (ed.), *Cognitive Models of Science*, Minnesota Studies in the Philosophy of Science, v. 15. Minneapolis, MN: University of Minnesota Press, pp. 251–273.

Lindley Darden<br>
College Park, Maryland 2005

# Introduction

This book discusses reasoning strategies for discovery that are exemplified in numerous biological cases. Scientific discovery should be viewed as an extended, piecemeal process with hypotheses undergoing iterative refinement. Construction, evaluation, and revision are tightly connected in ways that philosophers of science have often not recognized, given their neglect of reasoning in hypothesis construction and revision. Examination of historical cases from twentieth-century biology reveals reasoning strategies that could have produced the changes that did occur. Such critically examined reasoning strategies constitute compiled hindsight gleaned from these past episodes. Examples come from the fields of molecular biology, biochemistry, immunology, neuroscience, and evolutionary biology. Making reasoning strategies explicit shows that they are not merely descriptions of unique historical changes or unwarranted overly general prescriptions. They are advisory. They may be of use as metascientific hypotheses in philosophical and historical analyses of scientific reasoning, of use in future empirical and computational biological research, and of use in science education. Hence, one goal of this book is to make explicit reasoning strategies for construction, evaluation, and revision of scientific hypotheses.

Biologists often seek to discover mechanisms. Knowing what is to be discovered aids the extended process of discovery. The examination of the nature and means of representing biological theories aids analysis of reasoning in their discovery. What play the roles of theories in molecular biology, for example, are diagrammatically represented sets of mechanism schemas for such widely found mechanisms as DNA replication, protein synthesis, and many varieties of gene regulation. Hence, another goal of this book is to find reasoning strategies for discovering such mechanisms.

Sometimes scientific discoveries occur entirely within one field. In other cases, two or more scientific fields contribute to a scientific discovery in

1

various ways. Two fields may both seek to discover the same mechanism, investigating different modules of the mechanism using different techniques. Another field may supply items for the construction of an intrafield theory. Two fields may be bridged by an interfield theory. A multifield theory may integrate views of hierarchically nested mechanisms. An abstract mechanism schema from one field may be used analogically to construct a similar type of theory in another field. Hence, another goal of this book is to demonstrate the role of interfield relations in biological discoveries.

Much of the previous work in philosophy of science has been hampered by an overly sharp dichotomy between discovery and justification and by viewing scientific discovery as a mysterious process. Accounts of "aha" experiences, while entertaining, are not adequate descriptions of the reasoning in extended episodes of scientific innovation. Also, the prescriptions in philosophy of science of methods for confirmation or Karl Popper's (1965) of falsification have been shown to be too simplistic. Multiple factors play roles in the evaluation of scientific knowledge claims (both in assessing theories and credentialing empirical evidence). Some failures on some evaluative dimensions call not for complete rejection of a hypothesis but for its refinement. Surprisingly few philosophers of science have paid attention to reasoning in the revision of scientific hypotheses that face anomalies. Consequently, an alternative perspective, as I have also argued elsewhere (Darden 1991), needs to replace the simplistic dichotomy of irrational discovery followed by logically characterized justification (or falsification). Further, neglect of revision should be remedied. Science should be viewed as an error-correcting process and philosophers should seek to find reasoning strategies that constrain and guide that process. Hence, another goal of this book is to find strategies for anomaly resolution, viewed as diagnostic and redesign processes.

Some philosophers of science are what Tom Nickles (1980c) called "friends of discovery" (e.g., Hanson 1958, 1961; Schaffner 1974a; Buchanan 1982, 1985; Kleiner 1993), while others have recognized the importance of mechanisms in science, especially biology (Wimsatt 1972; Brandon 1985; Burian 1996a; Glennan 1996, 2002, 2005). Only a few have investigated reasoning in the discovery of mechanisms. Rom Harré was an early advocate: "Generally speaking, making models for unknown mechanisms is the creative process in science" (Harré 1970, p. 40). He emphasized the role of analogies in discovering mechanistic models. Harré also endeavored to find an analysis of causality compatible with his mechanistic view (Harré and Madden 1975). For numerous biological cases, William Bechtel and Robert Richardson showed how the heuristics of decomposition and localization aided the discovery of "mechanistic explanations" in the "dynamics of theory development" (Bechtel and Richardson 1993, pp. xii–xiii). Bechtel and Abrahamsen (2005) discussed a variety of experimental procedures for discovering the parts of mechanisms, their operations, and their organization. Paul Thagard, long interested in reasoning in discovery (e.g., Thagard 1992; Holyoak and Thagard 1995), turned his attention to the discovery of mechanisms in the biomedical sciences: "Discovery of a pathway provides a mechanism that describes the productive activity that enables the cell to perform tasks . . . ." (Thagard 2003, p. 239). Then, he continued, the discovery of such molecular cell mechanisms aids medical discovery: "Many diseases can be explained by defects in pathways, and new treatments often involve finding drugs that correct those defects" (Thagard 2003, p. 235). Some philosophers examined the dynamics of scientific change during anomaly characterization and error correction (e.g., Wimsatt 1987; Allchin 2002; Elliott 2004).

Hence, an important goal here is to integrate work on biological mechanisms and reasoning in discovery. The chapters of Part I, "Biological Mechanisms," focus on the characterization of biological mechanisms and the roles of reasoning strategies and techniques from different fields in their discovery, with examples from Mendelian genetics, molecular biology, and neurobiology. The chapters of Part II, "Reasoning Strategies: Relating Fields, Resolving Anomalies," discuss ways of representing biological theories, as well as reasoning strategies in finding interrelations between biological fields and in resolving anomalies for biological theories. These chapters add examples from evolutionary biology and immunology. Part III, "Discovering Mechanisms: Construction, Evaluation, Revision," integrates the earlier parts and expands the discussion of reasoning in discovering mechanisms. Chapter 12 responds to some of the criticisms of earlier work, elaborates the features of mechanisms that need to be discovered, and elaborates reasoning strategies for discovering mechanisms during construction, evaluation, and revision.

In our collaborative work, beginning in 1997, my colleagues Peter Machamer and Carl Craver and I analyzed aspects of mechanisms in biology. Peter Machamer brought his insights about seventeenth-century mechanisms (e.g., Machamer and Woody 1994; Machamer 1998). Carl Craver's ideas were informed by his work in neurobiology and mine by examination of molecular biology. Our collaborative chapters in Part I analyze these issues: characterization of biological mechanisms in molecular biology and neurobiology, constraints on an adequate description of a mechanism, some reasoning strategies for the discovery of mechanisms, and interfield integration in mechanism discovery.

Chapter 1, "Thinking About Mechanisms" with Peter Machamer and Carl F. Craver, was originally published in 2000. We refer to this paper as "MDC."

3

It provides a characterization of mechanisms that produce phenomena in terms of entities, activities, and their productive continuity. The activities in twentieth-century mechanisms are much more varied than the contact action of geometrico-mechanical seventeenth-century clockwork mechanisms. Especially important for molecular biological mechanisms are the activities of chemical bonding, especially weak hydrogen bonding. This chapter is programmatic, suggesting a new mechanistic approach for philosophy of biology that may be applicable to fields beyond molecular biology and neuroscience. It sketches analyses of mechanistic explanation, theory structure, theory change, causality, the unimportance of universal laws and derivational reduction, as well as interrelations among fields of biology. (Section 12.2 responds to critiques of this MDC characterization of mechanisms and I develop it further there; cf. Machamer 2004; Bogen 2005.)

Chapter 2, "Discovering Mechanisms in Neurobiology: The Case of Spatial Memory" with Carl F. Craver, elaborates constraints that an adequate description of a mechanism should satisfy, including componency, spatial, temporal, and hierarchical constraints. Another topic is finding experimental strategies for investigating mechanisms. Given an experimental setup with a running mechanism, one can intervene via inhibition or excitation and, then, detect the downstream effect. We argue that the neurobiological case study shows the hierarchical integration of work at different levels on mechanisms of spatial memory. The mechanistic approach in cases from neurobiology has been further developed by Craver (2001, 2002a, 2002b, 2003, 2005). (In Part III, I slightly expand what I there call "the features of mechanisms.")

Chapter 3, "Strategies in the Interfield Discovery of the Mechanism of Protein Synthesis" with Carl F. Craver, uses the mechanistic approach to illuminate the extended discovery of the mechanism of protein synthesis. Molecular biologists and biochemists worked on different ends of the mechanism. The molecular biologists began with the genetic material, DNA, while biochemists studied peptide bond formation between activated amino acids in proteins. Their work converged in the middle of the mechanism, with the discoveries of the types and roles of RNAs. In this case of interfield interaction, researchers in two different fields worked to understand the same mechanism. This case exemplifies two strategies for discovery: schema instantiation, and forward/backward chaining. (These strategies are put into a larger context of reasoning strategies of construction, evaluation, and revision in Part III.)

In Chapter 4, "Relations Among Fields: Mendelian, Cytological, and Molecular Mechanisms," I criticize earlier philosophical claims about relations between the fields of Mendelian genetics and molecular biology. I argue that these relations should not be analyzed in terms of reduction, replacement, or (one form of) explanatory extension. Instead, the two fields are shown to have investigated different, serially integrated, hereditary *mechanisms*. The mechanisms operate at different times and contain different working entities. Molecular biological mechanisms filled black boxes that were noted, but unilluminated, by Mendelian geneticists. (For another argument against reduction from his multilevel mechanistic perspective, see Craver 2005.)

The chapters in Part II extract additional reasoning strategies for problem solving from extended discovery episodes. Chapter 5, "Interfield Theories" with Nancy Maull, discusses cases in which interfield theories, bridging two fields, solve problems that could not be solved within a single field alone. Interrelations among fields, even if not developed into a full-fledged interfield theory, may provide new ideas for one or both of the bridged fields. I examine the chromosome theory of Mendelian heredity bridging genetics and cytology. Nancy Maull's cases are the operon theory of gene regulation, bridging genetics and biochemistry, and the theory of allosteric regulation bridging biochemistry and physical chemistry. (Compare additional analyses of interfield relations in, e.g., Bechtel 1984, 1986; Darden 1991; Craver 2005.)

Chapter 6, "Theory Construction in Genetics," contrasts the vague analogies used by William Bateson with the interfield relations that proved fruitful in T. H. Morgan's development of the theory of the gene. Analogies may serve as a source for new ideas; however, this chapter argues, interfield relations, if available, are likely to be more fruitful. (The extended discovery of the theory of the gene and reasoning strategies it exemplified were discussed in more detail in Darden 1991.)

Sometimes, hypotheses about mechanisms at several hierarchical levels can be integrated in a multilevel theory, as in the synthetic theory of natural selection. Chapter 7, "Relations Among Fields in the Evolutionary Synthesis," discusses the integration of mechanisms at three levels in the work of Theodosius Dobzhansky (1937). Genetics and cytology study mutations and chromosomal changes in organisms, which are the raw material for evolutionary change. Population genetics studies the impact of the environment on populational changes via, for example, selection or migration. Finally, evolutionary biology studies speciation mechanisms, investigating the study of isolating mechanisms that prevent interbreeding and thereby produce new species. (Compare the case of multifield integration in neuroscience in Craver 2005 and discussion of speciation mechanisms in Baker 2005.)

Once a new theory has been constructed, it may be seen as representative of a type. Joseph A. Cain and I discuss one prevalent type in Chapter 8, "Selection Type Theories." Abstracting (namely, eliminating details) from the theory of natural selection provides a schema for selection type theories. In contrast to the more usual analysis of selection in terms of replicators and interactors (Hull 1980), this selection schema relegates replication to a more minor role, a possible downstream benefit. The first step in a schema for a selection mechanism is the production of a population of variants. Even if this step is stochastic, as long as it provides variants for the next stage, it fulfills its appropriate role in the selection mechanism. (The related issue of possible mechanisms for producing adaptive mutations is the subject of Chapter 11.) The second step in the Darden and Cain schema is the crucial, difference-making step – the interaction of the variants with a critical environmental factor. The next step abstractly characterizes the result of the selective inter-action – some variants benefit and others suffer. These terms are sufficiently abstract that many different kinds of outcomes can count as benefits. In Dar-winian natural selection, a short-range benefit is survival and a longer range benefit is increased reproduction of the successful variants. Such a schema – variants, interaction, benefit – can be used to guide the construction of other selection type theories to solve adaptation problems. Once selection had been discovered in evolutionary biology, it became available as an analogy for other fields. Chapter 8 includes the historical examples of the clonal selection theory in immunology and the more speculative neural Darwinism. (Compare the further development of this schema as a mechanism schema in Skipper 1999, 2001; and critiques in Skipper and Millstein 2005. Also see the critique of the peppered moth example in Rudge 1999.)

Chapters 9, 10, and 11 categorize types of anomalies and refine strate-gies for anomaly resolution. Chapter 9, "Strategies for Anomaly Resolution: Diagnosis and Redesign," argues that anomaly-driven scientific change can be viewed as, first, a diagnostic reasoning process. A failure in a theory must be localized in a theoretical component. This analogy between localization of an anomaly and diagnostic reasoning is a fruitful one that allows philosophers of science to make use of the extensive work on reasoning in diagnosis. Sec-ond, fixing the failed component(s) of the system is a redesign process. Steps and strategies for localizing and fixing anomalies for scientific theories are outlined. A simulation model represents a Mendelian breeding experiment. The model is systematically debugged as an illustration of anomaly local-ization in a computational philosophy of science experiment. (For further development of the computational perspective for discovering mechanisms, see Darden 2001.)

Chapter 10, "Exemplars, Abstractions, and Anomalies: Representations and Theory Change in Mendelian and Molecular Genetics," shows that rep-resentations of scientific theories are closely tied to reasoning strategies for

6

theory change. One way a scientific theory may be represented is by a set of concrete exemplary problem solutions. Alternatively, a theory may be depicted in an abstract schema, which, when its variables are filled with constants, becomes a particular explanation. The exemplars and abstractions may be depicted diagrammatically, as they are in the cases from Mendelian genetics and molecular biology. One way that a theory grows is by adding new types of exemplars to its explanatory repertoire. Model anomalies show the need for a new wide-scope exemplar; they turn out to be examples of a typical, normal pattern that had not been included in the previous stage of theory development. The discovery of the linkage of genes resulted from the resolution of a model anomaly. In contrast, a special-case anomaly indicates the need for a new exemplar or abstraction with only a small scope of applicability. The discovery of reverse transcriptase provided a special-case anomaly for the central dogma of molecular biology, which is a mechanism schema for protein synthesis with wide scope: DNA −>RNA −>protein. For retroviruses, an additional step was added to the beginning of this mechanism schema: RNA −>DNA. Neither the linkage anomalies nor the reverse transcriptase anomaly could be barred as a monster that did not require theory change. (For more on Temin's discovery of reverse transcriptase, see Marcum 2002.)

Chapter 11, "Strategies for Anomaly Resolution in the Case of Adaptive Mutation," examines the controversial anomaly of directed or adaptive mutations. The anomaly has received many diverse responses, beginning in 1988 with radical challenges to the theory of natural selection and the central dogma of molecular biology. The hypothesized instructive mechanisms to produce directed mutation provide a contrast to selection mechanisms. As of 2003, this anomaly appears to be resolvable by appeal to operation of known types of mechanisms. Examination of this anomaly allows refinement of strategies for anomaly resolution.

Finally, Part III, Chapter 12, "Strategies for Discovering Mechanisms: Construction, Evaluation, Revision," summarizes and extends analyses in earlier chapters. A few criticisms of our MDC characterization of mechanism are briefly addressed. The list of features of mechanisms is expanded. Reasoning strategies for discovery of mechanisms via iterative refinement are discussed. The categories of strategies are construction, evaluation, and revision. Guidance in construction may be provided by the reasoning strategies of schema instantiation, modular subassembly, and forward/backward chaining. Evaluation strategies serve to assess adequacy. Evaluation detects the incompleteness in mechanism sketches. Proposed mechanism schemas are transformed as various evaluative strategies are employed, moving from how possibly, to how plausibly, to how actually the mechanism works. Anomaly resolution strategies guide diagnosis and repair during revision of mechanism schemas and sketches.

Thus, these chapters extract reasoning strategies for biological discovery from episodes in the history of biology. They may be useful for philosophers and historians of science interested in reasoning in discovery. They may serve as heuristics for scientists in lab meetings. They may serve as guides for educators who teach scientific reasoning. They may be of use for building computational systems to make biological discoveries.

REFERENCES

Allchin, Douglas (2002), "Error Types," *Perspectives on Science* 9: 38–58.

Baker, Jason M. (2005), "Adaptive Speciation: The Role of Natural Selection in Mechanisms of Geographic and Non-geographic Speciation," in Carl F. Craver and Lindley Darden (eds.), Special Issue: "Mechanisms in Biology," *Studies in History and Philosophy of Biological and Biomedical Sciences* 36: 303–326.

Bechtel, William (1984), "Reconceptualizations and Interfield Connections: The Discovery of the Link Between Vitamins and Coenzymes," *Philosophy of Science* 51: 265–292.

Bechtel, William (1986), "Introduction: The Nature of Scientific Integration," in W. Bechtel (ed.), *Integrating Scientific Disciplines*. Dordrecht: Nijhoff, pp. 3–52.

Bechtel, William and Adele Abrahamsen (2005), "Explanation: A Mechanist Alternative," in Carl F. Craver and Lindley Darden (eds.), Special Issue: "Mechanisms in Biology," *Studies in History and Philosophy of Biological and Biomedical Sciences* 36: 421–441.

Bechtel, William and Robert C. Richardson (1993), *Discovering Complexity: Decomposition and Localization as Strategies in Scientific Research*. Princeton, NJ: Princeton University Press.

Bogen, James (2005), "Regularities and Causality; Generalizations and Causal Explanations," in Carl F. Craver and Lindley Darden (eds.), Special Issue: "Mechanisms in Biology," *Studies in History and Philosophy of Biological and Biomedical Sciences* 36: 397–420.

Brandon, Robert (1985), "Grene on Mechanism and Reductionism: More Than Just a Side Issue," in Peter Asquith and Philip Kitcher (eds.), *PSA 1984*, v. 2. East Lansing, MI: Philosophy of Science Association, pp. 345–353.

Buchanan, Bruce (1982), "Mechanizing the Search for Explanatory Hypotheses," in Peter Asquith and Thomas Nickles (eds.), *PSA 1982*, v. 2. East Lansing, MI: Philosophy of Science Association, pp. 129–146.

Buchanan, Bruce (1985), "Steps Toward Mechanizing Discovery," in K. Schaffner (ed.), *Logic of Discovery and Diagnosis in Medicine*. Berkeley, CA: University of California Press, pp. 94–114.

Burian, Richard M. (1996a), "Underappreciated Pathways Toward Molecular Genetics as Illustrated by Jean Brachet's Cytochemical Embryology," in S. Sarkar (ed.), *The Philosophy and History of Molecular Biology: New Perspectives*. Dordrecht: Kluwer, pp. 67–85.

Craver, Carl F. (2001), "Role Functions, Mechanisms, and Hierarchy," *Philosophy of Science* 68: 53–74.

Craver, Carl F. (2002a), "Structures of Scientific Theories," in P. K. Machamer and M. Silberstein (eds.), *Blackwell Guide to the Philosophy of Science.* Oxford, UK: Blackwell, pp. 55–79.

Craver, Carl F. (2002b), "Interlevel Experiments, Multilevel Mechanisms in the Neuroscience of Memory," *Philosophy of Science (Supplement)* 69: S83–S97.

Craver, Carl F. (2003), "The Making of a Memory Mechanism," *Journal of the History of Biology* 36: 153–195.

Craver, Carl F. (2005), "Beyond Reduction: Mechanisms, Multifield Integration, and the Unity of Neuroscience," in Carl F. Craver and Lindley Darden (eds.), Special Issue: "Mechanisms in Biology," *Studies in History and Philosophy of Biological and Biomedical Sciences* 36: 373–397.

Darden, Lindley (1991), *Theory Change in Science: Strategies from Mendelian Genetics.* New York: Oxford University Press.

Darden, Lindley (2001), "Discovering Mechanisms: A Computational Philosophy of Science Perspective," in Klaus P. Jantke and Ayumi Shinohara (eds.), *Discovery Science* (Proceedings of the 4th International Conference, DS2001). New York: Springer-Verlag, pp. 3–15.

Dobzhansky, Theodosius (1937), *Genetics and the Origin of Species.* New York: Columbia University Press.

Elliott, Kevin (2004), "Error as Means to Discovery," *Philosophy of Science* 71: 174–197.

Glennan, Stuart S. (1996), "Mechanisms and the Nature of Causation," *Erkenntnis* 44: 49–71.

Glennan, Stuart S. (2002), "Rethinking Mechanistic Explanation," *Philosophy of Science (Supplement)* 69: S342–S353.

Glennan, Stuart S. (2005), "Modeling Mechanisms," in Carl F. Craver and Lindley Darden (eds.), Special Issue: "Mechanisms in Biology," *Studies in History and Philosophy of Biological and Biomedical Sciences* 36: 443–464.

Hanson, Norwood Russell (1958), *Patterns of Discovery.* Cambridge: Cambridge University Press.

Hanson, Norwood Russell ([1961] 1970), "Is There a Logic of Scientific Discovery?" in H. Feigl and G. Maxwell (eds.), *Current Issues in the Philosophy of Science.* New York: Holt, Rinehart and Winston, 1961. Reprinted in B. Brody (ed.), *Readings in the Philosophy of Science.* Englewood Cliffs, NJ: Prentice Hall, pp. 620–633.

Harré, Rom (1970), *The Principles of Scientific Thinking.* Chicago, IL: University of Chicago Press.

Harré, Rom and E. H. Madden (1975), *Causal Powers: A Theory of Natural Necessity.* Totowa, NJ: Rowman and Littlefield.

Holyoak, Keith J. and Paul Thagard (1995), *Mental Leaps: Analogy in Creative Thought.* Cambridge, MA: MIT Press.

Kleiner, Scott A. (1993), *The Logic of Discovery: A Theory of the Rationality of Scientific Research.* Dordrecht: Kluwer.

Machamer, Peter (1998), "Galileo's Machines, His Mathematics and His Experiments," in Peter Machamer (ed.), *Cambridge Companion to Galileo.* Cambridge: Cambridge University Press, pp. 53–79.

Machamer, Peter (2004), "Activities and Causation: The Metaphysics and Epistemology of Mechanisms," *International Studies in the Philosophy of Science* 18: 27–39.

Machamer, Peter, Lindley Darden, and Carl F. Craver (2000), "Thinking About Mechanisms," *Philosophy of Science* 67: 1–25.

Machamer, Peter and Andrea Woody (1994), "A Model of Intelligibility in Science: Using Galileo's Balance as a Model for Understanding the Motion of Bodies," *Science and Education* 3: 215–244.

Marcum, James A. (2002), "From Heresy to Dogma in Accounts of Opposition to Howard Temin's DNA Provirus Hypothesis," *History and Philosophy of the Life Sciences* 24: 165–192.

Nickles, Thomas (1980c), "Introductory Essay: Scientific Discovery and the Future of Philosophy of Science," in Thomas Nickles (ed.), *Scientific Discovery, Logic, and Rationality*. Dordrecht: Reidel, pp. 1–59.

Popper, Karl R. (1965), *The Logic of Scientific Discovery*. New York: Harper Torchbooks.

Rudge, David W. (1999), "Taking the Peppered Moth with a Grain of Salt," *Biology & Philosophy* 14: 9–37.

Schaffner, Kenneth (1974a), "Logic of Discovery and Justification in Regulatory Genetics," *Studies in the History and Philosophy of Science* 4: 349–385.

Skipper, Robert A., Jr. (1999), "Selection and the Extent of Explanatory Unification," *Philosophy of Science* 66 (Proceedings): S196–S209.

Skipper, Robert A., Jr. (2001), "The Causal Crux of Selection," in *Behavioral and Brain Sciences* 24: 556.

Skipper, Robert A., Jr., and Roberta L. Millstein (2005), "Thinking about Evolutionary Mechanisms: Natural Selection," in Carl F. Craver and Lindley Darden (eds.), Special Issue: "Mechanisms in Biology," *Studies in History and Philosophy of Biological and Biomedical Sciences* 36: 327–347.

Thagard, Paul (1992), *Conceptual Revolutions*. Princeton, NJ: Princeton University Press.

Thagard, Paul (2003), "Pathways to Biomedical Discovery," *Philosophy of Science* 70: 235–254.

Wimsatt, William (1972), "Complexity and Organization," in Kenneth F. Schaffner and Robert S. Cohen (eds.), *PSA 1972, Proceedings of the Philosophy of Science Association*. Dordrecht: Reidel, pp. 67–86.

Wimsatt, William (1987), "False Models as Means to Truer Theories," in Matthew Nitecki and Antoni Hoffman (eds.), *Neutral Models in Biology*. New York: Oxford University Press, pp. 23–55.

# I

# Biological Mechanisms

# 1

# Thinking About Mechanisms[1]

## 1.1  INTRODUCTION

In many fields of science what is taken to be a satisfactory explanation requires providing a description of a mechanism. So it is not surprising that much of the practice of science can be understood in terms of the discovery and description of mechanisms. Our goal is to sketch a mechanistic approach for analyzing neurobiology and molecular biology that is grounded in the details of scientific practice, an approach that may well apply to other scientific fields.

Mechanisms have been invoked many times and places in philosophy and science. A key word search on "mechanism" for 1992–1997 in titles and abstracts of *Nature* (including its subsidiary journals, such as *Nature Genetics*) found 597 hits. A search in the *Philosophers' Index* for the same period found 205 hits. Yet, in our view, there is no adequate analysis of what mechanisms are and how they work in science.

We begin (Section 1.2) with a dualistic analysis of the concept of mechanism in terms of both the *entities* and *activities* that compose them. Section 1.3 argues for the ontic adequacy of this dualistic approach and indicates some of its implications for analyses of functions, causality, and laws. Section 1.4

[1] This chapter was originally published as Machamer, Peter, Lindley Darden, and Carl F. Craver (2000), "Thinking About Mechanisms," *Philosophy of Science* 67: 1–25. We thank the following people for their help: D. Bailer-Jones, A. Baltas, J. Bogen, R. Burian, G. Carmadi, R. Clifton, N. Comfort, S. Culp, F. di Poppa, G. Gale, S. Glennan, N. Hall, L. Holmes, T. Iseda, J. Josephson, J. Lederberg, J. E. McGuire, G. Piccinini, P. Pietroski, H. Rheinberger, W. Salmon, S. Sastry, K. Schaffner, R. Skipper, P. Speh, D. Thaler, and N. Urban. Lindley Darden's work was supported by the General Research Board of the Graduate School of the University of Maryland and as a Fellow in the Center for Philosophy of Science at the University of Pittsburgh; Carl Craver's work was supported by a Cognitive Studies Postdoctoral Fellowship of the Department of Philosophy of the University of Maryland. Both Lindley Darden and Carl Craver were supported by the U.S. National Science Foundation Grant SBR-9817942.

uses the example of the mechanism of neuronal depolarization to demonstrate the adequacy of the mechanism definition. Section 1.5 characterizes the descriptions of mechanisms by elaborating such aspects as hierarchies, bottom-out activities, mechanism schemata, and sketches. This section also suggests a historiographic point to the effect that much of the history of science might be viewed as written with the notion of mechanism. Another example in Section 1.6, the mechanism of protein synthesis, shows how thinking about mechanisms illuminates aspects of discovery and scientific change. The final sections hint at new ways to approach and solve or dissolve some major philosophical problems (viz., explanation and intelligibility in Section 1.7 and reduction in Section 1.8). These arguments are not developed in detail but should suffice to show how thinking about mechanisms provides a distinctive approach to many problems in the philosophy of science.

Quickly, though, we issue a few caveats. First, we use "mechanism" because the word is commonly used in science. But as we shall detail more precisely, one should not think of mechanisms as exclusively mechanical (push-pull) systems. What counts as a mechanism in science has developed over time and presumably will continue to do so. Second, we will confine our attention to mechanisms in molecular biology and neurobiology. We do not claim that all scientists look for mechanisms or that all explanations are descriptions of mechanisms. We suspect that this analysis is applicable to many other sciences, and maybe even to cognitive or social mechanisms, but we leave this as an open question. Finally, many of our points are only provocatively and briefly stated. We believe there are full arguments for these points but detailing them here would obscure the overall vision.

## 1.2   MECHANISMS

Mechanisms are sought to explain how a phenomenon comes about or how some significant process works. Specifically:

> Mechanisms are entities and activities organized such that they are productive of regular changes from start or set-up to finish or termination conditions.

For example, in the mechanism of chemical neurotransmission, a presynaptic neuron transmits a signal to a post-synaptic neuron by releasing neurotransmitter molecules that diffuse across the synaptic cleft, bind to receptors, and so depolarize the post-synaptic cell. In the mechanism of DNA replication, the DNA double helix unwinds, exposing slightly charged bases to which

complementary bases bond, producing, after several more stages, two dupli-
cate helices. Descriptions of mechanisms show how the termination conditions are produced by the set-up conditions and intermediate stages. To give a
description of a mechanism for a phenomenon is to explain that phenomenon,
that is, to explain how it was produced.

Mechanisms are composed of both *entities* (with their properties) and
*activities*. Activities are the producers of change. Entities are the things that
engage in activities. Activities usually require that entities have specific types
of properties. The neurotransmitter and receptor (two entities) bind (an activity) by virtue of their structural properties and charge distributions. A DNA
base and a complementary base hydrogen bond because of their geometric
structures and weak charges. The organization of these entities and activities determines the ways in which they produce the phenomenon. Entities
often must be appropriately located, structured, and oriented, and the activities in which they engage must have a temporal order, rate, and duration. For
example, two neurons must be spatially proximate for diffusion of the neurotransmitter. Mechanisms are regular in that they work always or for the most
part in the same way under the same conditions. The regularity is exhibited in
the typical way that the mechanism runs from beginning to end; what makes
it regular is the *productive continuity* between stages. Complete descriptions
of mechanisms exhibit productive continuity without gaps from the set-up
to termination conditions. Productive continuities are what make the connections between stages intelligible. If a mechanism is represented schematically
by A –> B –> C, then the continuity lies in the arrows and their explication is
in terms of the activities that the arrows represent. A missing arrow, namely,
the inability to specify an activity, leaves an explanatory gap in the productive
continuity of the mechanism.

We are not alone in thinking that the concept of "mechanism" is central to
an adequate philosophical understanding of the biological sciences. Others have argued for the importance of mechanisms in biology (Bechtel and
Richardson 1993; Brandon 1985; Kauffman 1971; Wimsatt 1972) and molecular biology in particular (Burian 1996a; Crick 1988). Wimsatt, for example,
says that, "At least in biology, most scientists see their work as explaining
types of phenomena by discovering mechanisms ... " (Wimsatt 1972, p. 67).
Schaffner often gestures to the importance of mechanisms in biology and
medicine, but argues, following Mackie (1974), that talk of causal mechanisms is dependent on prior and more fundamental talk of "laws of working" (Schaffner 1993, p. 287, pp. 306–307). Elsewhere Schaffner claims that
"mechanism," as used by Wimsatt and others, is an "unanalyzed term" that
he wishes to avoid (Schaffner 1993, p. 287).

When the notion of a "mechanism" has been analyzed, it has typically been analyzed in terms of the decomposition of "systems" into their "parts" and "interactions" (Wimsatt 1976; Bechtel and Richardson 1993). Following in this "interactionist" tradition, Glennan (1992, 1996) defines a mechanism as follows:

> A mechanism underlying a behavior is a complex system which produces that behavior by . . . the interaction of a number of parts according to direct causal laws. (Glennan 1996, p. 52)

He claims that all causal laws are explicated by providing a lower level mechanism until one bottoms out in the fundamental, noncausal laws of physics. We find Glennan's reliance on the concept of a "law" problematic because, in our examples, there are rarely "direct causal laws" to characterize how activities operate. More important, as we argue in Section 1.3, the interactionist's reliance on laws and interactions seems to us to leave out the productive nature of activities.

Our way of thinking emphasizes the *activities* in mechanisms. The term "activity" brings with it appropriate connotations from its standard usage; however, it is intended as a technical term. An activity is usually designated by a verb or verb form (e.g., participles, gerundives). Activities are the producers of change. They are constitutive of the transformations that yield new states of affairs or new products. Reference to activities is motivated by ontic, descriptive, and epistemological concerns. We justify this break from parsimony, this dualism of entities and activities, by reference to these philosophical needs.

### 1.3  ONTIC STATUS OF MECHANISMS (ONTIC ADEQUACY)

Both activities and entities must be included in an adequate *ontic* account of mechanisms. Our analysis of the concept of mechanism is explicitly dualist. We are attempting to capture the healthy philosophical intuitions underlying both substantivalist and process ontologies. Substantivalists confine their attention to entities and properties, believing that it is possible to reduce talk of activities to talk of properties and their transitions. Substantivalists thus speak of entities with capacities (Cartwright 1989) or dispositions to act. However, to identify a capacity of an entity, one must first identify the activities in which that entity engages. One does not know that aspirin has the capacity to relieve a headache unless one knows that aspirin produces headache relief. Substantivalists also talk about interactions of entities (Glennan 1996) or their state transitions. We think state transitions have to be more completely described in terms of the activities of the entities and how those activities produce changes that constitute the next stage. The same is true of talk of interactions, which emphasizes spatio-temporal intersections and changes in properties without characterizing the productivity by which those changes are effected at those intersections.

Substantivalists appropriately focus attention on the entities and properties in mechanisms, for example, the neurotransmitter, the receptor, and their charge configurations or DNA bases and their weak polarities. It is the entities that engage in activities, and they do so by virtue of certain of their properties. This is why statistical relevance relations (cf. Salmon 1984) between the properties of entities at one time and the properties of entities at another (or generalizations stating "input-output" relations and state changes) are useful for describing mechanisms. Yet it is artificial and impoverished to describe mechanisms solely in terms of entities, properties, interactions, inputs-outputs, and state changes over time. Mechanisms do things. They are active and so ought to be described in terms of the activities of their entities, not merely in terms of changes in their properties.

In contrast to substantivalists, process ontologists reify activities and attempt to reduce entities to processes (cf. Rescher 1996). While process ontology does acknowledge the importance of active processes by taking them as fundamental ontological units, its program for entity reduction is problematic at best. As far as we know, there are no activities in neurobiology and molecular biology that are not activities *of* entities. Nonetheless, the process ontologists appropriately highlight the importance of active kinds of changing. There are kinds of changing just as there are kinds of entities. These different kinds are recognized by science and are basic to the ways that things work.

Activities are identified and individuated in much the same way as are entities. Traditionally, one identifies and individuates entities in terms of their properties and spatio-temporal location. Activities, likewise, may be identified and individuated by their spatio-temporal location. They also may be individuated by their rate, duration, types of entities, and types of properties that engage in them. More specific individuation conditions may include their mode of operation (e.g., contact action versus attraction at a distance), directionality (e.g., linear versus at right angles), polarity (e.g., attraction versus attraction and repulsion), energy requirements (e.g., how much energy is required to form or break a chemical bond), and the range of activity (e.g., electromagnetic forces have a wider influence than do the strong and weak forces in the nucleus). Often, generalizations or laws are statements whose predicates refer to the entities and properties that are important for the individuation of activities. Mechanisms are identified and individuated by the activities and entities that constitute them, by their start and finish conditions, and by their functional roles.

Functions are the roles played by entities and activities in a mechanism. To see an activity as a function is to see it as a component in some mechanism, that is, to see it in a context that is taken to be important, vital, or otherwise significant. It is common to speak of functions as properties "had by" entities, as when one says that the heart "has" the function of pumping blood or the channel "has" the function of gating the flow of sodium. This way of speaking reinforces the substantivalist tendency against which we have been arguing. Functions, rather, should be understood in terms of the activities by virtue of which entities contribute to the workings of a mechanism. It is more appropriate to say that the function of the heart is to pump blood and thereby deliver (with the aid of the rest of the circulatory system) oxygen and nutrients to the rest of the body. Likewise, a function of sodium channels is to gate sodium current in the production of action potentials. To the extent that the activity of a mechanism as a whole contributes to something in a context that is taken to be antecedently important, vital, or otherwise significant, that activity too can be thought of as the (or a) function of the mechanism as a whole (Craver 1998, 2001).

Entities and a specific subset of their properties determine the activities in which they are able to engage. Conversely, activities determine what types of entities (and what properties of those entities) are capable of being the basis for such acts. Put another way, entities having certain kinds of properties are necessary for the possibility of acting in certain specific ways, and certain kinds of activities are only possible when there are entities having certain kinds of properties. Entities and activities are correlatives. They are interdependent. An ontically adequate description of a mechanism includes both.

### 1.3.1  Activities and Causing

Activities are types of causes. Terms like "cause" and "interact" are abstract terms that need to be specified with a type of activity and are often so specified in typical scientific discourse. Anscombe ([1971], 1981, p. 137) noted that the word "cause" itself is highly general and only becomes meaningful when filled out by other more specific causal verbs, for example, scrape, push, dry, carry, eat, burn, knock over. An entity acts as a cause when it engages in a productive activity. This means that objects *simpliciter,* or even natural kinds, may be said to be causes only in a derivative sense. It is not the penicillin that causes the pneumonia to disappear but what the penicillin does.

Mackie's (1974) attempt to analyze the necessity of causality in terms of laws of working is similar to our analysis in many ways. He stresses that laws of working must be discovered empirically and are not found a priori (Mackie 1974, p. 213, p. 221). He also claims that counterfactuals are supported by the inductive evidence that such basic processes are at work (Mackie 1974, p. 229). However, he wants to analyze causality in terms of qualitative or structural continuity of processes (Mackie 1974, p. 224) and more vaguely in terms of "flowing from" or "extruding" (Mackie 1974, p. 226). It is unclear how to apply such concepts in our biological cases. But perhaps he is trying to use them to refer to what we call "activities" and to capture what we mean by "productivity."

Our emphasis on mechanisms is compatible, in some ways, with Salmon's mechanical philosophy, since mechanisms lie at the heart of the mechanical philosophy. Mechanisms, for Salmon, are composed of processes (things exhibiting consistency of characteristics over time) and interactions (spatio-temporal intersections involving persistent changes in those processes). It is appropriate to compare our talk of activities with Salmon's talk of interactions. Salmon identifies interactions in terms of transmitted marks and statistical relevance relations (Salmon 1984) and, more recently, in terms of exchanges of conserved quantities (Salmon 1997, 1998). Although we acknowledge the possibility that Salmon's analysis may be all there is to certain fundamental types of interactions in physics, his analysis is silent as to the character of the productivity in the activities investigated by many other sciences. Mere talk of transmission of a mark or exchange of a conserved quantity does not exhaust what these scientists know about productive activities and about how activities effect regular changes in mechanisms. As our examples will show, much of what neurobiologists and molecular biologists do should be seen as an effort to understand these diverse kinds of production and the ways that they work.

## 1.3.2   Activities and Laws

The traditional notion of a universal law of nature has few, if any, applications in neurobiology or molecular biology. Sometimes the regularities of activities can be described by laws. Sometimes they cannot. For example, Ohm's law is used to describe aspects of the activities in the mechanisms of neurotransmission. There is no law that describes the regularities of protein binding to regions of DNA. Nonetheless, the notion of activity carries with it some of the characteristic features associated with laws. Laws are taken to be determinate regularities. They describe something that acts in the same way under the same conditions, that is, same cause, same effect. (Schaffner 1993, p. 122, calls these "universal generalizations$_2$.") This is the same way we talk about mechanisms and their activities. A mechanism is the series of activities of entities that bring about the finish or termination conditions in a regular way. These regularities are nonaccidental and support counterfactuals to the extent that they describe activities. For example, if this single base in DNA were changed and the protein synthesis mechanism operated as usual, then the protein produced would have an active site that binds more tightly. This counterfactual justifies talking about mechanisms and their activities with some sort of necessity. No philosophical work is done by positing some further thing, a law, that underwrites the productivity of activities.

In sum, we are dualists: Both entities and activities constitute mechanisms. There are no activities without entities, and entities do not do anything without activities. We have argued for the ontic adequacy of this dualism by showing that it can capture insights of both substantivalists and process ontologists, by showing how activities are needed to specify the term "cause," and by an analysis of activities showing their regularity and necessity sometimes characterized by laws.

### 1.4   EXAMPLE OF A MECHANISM (DESCRIPTIVE ADEQUACY)

Consider the classic textbook account of the mechanisms of chemical transmission at synapses (Shepherd 1994). Chemical transmission can be understood abstractly as the activity of converting an electrical signal in one neuron, the relevant entity, into a chemical signal in the synapse. This chemical signal is then converted to an electrical signal in a second neuron. Consider the diagram in Figure 1.1.

The diagram is a two-dimensional spatial representation of the entities, properties, and activities that constitute these mechanisms. Mechanisms are often represented this way. Such diagrams exhibit spatial relations and structural features of the entities in the mechanism. Labeled arrows often represent the activities that produce changes. In these ways, diagrams represent features of mechanisms that could be described verbally but are more easily apprehended in visual form.

In the synaptic diagram, the entities are almost exclusively represented pictorially. These include the cell membrane, vesicles, microtubules, molecules, and ions. The activities are represented with labeled arrows. These include biosynthesis, transport, depolarization, insertion, storage, recycling, priming, diffusion, and modulation. The diagram is complicated in its attempt to

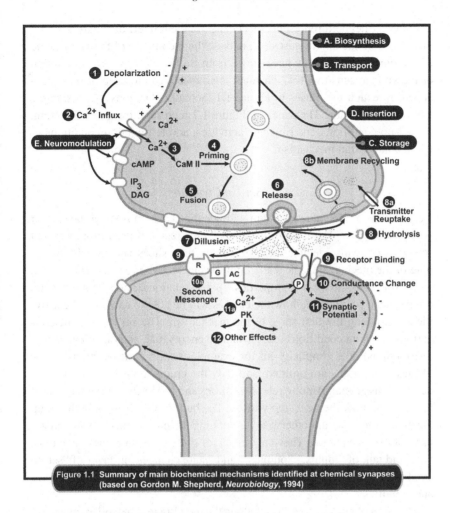

Figure 1.1 Summary of main biochemical mechanisms identified at chemical synapses (based on Gordon M. Shepherd, *Neurobiology*, 1994)

represent the many different mechanisms that can be found at chemical synapses. We use the first stage of this mechanism, *depolarization,* to exhibit the features of mechanisms in detail.

Neurons are electrically polarized in their resting state (i.e., their resting membrane potential, roughly –70 mV); the fluid inside the cell membrane is negatively charged with respect to the fluid outside of the cell. Depolarization is a positive change in the membrane potential. Neurons depolarize when sodium ($Na^+$) selective channels in the membrane open, allowing $Na^+$ to move into the cell by diffusion and electrical attraction. The resulting changes in ion distribution make the intracellular fluid progressively less negative and, eventually, more positive than the extracellular fluid (peaking at roughly

+50 mV). Shepherd represents this change in the top left of Figure 1.1 with pluses (+) inside and minuses (−) outside the membrane of the presynaptic cell. Figure 1.2, which we have drawn from Hall's (1992) verbal description of the voltage sensitive $Na^+$ channel, is an idealized close up of the mechanism by which the pluses in Figure 1.1 (actually $Na^+$ ions) get inside the neuronal membrane. The panels in Figure 1.2 represent, from top to bottom, the set-up conditions, intermediate activities, and termination conditions of the depolarization mechanism.

### 1.4.1  Set-Up Conditions

Descriptions of mechanisms begin with idealized descriptions of the start or set-up conditions. These conditions may be the result of prior processes, but scientists typically idealize them into static time slices taken as the beginning of the mechanism. The start conditions include the relevant entities and their properties. Structural properties, spatial relations, and orientations are often crucial for showing how the entities will be able to carry out the activities comprising the first stage of the mechanism. The set-up also includes various enabling conditions (e.g., available energy, pH, and electrical charge distributions). For simplicity in, for example, textbook descriptions, many of these conditions are omitted, and only the crucial entities and structural descriptions appear. Among relevant entities and properties, some are crucial for showing how the next step will go. The bulk of the features in the setup (spatial, structural, and otherwise) are not inputs into the mechanism but are parts of the mechanism. They are crucial for showing what comes next; thus, we avoid talk of "inputs," "outputs," and "state changes" in favor of "set-up conditions," "termination conditions," and "intermediate stages" of entities and activities.

The lines of pluses and minuses along the membrane at the top of Figure 1.1 represent the spreading depolarization of the axon, a crucial set-up condition for the depolarization of the axon terminal. This set-up condition is labeled in the top panel of Figure 1.2.

Also crucial are the locations, orientations, and charge distributions of the components of the $Na^+$ channel and the differential intra- and extra-cellular concentrations of $Na^+$. Two structural features of the $Na^+$ channel are crucial; each is depicted in the top panel of Figure 1.2. The first is the corkscrew-shaped portion of the protein (an alpha helix) known as the "voltage gate." It contains evenly spaced, positively charged amino acids. The second is a hairpin turn in the protein, known as the "pore lining," that has its own

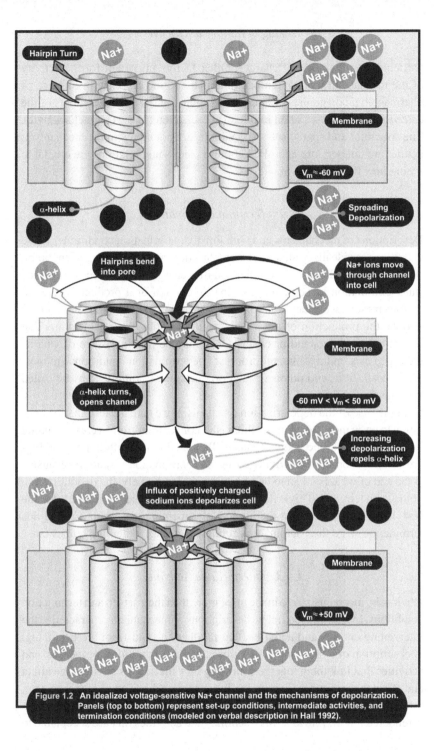

Figure 1.2 An idealized voltage-sensitive Na+ channel and the mechanisms of depolarization. Panels (top to bottom) represent set-up conditions, intermediate activities, and termination conditions (modeled on verbal description in Hall 1992).

particular configuration of charges. Other factors important for the activity of the mechanism include temperature, pH, and the presence or absence of pharmacological agonists or antagonists; such factors are the contents of the *ceteris paribus* clauses often implicit in descriptions of the channel's activity. The structural and spatial set-up conditions are not inputs to the mechanism; neither are temperature and pH. Yet these factors and relations are crucial to seeing how the mechanism will go.

### 1.4.2 Termination Conditions

Descriptions of mechanisms end with finish or termination conditions. These conditions are idealized states or parameters describing a privileged endpoint, such as rest, equilibrium, neutralization of a charge, a repressed or activated state, elimination of something, or the production of a product. There are various reasons why such states are privileged. For example, the end product may be the production of a particular kind of entity or state of affairs that we set out to understand or create. Or, it may be the final stage of what is identified as a unitary, integral process. The termination conditions are most often idealized as end points or final products; misleadingly, they are called "outputs."

In the case of the depolarization mechanism, we take the termination condition to be an increase in intracellular $Na^+$ concentration and a corresponding increase in membrane voltage. This is illustrated in the bottom panel of Figure 1.2. This condition is privileged and so a termination condition because it is the end of what is taken to be a unitary process, namely, the depolarization of the axon terminal. This is illustrated in the bottom panel of Figure 1.2 as the $Na^+$ ions lining up against the intracellular membrane surface. Calling this termination stage the "output" inaccurately suggests something comes out.

### 1.4.3 Intermediate Activities

Obviously, mechanisms are made up of more than their set-up and termination conditions. In addition, complete descriptions of mechanisms characterize the intervening entities and activities that produce the end from the beginning. A description of a mechanism describes the relevant entities, properties, and activities that link them together, showing how the actions at one stage affect and effect those at successive stages. In a complete description of mechanism, there are no gaps that leave specific steps unintelligible; the process as a whole is rendered intelligible in terms of entities and activities that are acceptable to a field at a time. In the simplest case, the stages of a mechanism are organized linearly, but they also may be forks, joins, or cycles. Often, mechanisms are continuous processes that may be treated for convenience as a series of discrete stages or steps.

Look again at the depolarization example. The activities by which the cell will depolarize are presaged in the set-up conditions. These intermediate activities are presented in the central panel of Figure 1.2. The spreading depolarization from the axonal action potential (1) repels the positive charges in the alpha helix voltage gates, and (2) rotates them about their central axis and opens a pore or channel through the membrane. The resulting conformation change in (or bending of) the protein (3) moves the extra-cellular hairpins into the pore. The particular configuration of charges on this pore lining makes the channel selective for $Na^+$. As a result, (4) $Na^+$ ions move through the pore and into the cell. This increase in $Na^+$ concentration depolarizes the axon terminal (see the final panel, Figure 1.2). Although we may describe or represent these intermediate activities as stages in the operation of the mechanism, they are more accurately viewed as continuous processes. As the axonal depolarization spreads, the repulsive forces acting on the positive charges in the corkscrew are increasingly pushed outward, rotating the helix and opening the $Na^+$ selective channel pore.

### 1.5   HIERARCHIES, BOTTOMING OUT, MECHANISM SCHEMATA AND SKETCHES

Mechanisms occur in nested hierarchies and the descriptions of mechanisms in neurobiology and molecular biology are frequently multilevel. The levels in these hierarchies should be thought of as part-whole hierarchies with the additional restriction that lower level entities, properties, and activities are components in mechanisms that produce higher level phenomena (Craver 1998; Craver and Darden 2001; see Chapter 2, this book). For example, the activation of the sodium channel is a component of the mechanism of depolarization, which is a component of the mechanism of chemical neurotransmission, which is a component of most higher level mechanisms in the central nervous system. Similar hierarchies can be found in molecular biology. James Watson (1965) discussed mechanisms for forming strong and weak chemical bonds, which are components of the mechanisms of replication, transcription, and translation of DNA and RNA, respectively, which are components of the mechanisms of numerous cell activities.

### 1.5.1  Bottoming Out

Nested hierarchical descriptions of mechanisms typically *bottom out* in lowest level mechanisms. These are the components that are accepted as relatively fundamental or taken to be unproblematic for the purposes of a given scientist, research group, or field. Bottoming out is relative: Different types of entities and activities are where a given field stops when constructing mechanisms. The explanation comes to an end and description of lower level mechanisms would be irrelevant to their interests. Also, scientific training is often concentrated at or around certain levels of mechanisms. Neurobiologists with different theoretical or experimental interests bottom out in different types of entities and activities. Some neurobiologists are primarily interested in behaviors of organisms, some are primarily interested in the activities of molecules composing nerves cells, and others devote their attention to phenomena in between. The fields of molecular biology and neurobiology do not typically regress to the quantum level to talk about the activities of, for example, chemical bonding. Rarely are biologists driven by anomalies or any other reason to go to such lower levels, although some problem might require it. Levels below molecules and chemical bonding are not fundamental for the fields of molecular biology and molecular neurobiology. But remember, what is considered the bottom-out level may change.

In molecular biology and molecular neurobiology, hierarchies of mechanisms bottom out in descriptions of the activities of macromolecules, smaller molecules, and ions. These are commonly recognized as bottom-out entities; we believe that we have identified the most important types of bottom-out activities. These bottom-out activities in molecular biology and molecular neurobiology can be categorized into the following four types:

   (i) geometrico-mechanical
  (ii) electro-chemical
 (iii) energetic
 (iv) electromagnetic

   (i) Geometrico-mechanical activities are those familiar from the seventeenth-century mechanical philosophy. They include fitting, turning, opening, colliding, bending, and pushing. The rotation of the alpha helix in the sodium channel and the geometrical fitting of a neurotransmitter and a post-synaptic receptor are examples of geometrico-mechanical activities.

   (ii) Attracting, repelling, bonding, and breaking are electro-chemical kinds of activity. Chemical bonding, such as the formation of strong covalent bonds between amino acids in proteins, is a more specific example. The lock-and-key docking of an enzyme and its substrate involves geometrical shape and mechanical stresses and chemical attractions. As we will see, the historical development of the mechanism of protein synthesis required finding an activity to order linearly the constituents of the protein, its amino acids; an early idea using primarily geometrico-mechanical activities was replaced by one involving, primarily, the weak electro-chemical activities of hydrogen bonding.

(iii) Energetic activities have thermodynamics as their source. A kind of energetic activity involves simple diffusion of a substance as, for example, when concentrations on different sides of a membrane lead to movement of substances across the membrane.

(iv) Electromagnetic activities are occasionally used to bottom out mechanisms in these sciences. The conduction of electrical impulses by nerve cells and the navigational mechanisms of certain marine species are examples.

## 1.5.2 An Historical Aside

These categories of relatively fundamental activities suggest an historical strategy for examining the history of mechanisms. The discovery and individuation of different entities and activities are important parts of scientific practice. In fact, much of the history of science has been well written, albeit unwittingly, by tracing the discoveries of new entities and activities that mark the changes in a discipline.

The modern idea of explaining with mechanisms became current in the seventeenth century when Galileo articulated a geometrico-mechanical form of explanation based on Archimedes's simple machines (Machamer 1998). Soon an expanded version of this geometrico-mechanical way of describing and thinking about the world became widespread across Europe (and the New World) and was called the "mechanical philosophy."

In the eighteenth and nineteenth centuries, chemists and electricians began to discover and describe other entities and activities that they took as fundamental to the structure of the world, and so expanded the concept of what could occur in mechanisms. The nineteenth century also saw an emerging emphasis on the concept of energy and electromagnetism. These different kinds of forces acting were new and different kinds of activities.

In every case, scientists were compelled to add new entities and new forms of activity in order to better explain how the world works. To do this, they would postulate an entity or activity, present criteria for its identification and recognition, and display the patterns by which these formed a unity that constituted a mechanism. These became the new laws or ways of working of the various sciences. Documenting such new entities and activities allows us to map out the changes that become the substance of the history of science.

This pastiche of history is a quick and simplistic way to show that the discovery of different kinds of mechanisms with their kinds of entities and different activities is an important part of scientific development. Contemporary sciences such as neurobiology and molecular biology are in this tradition and draw on the entities and activities made available through some of these historical discoveries.

The history of these changes implies that what count as acceptable types of entities, activities, and mechanisms change with time. At different historical moments, in different fields, different mechanisms, entities, and activities have been discovered and accepted. The set of types of entities and activities so far discovered likely is not complete. Further developments in science will lead to the discovery of additional ones.

### 1.5.3    Mechanism Schemata and Sketches

Scientists do not always provide complete descriptions of mechanisms at all levels in a nested hierarchy. Also, they are typically interested in types of mechanisms, not all the details needed to describe a specific instance of a mechanism. We introduce the term "mechanism schema" for an abstract description of a type of mechanism. A mechanism schema is a truncated abstract description of a mechanism that can be filled with descriptions of known component parts and activities. An example is represented in Watson's (1965) diagram of the central dogma of molecular biology (Figure 1.3).

Schemata exhibit varying degrees of abstraction, depending on how much detail is included. Abstractions may be constructed by taking an exemplary case or instance and removing detail. For example, a constant can be made into a variable (Darden 1995; see Chapter 10, this book). A particular DNA sequence may be abstracted to any DNA sequence. Often, scientists use schema terms, such as "transcription" and "translation," to capture compactly many aspects of the underlying mechanism. These may be characterized as activities in higher level mechanisms.

Degrees of abstraction should not be confused with degrees of generality or scope (Darden 1996). Abstraction is an issue of the amount of detail included in the description of one or more mechanism instances. The generality of a schema is the scope (small or large) of the domain in which it can be instantiated. One can describe a single instance of a mechanism more or less abstractly. Alternatively, the schema, at whatever degree of abstraction, may have a quite general scope. The schema for the central dogma is nearly

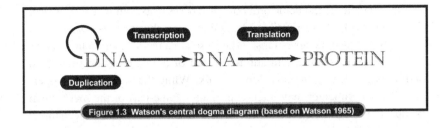

Figure 1.3 Watson's central dogma diagram (based on Watson 1965)

terrestrially universal, holding for most instances of protein synthesis in most species. However, the schema for protein synthesis in some RNA viruses is just

RNA –> protein

In other RNA retroviruses, it is

RNA –> DNA –> RNA –> protein

These schemata are just as abstract as Watson's schema of the central dogma (Figure 1.3) but they are much more limited in scope.

Neurobiologists and molecular biologists sometimes use the term "theory" to refer to hierarchically organized mechanism schemata of variable, though generally less than universal, scope. Mechanism schemata, as well as descriptions of particular mechanisms, play many of the roles attributed to theories. They are constructed, evaluated, and revised in cycles as science proceeds. They are used to describe, predict, and explain phenomena; to design experiments; and to interpret experimental results.

Thinking about mechanisms as composed of entities and activities provides resources for thinking about strategies for scientific change. Known types of entities and activities in a field provide the intelligible building blocks from which to construct hypothesized mechanism schemata. If one knows what kind of activity is needed to do something, then one seeks kinds of entities that can do it, and vice versa. Scientists in the field often recognize whether there are known types of entities and activities that can possibly accomplish the hypothesized changes and whether there is empirical evidence that a possible schemata is plausible.

When instantiated, mechanism schemata yield mechanistic explanations of the phenomenon that the mechanism produces. For example, the schema for the $Na^+$ channel depicted in Figure 1.2, when instantiated, can be used to explain the depolarization of a specific nerve cell. Mechanism schemata can also be specified to yield predictions. For example, the order of the amino acids in a protein can be predicted from specification of the central dogma schema that includes a specific order of DNA bases in its coding region. Third, schemata provide "blueprints" for designing research protocols (Darden and Cook 1994). A technician can instantiate a schema in an experiment by actually choosing physical instantiations of each of the entities and the set-up conditions and letting the mechanism work. While the mechanism is operating, the experimenter may intervene to alter some part of the mechanism and observe the changes in a termination condition or what the mechanism does. Changes produced by such interventions can provide evidence for the hypothesized schema (Craver and Darden 2001; see Chapter 2, this book).

When a prediction made on the basis of a hypothesized mechanism fails, then one has an anomaly, and a number of responses are possible. If the experiment was conducted properly and the anomaly is reproducible, then perhaps something other than the hypothesized mechanism schema is at fault, such as hypotheses about the set-up conditions. If the anomaly cannot be resolved otherwise, then the hypothesized schema may need to be revised. One might abandon the entire mechanism schema and propose a new one. Alternatively, one can revise a portion of the failed schema. Reasoning in the light of failed predictions involves, first, a diagnostic process to isolate where the mechanism schema is failing, and, second, a redesign process to change one or more entities or activities or stages to improve the hypothesized schema (Darden 1991, 1995; see Chapter 10, this book).

Mechanism schemata can be instantiated in biological wet-ware (as in the experimental case discussed above) or represented in the hardware of a machine. For example, a computational biologist can write an algorithm that depicts the relations among the order of DNA bases, RNA bases, and amino acids in proteins. This algorithm represents the mechanism schema of the central dogma. Yet the algorithm itself becomes an actual mechanism of a very different kind when written in a programming language and instantiated in hardware that can run it as a simulation.

For epistemic purposes, a *mechanism sketch* may be contrasted with a schema. A sketch is an abstraction for which bottom-out entities and activities cannot (yet) be supplied or which contains gaps in its stages. The productive continuity from one stage to the next has missing pieces, black boxes, which we do not yet know how to fill in. A sketch thus serves to indicate what further work needs to be done in order to have a mechanism schema. Sometimes a sketch has to be abandoned in the light of new findings. In other cases, it may become a schema, serving as an abstraction that can be instantiated as needed for the tasks mentioned above (e.g., explanation, prediction, and experimental design).

## 1.6   CASE STUDY: DISCOVERING THE MECHANISM
## OF PROTEIN SYNTHESIS

The discovery of the mechanism of protein synthesis illustrates piecemeal discovery of a mechanism schema, with different components discovered by different fields. It also emphasizes the importance of finding the activities, as well as the entities, during mechanism discovery.

Prior to the discovery of messenger RNA (mRNA), biochemists and molecular biologists proposed mechanisms for protein synthesis focusing on different entities and activities. The contrasting mechanism schemata are vividly illustrated in two diagrams (Figure 1.4): one by Zamecnik, a biochemist, and the other by Watson, a molecular biologist. Zamecnik's 1953 diagram focuses on energy production (formation of ATP) and the activation of amino acids prior to their incorporation into the protein's polypeptide chain. It

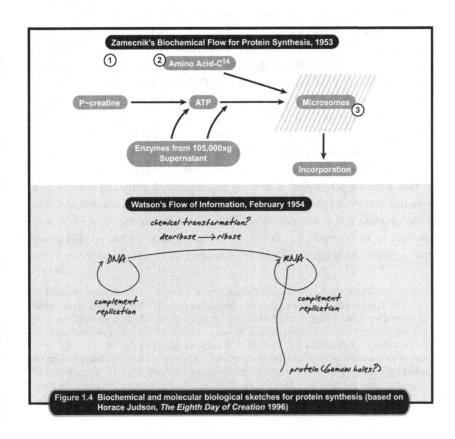

Figure 1.4  Biochemical and molecular biological sketches for protein synthesis (based on Horace Judson, *The Eighth Day of Creation* 1996)

31

depicts the microsomes (labeled 3 in the diagram) as the site of protein synthesis. (Microsomes were later shown to be ribosomes associated with other cellular components; see Zamecnik 1969; discussed in Rheinberger 1997.) This diagram clearly lacks any step for ordering the amino acids as they are incorporated into the protein. Although the nucleic acid RNA was known to be part of the microsomes, Zamecnik does not explicitly represent any nucleic acids as component entities of the mechanism. The biochemist's diagram is therefore an incomplete sketch; it lacks crucial entities and, more important, any reference to activities capable of ordering the amino acids.[2]

Watson's 1954 diagram (Figure 1.4) exhibits the molecular biological focus on the activities of the nucleic acids, DNA and RNA. It depicts an early geometrico-mechanical schema for determining the order of the amino acids. George Gamow (1954), a physicist, had proposed that proteins were synthesized directly on the DNA double helix by fitting into "holes" in the helix (more technically, the major and minor grooves of the helix). Watson was aware of biochemical evidence that proteins do not form directly on DNA but instead are associated with RNA. Modifying Gamow's idea in light of this evidence, Watson proposed that RNA had "Gamow holes" whose shapes were determined by the surrounding bases. Different amino acids would then fit into different holes. The ordering of the RNA bases determined the shape of the sequential holes and, therefore, the ordering of the amino acids (via a geometrico-mechanical activity). After amino acids fell into the holes, adjacent amino acids would covalently bond (an electro-chemical activity) to one another, forming the protein (discussed in Watson [1962] 1977).

This geometrical "holes" schema was plausible: It provided entities and activities that could produce the end product (i.e., the ordered amino acids in the protein), and it was consistent with available evidence that RNA was involved in the mechanism. However, evidence soon disproved this plausible schema. Although the DNA-base sequences in different species were

---

[2] In a letter of December 8, 1999 to LD, Zamecnik recalls that they were aware of the need to include a role for DNA, beginning in 1944, because of Avery's work. Sanger's presentation in 1949 at the Cold Spring Harbor Symposium at which Zamecnik spoke showed that protein sequences did not have simple repeats. Watson and Zamecnik were discussing connections between their work, beginning with a visit in 1954 and subsequent contacts. The role of RNA was also being considered as an intermediary because of work by others. Zamecnik concludes with an apt metaphor showing how the two lines of investigation were joined: "As in building a transcontinental railroad, one team starts from San Francisco and the other from the mid-continent. They are both conscious of the way the compass is pointing, if they are to meet somewhere in the middle."

very different, the base sequences of ribosomal RNA (where most RNA was concentrated) were very similar across species (Belozersky and Spirin 1958; discussed in Crick 1959). If ribosomes were similar from species to species, then it was unlikely that they had sufficiently differently shaped holes to produce the different orderings of amino acids in different proteins.

Thus, both the biochemical and molecular biological schemata proved problematic. Although the biochemical schema clearly indicated the source of energy for the formation of covalent bonds (ATP) and identified microsomes as the site of protein synthesis, it had no activity to order the amino acids. The hypothesized molecular biological mechanism proved to be wrong because the ordering of amino acids is not accomplished by geometrically arranging them in holes in RNA. Additional theoretical and empirical work was required to discover the additional entities and activities necessary for protein synthesis. These include transfer RNAs (Crick 1958), which deliver each of the twenty amino acids to the ribosome, and messenger RNA. Messenger RNA is the linear copy of DNA that provides the ordering of the amino acids via the activity of hydrogen bonding between its bases and the complementary ones in the transfer RNAs. The ribosome turned out to be the nonspecific site where mRNA and transfer RNAs come together to properly orient the amino acids in space for covalent bonding in the proper order. (For more on the discovery of transfer and messenger RNA, see Judson 1996; Morange 1998; Olby 1970; Rheinberger 1997). The discovery of the mechanism of protein synthesis required entities and activities from both fields to correct and elaborate hypotheses about the RNA stage of the mechanism and to find the appropriate activity, hydrogen bonding, for ordering amino acids during protein synthesis.

The theories in the field of molecular biology can be viewed as sets of mechanism schemata. The primary ones are DNA replication, the mechanism of protein synthesis, and the many mechanisms of gene regulation. A complete history of their development would emphasize the importance of the discovery of weak chemical bonding by Linus Pauling and the critical role of this activity in these discoveries by Francis Crick (1988, 1996) and others. Thus, descriptively adequate historical accounts need to discuss the discovery of new kinds of activities, such as hydrogen bonding, as well as the discovery of new entities (which is where the focus usually lies). This example also illustrates how thinking about a kind of activity can guide the construction of a mechanism, when Crick reasoned that nucleic acid bases were particularly suited to hydrogen bonding and used that activity to postulate transfer RNAs and their action. Further, the example shows how incomplete sketches point to black boxes that need to be filled and how incorrect schemata can be changed by substituting another kind of activity. Explicit knowledge of kinds of activities is thus crucial when resolving anomalies and constructing new mechanisms.

## 1.7 ACTIVITIES, INTELLIGIBILITY, AND EXPLANATION (EPISTEMIC ADEQUACY)

Yet another justification (our third, along with the ontic and descriptive) for thinking about mechanisms in terms of activities and entities is epistemic: As we have illustrated, both are integral to giving mechanistic explanations. The contemporary mechanical worldview, among other things, is a conviction about how phenomena are to be understood. Activities are essential for rendering phenomena intelligible (Machamer 2000). The intelligibility consists in the mechanisms being portrayed in terms of a field's bottom-out entities and activities.

Let us briefly, and incompletely, sketch some of the implications of this claim. The understanding provided by a mechanistic explanation may be correct or incorrect. Either way, the explanation renders a phenomenon intelligible. Mechanism descriptions show *how possibly*, *how plausibly*, or *how actually* things work. Intelligibility arises not from an explanation's correctness but rather from an elucidative relation between the explanans (i.e., the set-up conditions and intermediate entities and activities) and the explanandum (i.e., the termination condition or the phenomenon to be explained). Protein synthesis can be elucidated by reference to Gamow holes. The ability of nerves to conduct signals can be rendered intelligible by reference to their internal vibrations. Neither of these explanations is correct, yet each provides intelligibility by showing how the phenomena might possibly be produced.

We should not be tempted to follow Hume and later logical empiricists into thinking that the intelligibility of activities (or mechanisms) is reducible to their regularity. Descriptions of mechanisms render the end stage intelligible by showing how it is produced by bottom-out entities and activities. To explain is not merely to redescribe one regularity as a series of several. Rather, explanation involves revealing the *productive* relation. It is the unwinding, bonding, and breaking that explain protein synthesis; it is the binding, bending, and opening that explain the activity of $Na^+$ channels. It is not the regularities that explain but the activities that sustain the regularities.

This discussion brings us back to our four bottom-out kinds of activities: geometrico-mechanical, electro-chemical, electromagnetic, and energetic.

34

These bottom-out activities are quite general kinds of abstract means of production that can fruitfully be applied in particular cases to explain phenomena. (For a discussion of how this works in the case of balancing, a geometrico-mechanical kind of activity, see Machamer and Woody 1994.) Mechanistic explanation in neurobiology and molecular biology involves showing or demonstrating that the phenomenon to be explained is a product of one or more of these abstract and recurring types of activity or the result of higher level, productive activities.

There is no logical story to be told about how these bottom-out activities, these kinds of production, come to inhabit a privileged explanatory position. What is taken to be intelligible (and the different ways of making things intelligible) changes over time as different fields within science bottom out their descriptions of mechanisms in different entities and activities that are taken as, or have come to be, unproblematic. This suggests quite plausibly that intelligibility is historically constituted and disciplinarily relative (which is nonetheless consistent with there being universal general characteristics of intelligibility).

We also believe it to be likely, although we cannot argue for it here, that what we take to be intelligible is a product of the ontogenic and phylogenetic development of human beings in a world such as ours. Briefly, sight is an important source for what we take to be intelligible; we directly see many activities, such as movement and collision (Cutting 1986; Schaffner 1993). But seeing is not our only means of access to activities. Importantly, our kinesthetic and proprioceptive senses also provide us with experience of activities (e.g., pushing, pulling, and rotating). Emotional experiences also are likely experiential grounds of intelligibility for activities of attraction, repulsion, hydrophobicity, and hydrophilicity. These activities give meanings that are then extended to areas beyond primitive sense perception. The use of basic perceptual verbs, such as "see" or "show," are extended to wider forms of intelligibility, such as proof or demonstration.

Intelligibility, at least in molecular biology and neurobiology, is provided by descriptions of mechanisms – that is, through the elaboration of constituent entities and activities that, by an extension of sensory experience with ways of working, provide an understanding of how some phenomenon is produced.

## 1.8 REDUCTION

Philosophical discussions of reduction have attempted to shed light on issues in ontology, scientific change, and explanation. Because we have introduced the notion of relative bottoming out, we do not address issues about ultimate ontology. Instead, our focus, vis-à-vis reduction, is on scientific change and explanation.

Models of reduction, including deductive models (e.g., Nagel 1961; Schaffner 1993), have been claimed to be ways to characterize scientific change and scientific explanation. These models do not fit neuroscience and molecular biology. Instead, we suggest the language of mechanisms.

Theory change in neuroscience and molecular biology is most accurately characterized in terms of the gradual and piecemeal construction, evaluation, and revision of multilevel mechanism schemata (Craver 1998; Craver and Darden 2001; see Chapter 2, this book). Elimination or replacement should be understood in terms of the reconceptualization or abandonment of the phenomenon to be explained, of a proposed mechanism schema, or of its purported components. This contrasts with the static two-place relations between different theories (or levels) and with the case of logical deduction.

Deductive models have also been taken to provide an analysis of explanation, with lower levels explaining higher levels through the identification of terms and the derivation of the higher level laws from the lower level (for the details, see Schaffner 1993). Aside from the fact that identification and derivation are peripheral to the examples we have discussed (as Schaffner admits), this model cannot accommodate the prevalent multilevel character of explanations in our sciences. In these cases, entities and activities at multiple levels are required to make the explanation intelligible. The entities and activities in the mechanism must be understood in their important, vital, or otherwise significant context, and this requires an understanding of the working of the mechanism at multiple levels. The activity of the $Na^+$ channel cannot be properly understood in isolation from its role in the generation of action potentials, the release of neurotransmitters, and the transmission of signals from neuron to neuron. Higher level entities and activities are thus essential to the intelligibility of those at lower levels, just as much as those at lower levels are essential for understanding those at higher levels. It is the integration of different levels into productive relations that renders the phenomenon intelligible and thereby explains it.

## 1.9   CONCLUSION

Thinking about mechanisms gives a better way to think about one's ontic commitments. Thinking about mechanisms offers an interesting and good way to look at the history of science. Thinking about mechanisms provides a descriptively adequate way of talking about science and scientific discovery. Thinking about mechanisms presages new ways to handle some important philosophical concepts and problems. In fact, if one does not think about mechanisms, one cannot understand neurobiology and molecular biology.

### REFERENCES

Anscombe, Gertrude and Elizabeth Margaret ([1971] 1981), "Causality and Determination," in *Metaphysics and the Philosophy of Mind, The Collected Philosophical Papers of G. E. M. Anscombe*, v. 2. Minneapolis, MN: University of Minnesota Press, pp. 133–147.

Bechtel, William and Robert C. Richardson (1993), *Discovering Complexity: Decomposition and Localization as Strategies in Scientific Research*. Princeton, NJ: Princeton University Press.

Belozersky, Andrei N. and Alexander S. Spirin (1958), "A Correlation between the Compositions of Deoxyribonucleic and Ribonucleic Acids," *Nature* 182: 111–112.

Brandon, Robert (1985), "Grene on Mechanism and Reductionism: More Than Just a Side Issue," in Peter Asquith and Philip Kitcher (eds.), *PSA 1984*, v. 2. East Lansing, MI: Philosophy of Science Association, pp. 345–353.

Burian, Richard M. (1996a), "Underappreciated Pathways Toward Molecular Genetics as Illustrated by Jean Brachet's Cytochemical Embryology," in Sahotra Sarkar (ed.), *The Philosophy and History of Molecular Biology: New Perspectives*. Dordrecht: Kluwer, pp. 67–85.

Cartwright, Nancy (1989), *Nature's Capacities and Their Measurement*. Oxford: Oxford University Press.

Craver, Carl F. (1998), *Neural Mechanisms: On the Structure, Function, and Development of Theories in Neurobiology*. Ph.D. Dissertation. Pittsburgh, PA: University of Pittsburgh.

Craver, Carl F. (2001), "Role Functions, Mechanisms, and Hierarchy," *Philosophy of Science* 68: 53–74.

Craver, Carl F. and Lindley Darden (2001), "Discovering Mechanisms in Neurobiology: The Case of Spatial Memory," in Peter Machamer, R. Grush, and P. McLaughlin (eds.), *Theory and Method in the Neurosciences*. Pittsburgh, PA: University of Pittsburgh Press, pp. 112–137.

Crick, Francis (1958), "On Protein Synthesis," *Symposium of the Society of Experimental Biology* 12: 138–167.

Crick, Francis (1959), "The Present Position of the Coding Problem," *Structure and Function of Genetic Elements: Brookhaven Symposia in Biology* 12: 35–39.

Crick, Francis (1988), *What Mad Pursuit: A Personal View of Scientific Discovery*. New York: Basic Books.

Crick, Francis (1996), "The Impact of Linus Pauling on Molecular Biology," in Ramesh S. Krishnamurthy (ed.), *The Pauling Symposium: A Discourse on the Art of Biography*. Corvallis, OR: Oregon State University Libraries Special Collections, pp. 3–18.

Cutting, James E. (1986), *Perception with an Eye for Motion*. Cambridge, MA: MIT Press.

Darden, Lindley (1991), *Theory Change in Science: Strategies from Mendelian Genetics*. New York: Oxford University Press.

Darden, Lindley (1995), "Exemplars, Abstractions, and Anomalies: Representations and Theory Change in Mendelian and Molecular Genetics," in James G. Lennox and Gereon Wolters (eds.), *Concepts, Theories, and Rationality in the Biological Sciences*. Pittsburgh, PA: University of Pittsburgh Press, pp. 137–158.

Darden, Lindley (1996), "Generalizations in Biology: Essay Review of K. Schaffner's *Discovery and Explanation in Biology and Medicine*," *Studies in History and Philosophy of Science* 27: 409–419.

Darden, Lindley and Michael Cook (1994), "Reasoning Strategies in Molecular Biology: Abstractions, Scans and Anomalies," in David Hull, Micky Forbes, and Richard M. Burian (eds.), *PSA 1994*, v. 2. East Lansing, MI: Philosophy of Science Association, pp. 179–191.

Gamow, George (1954), "Possible Relation between Deoxyribonucleic Acid and Protein Structures," *Nature* 173: 318.

Glennan, Stuart S. (1992), *Mechanisms, Models, and Causation*. Ph.D. Dissertation. Chicago, IL: University of Chicago.

Glennan, Stuart S. (1996), "Mechanisms and The Nature of Causation," *Erkenntnis* 44: 49–71.

Hall, Zach W. (ed.) (1992), *An Introduction to Molecular Neurobiology*. Sunderland, MA: Sinauer Associates.

Judson, Horace F. (1996), *The Eighth Day of Creation: The Makers of the Revolution in Biology*. Expanded Edition. Cold Spring Harbor, NY: Cold Spring Harbor Laboratory Press.

Kauffman, Stuart A. (1971), "Articulation of Parts Explanation in Biology and the Rational Search for Them," in Roger C. Buck and Robert S. Cohen (eds.), *PSA 1970*. Dordrecht: Reidel, pp. 257–272.

Machamer, Peter (1998), "Galileo's Machines, His Mathematics and His Experiments," in Peter Machamer (ed.), *Cambridge Companion to Galileo*. New York: Cambridge University Press, pp. 27–52.

Machamer, Peter (2000), "The Nature of Metaphor and Scientific Description," in Fernand Hallyn (ed.), *Metaphor and Analogy in the Sciences*. Dordrecht: Kluwer, pp. 35–52.

Machamer, Peter and Andrea Woody (1994), "A Model of Intelligibility in Science: Using Galileo's Balance as a Model for Understanding the Motion of Bodies," *Science and Education* 3: 215–244.

Mackie, John Leslie (1974), *The Cement of the Universe: A Study of Causation*. Oxford: Oxford University Press.

Morange, Michel (1998), *A History of Molecular Biology*. Translated by Matthew Cobb. Cambridge, MA: Harvard University Press.

Nagel, Ernest (1961), *The Structure of Science*. New York: Harcourt, Brace and World.

Olby, Robert (1970), "Francis Crick, DNA, and the Central Dogma," in Gerald Holton (ed.), *The Twentieth Century Sciences*. New York: W. W. Norton, pp. 227–280.

Rescher, Nicholas (1996), *Process Metaphysics: An Introduction to Process Philosophy*. Albany, NY: State University of New York Press.

Rheinberger, Hans-Jörg (1997), *Experimental Systems: Towards a History of Epistemic Things. Synthesizing Proteins in the Test Tube*. Stanford, CA: Stanford University Press.

Salmon, Wesley (1984), *Scientific Explanation and the Causal Structure of the World*. Princeton, NJ: Princeton University Press.

Salmon, Wesley (1997), "Causality and Explanation: A Reply to Two Critiques," *Philosophy of Science* 64: 461–477.

Salmon, Wesley (1998), *Causality and Explanation*. New York: Oxford University Press.

Schaffner, Kenneth (1993), *Discovery and Explanation in Biology and Medicine*. Chicago, IL: University of Chicago Press.

Shepherd, Gordon M. (1994), *Neurobiology*. 3rd ed. New York: Oxford University Press.

Watson, James D. ([1962] 1977), "The Involvement of RNA in the Synthesis of Proteins," in *Nobel Lectures in Molecular Biology 1933–1975*. New York: Elsevier, pp. 179–203.

Watson, James D. (1965) *Molecular Biology of the Gene*. New York: W. A. Benjamin.

Wimsatt, William (1972), "Complexity and Organization," in Kenneth F. Schaffner and Robert S. Cohen (eds.), *PSA 1972, Proceedings of the Philosophy of Science Association*. Dordrecht: Reidel, pp. 67–86.

Wimsatt, William (1976), "Reductive Explanation: A Functional Account," in Robert S. Cohen (ed.), *PSA 1974*. Dordrecht: Reidel, pp. 671–710. Reprinted in Elliott Sober (ed.) (1984), *Conceptual Issues in Evolutionary Biology: An Anthology*. 1st ed. Cambridge, MA: MIT Press, pp. 477–508.

Zamecnik, Paul C. (1969), "An Historical Account of Protein Synthesis, with Current Overtones – A Personalized View," *Cold Spring Harbor Symposia on Quantitative Biology* 34: 1–16.

# 2

# Discovering Mechanisms in Neurobiology

## The Case of Spatial Memory[1]

### 2.1  INTRODUCTION

This chapter is about discovery in neurobiology; more specifically, it is about the discovery of mechanisms. The search for mechanisms is widespread in contemporary neurobiology, and, understandably, the character of this product shapes the process by which mechanisms are discovered. Analyzing mechanisms and their characteristic organization reveals constraints on their discovery. These constraints reflect, at least in part, what it is to have a plausible description of a mechanism. These constraints also highlight varieties of evidence that both guide and delimit the construction, evaluation, and revision of such plausible descriptions.

The central example in the following discussion is the continuing discovery of the mechanism of spatial memory. Spatial memory, roughly speaking, is the ability to learn to navigate through a novel environment. The mechanism of spatial memory is multilevel, and recently an integrated sketch of the mechanism at each of these levels has started to emerge. Even though this sketch is far from complete at this time, the example offers a glimpse at the kinds of constraints that are delimiting and guiding this gradual and piecemeal discovery process.

[1]  This chapter was originally published as Craver, Carl F. and Lindley Darden, "Discovering Mechanisms in Neurobiology: The Case of Spatial Memory," in Peter K. Machamer, R. Grush, and P. McLaughlin (eds.), *Theory and Method in the Neurosciences*, © 2001 University of Pittsburgh Press, pp. 112–137. Reprinted by permission of the University of Pittsburgh Press. This work was supported by the National Science Foundation under grant SBR-9817942. Craver's work was also partly funded by the Committee on Cognitive Studies Postdoctoral Fellowship in the Department of Philosophy at the University of Maryland, College Park. We thank German Barrionuevo, Nancy Hall, Tetsuji Iseda, Peter Machamer, Gualtiero Piccinini, Rob Skipper, and Nathan Urban for useful comments on earlier drafts and Payman Farsaii for help as an undergraduate research assistant.

This chapter opens with an analysis of mechanisms, discussing their components and their characteristic spatial, temporal, and multilevel organization. Mechanisms are often discovered gradually and piecemeal. The second section introduces conventions for constructing incomplete and abstract descriptions of mechanisms (namely, the mechanism sketch and the mechanism schema) and for describing the constraints under which the gradual and piecemeal discovery of these sketches and schemata proceeds. The third section uses the case study of spatial memory to illustrate how constraints on the organization of mechanisms have guided and delimited their discovery. The final section focuses on hierarchical constraints in particular and discusses the use of multilevel experiments to integrate the different levels in such a description. Throughout, the goal is to show that the products, multilevel descriptions of mechanisms, shape the process by which they are discovered.

### 2.2   MECHANISMS AND THEIR ORGANIZATION

#### 2.2.1   Mechanisms

Neurobiologists often speak of "systems" and "cascades" to describe what we call, also consistently with the field's language, "mechanisms." Through our collaboration with Peter Machamer, we have come to think about mechanisms as follows. Mechanisms are collections of entities and activities organized in the production of regular changes from start or set-up conditions to finish or termination conditions (Machamer, Darden, Craver 2000; see Chapter 1, this book). Entities in neurobiology include such things as pyramidal cells, neurotransmitters, brain regions, and mice. Activities are the various doings in which these entities engage: pyramidal cells *fire*, neurotransmitters *bind*, brain regions *process*, and mice *swim* in water while eagerly *searching* for a means of escape. When neurobiologists speak generally about activities, they use a variety of terms; activities are often called "processes," "functions," and "interactions." Activities are the things that entities do; they are the productive components of mechanisms, and they constitute the stages of mechanisms.

#### 2.2.2   Organization

The entities and activities composing mechanisms are organized; they are organized such that they *do* something, *carry out* some process, *exercise* some faculty, *perform* some function, or *produce* some end product. We refer to this activity or behavior of the mechanism as a whole as the *phenomenon* to be explained by the description of the mechanism. This is the activity at

41

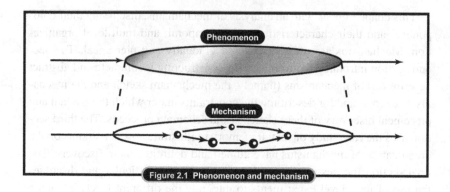

Figure 2.1 Phenomenon and mechanism

the top of Figure 2.1. Below it are the entities and activities composing the mechanism for that phenomenon.

The phenomena to be explained by descriptions of mechanisms can be understood in the spirit of Bogen and Woodward (1988, p. 317). We think of phenomena as relatively stable and repeatable properties or activities that can be produced, manipulated, or detected in a variety of experimental arrangements. Examples of phenomena in neurobiology include the acquisition, storage, and retrieval of spatial memories; the release of neurotransmitters; and the generation of action potentials.

Mechanisms are *organized* in the production of phenomena. One aspect of this organization is *temporal*. The stages of mechanisms have a productive order from beginning to end, with earlier stages giving rise to later stages. The stages of mechanisms also have characteristic rates and durations that can be crucial to their operation. Order, rate, and duration are crucial, for example, for the generation of action potentials in neurons: sodium channels open before potassium channels, and the respective timing and duration of their opening account for the characteristic waveform of the action potential.

A second aspect of the organization of mechanisms is *spatial*. Different stages of the mechanism may be *compartmentalized* within some boundary or otherwise *localized* within some more or less well-defined region. These stages are *connected* with one another by, for example, motion and contact. Often, the connection between stages depends crucially on the *structures* of the entities in the mechanism and on those structured entities being *oriented* with respect to one another in particular ways.

These temporal and spatial aspects of the organization of mechanisms trace the productive relationships among the component stages – the relationship of one stage giving rise to, driving, making, or allowing its successor. Importantly, mechanisms exhibit a productive continuity, without gaps, from

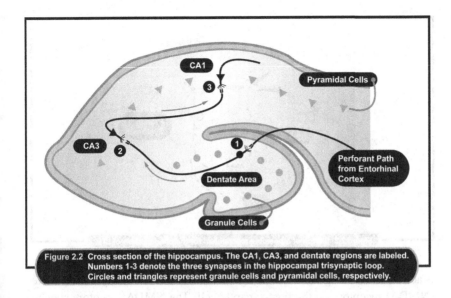

**Figure 2.2 Cross section of the hippocampus. The CA1, CA3, and dentate regions are labeled. Numbers 1-3 denote the three synapses in the hippocampal trisynaptic loop. Circles and triangles represent granule cells and pyramidal cells, respectively.**

setup to termination. Mechanisms require productive continuity to work, and accordingly our understanding of mechanisms turns on our ability to establish a seamless continuity between setup and termination. The discovery of mechanisms is often driven by the goal of eliminating gaps in this productive continuity.

Consider an example from contemporary neurobiology: the mechanism of long-term potentiation (LTP). LTP is a means of strengthening synapses, and many think that LTP, or something like it, is a crucial activity in the mechanism of spatial memory. The idea is essentially Hebb's (1949): when the presynaptic and the post-synaptic neurons are simultaneously active, the synapse is strengthened. LTP is commonly studied in the synapses of the mammalian hippocampus, a medial temporal lobe structure that is thought to be an important entity in the mechanism of spatial memory. A cross section of the hippocampus highlighting some of its major anatomical regions is shown in Figure 2.2. Spatial memories are thought to be formed through the changing strengths of synapses between neurons in the hippocampus, and this is how LTP is thought to fit into the context of the mechanism of spatial memory.

The mechanisms of LTP are not yet completely understood. Robert Malinow, an LTP researcher, has worried in print that the LTC (long-term controversy) over LTP is becoming an LTTP (long-term tar pit) for neurobiologists (Malinow 1998, p. 1226). Nonetheless, one popular sketch of the mechanisms of LTP, visually represented in Figure 2.3, includes the following organized collection of entities and activities. When the presynaptic neuron

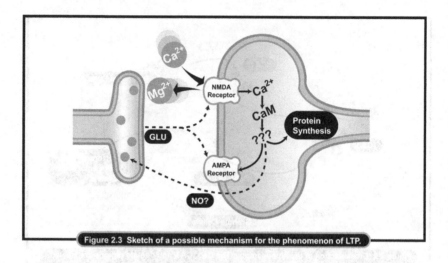

Figure 2.3 Sketch of a possible mechanism for the phenomenon of LTP.

is active, it releases glutamate. This glutamate binds to $N$-methyl-D-aspartate (NMDA) receptors on the post-synaptic cell. The NMDA receptors change their conformation, exposing a pore in the cell membrane. If the post-synaptic cell is inactive, the channel remains blocked by large $Mg^{2+}$ ions. But if the post-synaptic cell is depolarized, these $Mg^{2+}$ ions float out of the channel, allowing $Ca^{2+}$ to diffuse into the cell. The rising intracellular $Ca^{2+}$ concentration sets in motion a long chain of biochemical activities terminating in the question marks of Figure 2.3.

A number of gaps arise in the story at this point, but three things are thought to happen. In the short term, it is thought that this cascade leads to an increase in the number or sensitivity of $\alpha$-amino-3-hydroxyl-5-methyl-4-isoxazolepropionic acid (AMPA) receptors (perhaps by phosphorylation). These changes account for the rapid induction of LTP. In the long term, the cascade leads to the production of proteins in the post-synaptic cell body. These proteins are then used to alter the structure of the dendritic spines at that synapse. Some suspect that there is also a presynaptic component of the LTP mechanism whereby, for example, the presynaptic cell releases more glutamate. Although incomplete, this description will suffice as a sketch of the mechanism of LTP. (More detail concerning this mechanism and the evidence that supports it can be found in Kuno 1995; Frey and Morris 1998.)

The entities in this mechanism are glutamate molecules, NMDA receptors, $Ca^{2+}$ ions, and the like. The activities include binding, diffusing, phosphorylating, and changing conformation. These entities and activities are organized in the production of LTP. The components exhibit a temporal organization that begins with the release of glutamate and terminates in structural changes that strengthen the synapse. The rates and durations of the different stages are crucial for the working of the mechanism; for example, there are the short-term modification of the AMPA receptors and the long-term structural changes to the dendritic spine. Stages of the mechanism are compartmentalized or localized in cells, membranes, and pores. Ignoring the question marks, the early stages of the mechanism are connected with one another, mostly through the motion, binding, and breaking of molecules. These molecular activities depend crucially on the structures and orientations of the entities involved; the size of the pore and the complementary shapes of glutamate and the NMDA receptor allow these entities to engage in the activities that produce the later stages of the mechanism.

### 2.2.3 Levels

There is often a third aspect to the organization of mechanisms in addition to the spatial and temporal; this is a hierarchical aspect.[2] Mechanisms in contemporary neurobiology are organized into multilevel hierarchies (Figure 2.4). The mechanism of spatial memory is a good illustration. The description of this mechanism includes mice learning to navigate, hippocampi generating spatial maps, synapses inducing LTP, and macromolecules (like the NMDA receptor) binding and bending.

The levels in this sort of hierarchy stand in part-whole relations to one another with the important additional restriction that the lower-level entities and activities are components of the higher-level mechanism. The binding of glutamate to the NMDA receptor is a lower-level activity in the mechanism of LTP, and LTP is thought to be a lower-level activity in spatial map formation, which, of course, is thought to be an activity in the mechanism of spatial memory.

The elaboration and refinement of these hierarchical descriptions typically proceeds piecemeal with the goal of *integrating* the entities and activities at different levels. Integrating a component of a mechanism into such a hierarchy involves, first, contextualizing the item within the mechanism of the phenomenon to be explained. This involves "looking up" a level and finding a functional role for the item in that higher-level mechanism. There is some question, for example, over the role of LTP in the mechanism of spatial memory; not only is there a debate over the correct role for LTP in this mechanism, but there is even some debate as to whether it has a role at all. Contextualizing

---

[2] Although the notion of "hierarchy" is often associated with the control or governance of things at lower levels by those at higher levels, this should in no way be associated with the notions of "level" and "hierarchy" explicated here.

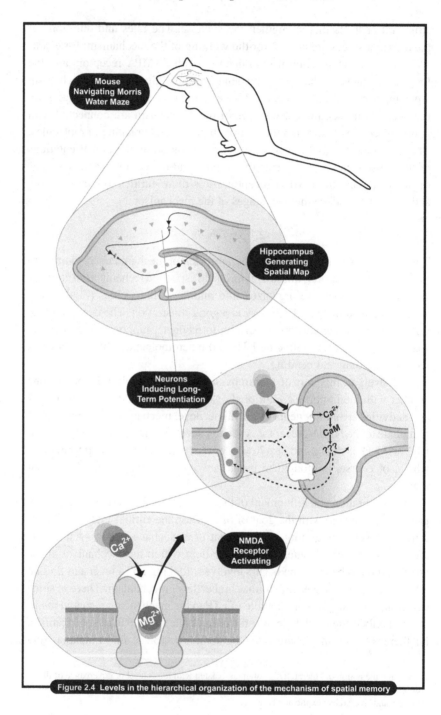

Figure 2.4  Levels in the hierarchical organization of the mechanism of spatial memory

an item within a relevant mechanism for the phenomenon to be explained is one step in the integration of the different levels in a hierarchy.

The second means of integrating a component into such a hierarchical description is downward looking. This involves showing that the properties or activities of an entity can be explicated in terms of a lower-level mechanism. The persistent failure to find a mechanism for a postulated property or activity signals that there is an irremediable gap in the productive continuity of the mechanism. The activity or property cannot be integrated with those at lower levels. Thus, integrating multilevel mechanisms involves both contextualizing an item within higher-level mechanisms and explicating that item in terms of lower-level entities and activities. (For more on the relationship among functions, mechanisms, and levels, see Craver 1998, 2001.)

## 2.3   DESCRIBING MECHANISMS

The preceding discussion of mechanisms and their organization is motivated by the idea that thinking carefully about the abstract structure of mechanisms can provide insight into how they are discovered. Philosophical discussions of discovery should be sensitive to the fact that there are many different kinds of things to be discovered and that these different kinds of things are not all discovered in the same way. The product shapes the process of discovery.

The mechanism for a given phenomenon, for example, is typically not discovered all at once. Instead, descriptions of mechanisms are typically constructed piecemeal. Often, neurobiologists understand some stages of the mechanism quite well and have only the sketchiest understanding of other stages. The question marks in the LTP diagram in Figure 2.3 make these gaps in the mechanism explicit. The descriptions of the various components of a mechanism are often evaluated, and so revised, independently of others. In order to capture this feature of the discovery of mechanisms, it is useful to distinguish mechanism schemata from mechanism sketches.

Mechanism schemata are abstract descriptions of mechanisms that can be instantiated to yield descriptions of particular mechanisms. The term "mechanism schemata" is fitting because their components are placeholders that can be filled in with detailed stages between the setup and termination.[3] Schemata are thus complete in the sense that they can be filled in without gaps in the productive continuity of the mechanism. Schemata and their component

---

[3] Skipper (1999) develops a mechanism schema for selection mechanisms.

placeholders typically have less than universal scope, and their scope can vary considerably.[4]

Mechanism *sketches*, in contrast to mechanism schemata, are abstract descriptions of mechanisms that cannot yet be filled in. Mechanism sketches have black boxes – they leave gaps in the productive continuity of the mechanism, such as the question marks in the LTP diagram. Such black boxes in mechanism sketches are useful in providing guidance about where further elaboration is needed. This role is especially important in the discovery of multilevel mechanisms.

The discovery of mechanisms unfolds gradually and piecemeal through the addition of constraints on plausible mechanism schemata and sketches. These constraints are used to construct plausible descriptions of mechanisms and to revise these plausible descriptions as constraints are added, deleted, or modified. Constraints determine the shape of the space of hypothesized mechanisms. Most simplistically, this space can be understood as a tree with terminal nodes representing possible mechanism schemata for the phenomenon to be explained. The addition of constraints prunes the tree or changes the weights on different branches. The removal of constraints, likewise, can add new branches to the hypothesis space. Understanding the discovery of mechanisms requires an understanding of these different constraints. This is the subject of the remainder of this chapter.

## 2.4   CONSTRAINTS ON THE ORGANIZATION OF MECHANISMS

Bechtel and Richardson's 1993 book *Discovering Complexity* discusses the use of localization and decomposition as research strategies in the discovery of

---

[4] Schaffner (1993) suggested that the "bulk" of theories in the biomedical sciences be seen as "overlapping interlevel temporal models of varying scope." Although we are sympathetic with the direction of Schaffner's thinking here, we prefer to think of such theories as schematic multilevel descriptions of mechanisms. Our discussion of spatial, temporal, and hierarchical constraints on descriptions of mechanisms is intended to exhibit the additional content of an explicit emphasis on mechanisms over and above less specific talk of "theories" or "models." Both mechanism schemata and their components (the placeholders for entities and activities) can have widely varying scope, from near-terrestrial universality (e.g., the mechanism of protein synthesis) to mechanisms, entities, or activities that are found only in some parts of some highly specific strains of organisms. (More on scope can be found in Darden's (1996) review of Schaffner (1993).) Because the sense of "level" articulated in Section 2.2.3 is explicitly defined in terms of "componency" and hence "mechanism," emphasis on mechanisms in the structure of these theories also brings with it a sensible interpretation of their multilevel character. For discussions of multilevel mechanism schemata as theories in neurobiology, see Craver (1998, pp. 10–48; 2001).

mechanistic explanations and the conditions under which these strategies are prone to fail. Their book is an important contribution to the relatively sparse discovery literature in the philosophy of biology (see, e.g., Darden 1991; Darden and Cook 1994). Yet the contribution remains incomplete without a careful look at the products of this discovery process. Thinking carefully about mechanisms and especially their organization highlights a broad variety of constraints on their discovery in addition to those that come from localizing and decomposing. Bechtel and Richardson's discussion of constraints on "causal and explanatory models" is in some ways complementary to our treatment (Bechtel and Richardson 1993, p. 235). For the remainder of the chapter, we focus on five varieties of constraint, including the character of the phenomenon, componency constraints, spatial constraints, temporal constraints, and hierarchical constraints. Localization provides one kind of spatial constraint.

### 2.4.1 Characterizing the Phenomenon

Mechanism schemata and sketches are constrained by the character of the phenomenon for which a mechanism is sought. How one characterizes the phenomenon determines what will count as an adequate description of the mechanism that produces it; the complete description of the mechanism shows how that phenomenon is produced. Spatial memory is the phenomenon to be explained in our working example. But it is not at all obvious in advance of considerable empirical inquiry that there is any such individuable phenomenon – spatial memory – for which there exists an individuable mechanism; and, given that there is such a faculty or phenomenon, it is not at all obvious in advance of considerable empirical inquiry how that phenomenon is properly to be characterized. Debates over the taxonomy of memory can be understood as debates about how to characterize and individuate different memory phenomena. The character of the phenomenon, like the description of the mechanism, is open to revision in light of evidence. This is a prevalent feature of the discovery of mechanisms.

Tolman's famous experiments on maze learning are a good example (Tolman and Honzick 1930; Tolman 1948). Tolman's work was instrumental in shaping the way in which contemporary neurobiologists think about spatial memory. Rats trained to navigate a circuitous route through a maze successfully were subsequently placed into the same maze with a new, more direct route from start to reward. If spatial memory were a simple association between stimulus and response, the rats would be expected to take the circuitous route for which they had been reinforced. But they did not; they preferred the more direct route. The rats could also construct efficient detours, shortcuts, and novel routes to the reward (see, e.g., Olton and Samuelson 1976; Chapuis et al. 1987). Importantly, these experiments and others like them suggest that spatial memory involves the formation of an internal spatial representation – a cognitive map – by which different locations and directions in the environment can be assessed. This characterization of the phenomenon guides the neurobiologist to seek out some entity, property, or activity in the central nervous system that could serve as a representation of space.

The characterization of the phenomenon is also shaped crucially by the accepted experimental protocols for producing, manipulating, and detecting it. As Bogen and Woodward (1988) argued, phenomena should not be confused with data, which are the evidence for phenomena. Data, among other things, are idiosyncratic to particular experimental arrangements; phenomena, as we think of them, are the stable and repeatable properties or activities that can be detected, produced, and manipulated in a variety of experimental arrangements. For our present purposes, it is important to note that different experimental arrangements reveal different aspects of the phenomenon.

So it was mazes of differing complexity that led Tolman to think of spatial memory in terms of the formation of spatial maps. Spatial memory is also tested in radial-arm mazes, three-table problems, and, most important for our purposes, the Morris water maze. The Morris water maze is a circular pool filled with an opaque liquid that covers a hidden platform. Mice are trained to escape over repeated trials. They do not like to swim, and so they learn quickly. The aquatic nature of the task also eliminates smell as a sensory cue. So the maze isolates the place of *visual information* in spatial memory. Sherry and Healy (1998, p. 133) underscore the importance of different experimental protocols for understanding the phenomenon for which a mechanism is sought: "The idea of the cognitive map, first proposed by Tolman (1948), has been an important and influential stimulus to research. But it is really more of a metaphor than a theory. Research on path integration, landmark use, the sun compass and snapshot orientation . . . attempts to specify more concretely exactly what makes up a 'cognitive map' of space." Different experimental assemblies accentuate different features of the phenomenon to be explained. Scientific debates often turn on the appropriateness of a given experimental arrangement for producing, manipulating, or detecting a given phenomenon. Debates over the ecological validity or ethological appropriateness of a task, for example, are debates over the character of the phenomenon. Experimental arrangements are often revised and adjusted over the course of the discovery of a mechanism.

Characterizing the higher-level phenomenon to be explained is a vital step in the discovery of mechanisms. Characterizing the phenomenon prunes the hypothesis space (since the mechanism must produce the phenomenon) and loosely guides its construction (since certain phenomena are suggestive of possible mechanisms). Yet, such a top-down approach, as Bechtel and Richardson (1993, p. 237) agree, cannot itself exhaust the discovery of a mechanism. One also must know the components of the mechanism and how they are organized.

### 2.4.2 Componency Constraints

Mechanisms, remember, are composed of both entities and activities. For a given field at a given time, there is typically a store of established or accepted components out of which mechanisms can be constructed and a set of components that have been excluded from the shelves. The store also contains accepted modules: organized composites of the established entities and activities. In contemporary neurobiology, for example, brain mechanisms will be composed of discrete neurons rather than a "reticulum." These neurons are connected by synapses, which may be electrical, chemical, or both. If they are chemical, then the mechanism will most likely involve action potentials, quantal release, and allosteric interactions. Modules in neurobiology include different second-messenger cascades, ionophore complexes, and cytoarchitectural structures, such as ocular dominance columns and glomeruli.

The store of entities, activities, and modules out of which mechanisms are constructed expands and contracts with the addition and removal of established entities and activities over time. Contracting the store adds constraints on plausible mechanisms by pruning those branches of the hypothesis space that represent mechanisms with such unestablished or unaccepted components. One commentator recently praised a hypothesized mechanism of LTP by saying, "If nothing else, this model is attractive because it requires only established intracellular signaling mechanisms" (Malinow 1998, p. 1226). Expanding the store of components loosens constraints by adding branches to the hypothesis space. The addition of nitric oxide (NO) to the store in the 1980s opened the hypothesis space to mechanisms involving retrograde transmission from the post-synaptic to the presynaptic neuron. It had previously been assumed that chemical neurotransmission was unidirectional, but this entity, which can diffuse freely through neuronal membranes, expanded the space of possibilities (Figure 2.3). Importantly, the store of mechanism components provides guidance in the construction of mechanisms by supplying a set of ingredients out of which mechanisms might be concocted. Introductory neurobiology textbooks acquaint students with this store of entities, activities, and modules.

These textbooks also introduce students to various limitations on the activities in which these entities can engage. These features are important componency constraints on plausible mechanisms. For example, action potentials do not travel at the speed of light. The fastest move at 120 meters per second, and that is in the squid, with its appropriately named giant axons. Human action potentials propagate at roughly 1 meter per second, and our neurons can only fire five hundred times in a second. Constraints can also be found in metabolic requirements, computational resources, temperature limits, rates of protein synthesis, and similar facts of carbon-based life on earth. A pair of authors recently dismissed the hypothesis that the proteins for altering the structure of the synapse in LTP were synthesized within each individual synaptic spine. They rejected this hypothesized mechanism as metabolically too demanding (Frey and Morris 1998, p. 182; they also discuss experimental evidence against the hypothesis). Although these componency constraints are always open to revision in the light of new evidence, they can be decisive in determining the fate of a proposed mechanism schema.

### 2.4.3 Spatial Constraints

Componency constraints shape the hypothesis space by delimiting the store of entities, activities, and modules that can be included in a mechanism and by limiting the possible activities in which the entities can engage. More specific constraints arise from empirical discoveries concerning the spatial organization of the mechanism. The components of mechanisms are often compartmentalized, localized, connected, structured, and oriented with respect to one another. Evidence concerning these sorts of spatial relationships among the components of a mechanism also constrains and guides the discovery process.

Often, the components or stages of mechanisms are *compartmentalized* within reasonably well-defined regions. As the term suggests, these regions are often sectioned off by physical boundaries, like a nuclear membrane, a cell membrane, or skin. Compartmentalization often provides a natural way to individuate the stages of a mechanism. Transcription happens in the nucleus and translation happens in the cytoplasm; there are pre- and post-synaptic

52

components of the mechanism of LTP. Compartmentalization also guides one to seek activities capable of linking the components inside this boundary with those outside – activities such as diffusion, active transport, and second messenger systems.

Closely related to compartmentalization is *localization*. Localizing components, the major focus of Bechtel and Richardson (1993), is often essential for understanding the spatial layout of a mechanism. For reasons to be discussed shortly, researchers are now reasonably confident that some components of the mechanism of spatial memory are to be found in the hippocampus. This finding opens up two new sets of research questions grounded in the spatial organization of the mechanism. For instance, one can look inside the hippocampus to see what makes it work. Such investigation allows one to restrict the store of components to just those that can be identified within this spatial region. One is constrained to understand the activity of the hippocampus in terms of the cells, synapses, neurotransmitters, and circuits that can be found there.

One can then begin to describe the connections of these hippocampal components. *Connectivity* is yet a third variety of spatial constraint; the productive continuity of mechanisms relies on the spatial connections among the components. Early research on the hippocampus has operated under the assumption that the anatomical connectivity of the hippocampal regions exhibits a characteristic clockwise "trisynaptic" loop (Figure 2.2). Perforant path fibers from the entorhinal cortex make the first synapse onto granule cells in the dentate gyrus. These in turn project their axons to the pyramidal cells of the CA3 region, which in turn project to CA1 pyramidal cells. Revising this simple wiring diagram by adding new connections or new types of cells alters the space of plausible mechanisms by changing the scaffolding on which the mechanism can be constructed. More recent research is beginning to emphasize the recurrent connections within these different regions.

Knowing that part of the spatial memory mechanism is localized to the hippocampus also constrains the description of the mechanistic context of the hippocampus. The anatomical connections into and out of the hippocampus become further constraints on the mechanism of spatial memory. This connected anatomy is the spatial scaffolding of the components of the mechanism. Localization is thus an important tool for revealing the connectivity of the mechanism.

Details concerning the geometrical *structure* and *orientation* of the entities in a mechanism, on the other hand, are important for understanding the productivity of mechanisms. As we noted earlier, the stages of mechanisms often depend crucially on entities with appropriate structures having appropriate orientations with respect to one another. Discovering these structures and orientations can place important constraints on the hypothesis space. Indeed, structural aspects of the LTP mechanism are a major focus of recent research on LTP. Articles by Engert and Bonhoeffer (1999) and Maletic-Savatic et al. (1999) present evidence for the addition of new dendritic spines to recently potentiated synapses. Although far from conclusive, such evidence argues for the existence of the structural basis for one plausible mechanism sketch for LTP.[5] The idea, yet to be confirmed, is that the addition of new dendritic spines makes the post-synaptic cell more responsive to glutamate. Structure thus provides clues to the activities that sustain the productive continuity of the mechanism. What remains to be shown, if this mechanism sketch is to be viable, is that the new post-synaptic spines are oriented properly with respect to the presynaptic axons.

The general point of this discussion is that mechanisms are organized spatially in the production of a phenomenon. Identifying aspects of that spatial organization guides and constrains the search through the hypothesis space in a number of ways that go beyond the strategy of localizing. Locations, boundaries, connections, positions, shapes, and orientations are especially important characteristics of or relations among the entities in mechanisms; these characteristics are especially important because they constrain the activities in which those entities can engage and so constrain the way that the mechanism can work. It is for this reason that characterizing these spatial aspects of the organization of mechanisms contributes to our understanding. But details of the spatial organization alone do not allow one to understand what mechanisms do. This spatial organization must be set in motion.

### 2.4.4  Temporal Constraints

Our efforts to understand just how a given mechanism moves are delimited and guided by knowledge of the mechanism's temporal organization. Knowledge of the order, rate, duration, and frequency of the activities in which the component entities of the mechanism engage provides clues to how the mechanism works.

Consider the temporal order of the activities of the entities composing a mechanism, that is, their relative position in the series, forks, and cycles that make up the mechanism. Spatial organization, in and of itself, does not reveal the direction of the productivity in the mechanism – the idea that the circuit of

---

[5] Additional details concerning this hypothesized mechanism continue to emerge. See Barinaga (1999) and Shi et al. (1999) on the delivery and activation of AMPA receptors in LTP.

neurons in the hippocampus goes clockwise or that NMDA receptors allow $Ca^{2+}$ influx into the post-synaptic cell, thereby initiating protein synthesis. Of course, temporal sequence is not by itself sufficient to establish these productive relationships, but given the temporal asymmetry of causality, temporal relations can place constraints on which entities and activities can be seen as productive of which others.

Temporal constraints on the discovery of mechanisms also include constraints imposed by the rates and durations of both the phenomenon to be explained and the stages of the mechanism. The speed limit for generating and propagating action potentials and for transmission at chemical synapses, for example, places limits on the number of sequential steps involved in a phenomenon of a given duration. Temporal constraints have been important in the discovery of the mechanisms of LTP. Researchers who believe that enduring LTP may be sustained by the addition of dendritic spines to the post-synaptic cell, for example, cannot use this mechanism to explain the initial induction of LTP because it takes around 30 minutes to produce the required proteins, distribute them, and insert them into the membrane. Short-term induction of LTP must rely on some faster mechanism, like the phosphorylation of AMPA receptors. Possible mechanisms are pruned from the hypothesis space on the grounds that the stages or steps take too long or happen too slowly to produce a phenomenon with a given rate or duration. Because mechanisms are active – because mechanisms do things – they take time to work, and the order, rate, and duration of the stages in a mechanism are therefore important tools for discovering their organization and culling the hypothesis space.

## 2.5 EXPERIMENTS FOR TESTING HIERARCHICAL MECHANISMS

The final set of constraints on the discovery of mechanisms is grounded in the hierarchical organization of mechanisms. Neurobiologists conduct experiments to reveal this hierarchical organization. Although there is a great deal to be said about how experiments are used to test mechanisms, the focus here is on the role of experiments in the integration of levels in multilevel schemata and sketches.

Experiments can be understood in terms of an abstract experimental protocol, which clearly has some affinities with Hacking's (1988, 1992) discussion of experimentation.[6] Hacking does not discuss experimentation in

---

[6] For an alternative schematic account of experiments, see Lederberg (1995).

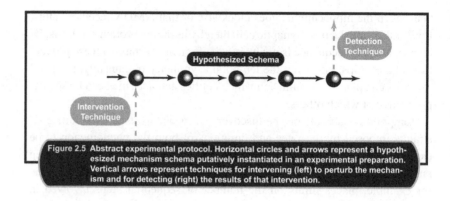

Figure 2.5 Abstract experimental protocol. Horizontal circles and arrows represent a hypothesized mechanism schema putatively instantiated in an experimental preparation. Vertical arrows represent techniques for intervening (left) to perturb the mechanism and for detecting (right) the results of that intervention.

the discovery of mechanisms per se, nor does he address their role in the integration of levels. We can achieve a more detailed understanding of experimentation by attending to the mechanistic organization that these experiments are used to reveal.

The abstract protocol in Figure 2.5 is most easily described in the case of a unilevel mechanism; the protocol can then be easily extended to a multi-level case. The connected circles and arrows represent a hypothesized mechanism schema putatively instantiated in some experimental preparation. The schema may be instantiated in vivo, in vitro, or in silico (i.e., in a mouse, in a petri dish, or in a computer). Having found or created such an experimental preparation, one then typically uses some *intervention technique* to perturb some component. The perturbation produced in the experimental preparation presumably has downstream results that are detected or amplified with the help of a *detection technique*. There is much to be said about this idealized protocol; we introduce it here simply to extend it to the multilevel case. And this is easy to do. Experiments designed to test the hierarchical organization of a mechanism typically involve intervening at one level and detecting at another. Sometimes, a single set of experiments involves intervening and detecting at multiple levels at once; we will get to such a case shortly.

For simplicity, though, we start with two-level experiments. The left-hand side of Figure 2.6 exhibits a case of intervening at the lower level and detecting at the higher level; we call these "bottom-up" experiments. The right-hand side of that figure shows the opposite: intervention at the higher level and detection at the lower level. These can be thought of as "top-down" experiments. Interventions, in either of these cases, may be either stimulatory or inhibitory. Our first set of examples is inhibitory, bottom-up experiments. Our second set of examples is top-down and stimulatory. Both sets of examples

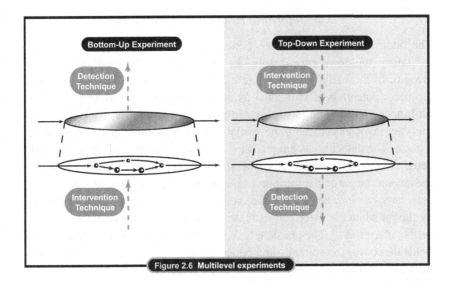

Figure 2.6 Multilevel experiments

are drawn from experiments that, taken together, forcefully suggest that the hippocampus is involved in the formation of spatial maps.

### 2.5.1 Bottom-Up Inhibitory Experiments

The first example is the now-familiar case study H.M., as reported by Scoville and Millner (1957). Because H.M. was plagued by incapacitating epileptic seizures, he consented to an experimental surgical procedure to remove portions of his medial temporal lobes, including the hippocampus. After the surgery, it quickly became apparent that H.M. had lost the ability to remember recent facts, even though he retained the ability to learn new skills. H.M.'s case famously suggested to researchers that the human hippocampus is a crucial entity in the mechanisms of what has since been called "declarative" memory (Zola-Morgan and Squire 1993).

Subsequent experiments in rats and mice have shown that bilateral removal of the hippocampus leads to profound deficits in spatial memory. For example, although rats with intact hippocampi learn very quickly, over repeated trials, to swim directly to the hidden platform in the Morris water maze, rats with bilateral hippocampal lesions continue over multiple trials to swim randomly through the pool, stopping only when they stumble onto the platform (Morris et al. 1982).

Both the case of H.M. and the subsequent ablation experiments in mice are examples of the two-tiered experimental structure exhibited in the left-hand side of Figure 2.6.

## 2.5.2   Top-Down Excitatory Experiments

The findings of these inhibitory, bottom-up experiments are reinforced by excitatory, top-down experiments like those on the right-hand side of Figure 2.6. In the early 1970s, O'Keefe and Dostrovsky (1971) recorded the electrical potentials of individual pyramidal cells in the CA1 region of the rat hippocampus while rats navigated a standard maze. The intervention in this case involves activating the spatial-memory system by putting the rat in a maze. The detection technique is the electrical recording. They found that certain of those pyramidal cells generate bursts of action potentials whenever the rat enters a particular location while facing in a particular direction. These cells have come to be called "place cells," and the region of space occupied by the rat when the place cell increases its activity is likewise known as the cell's "place field." The place cells of CA1 have slightly overlapping place fields that cover the animal's immediate spatial environment. The pattern of activity across this subpopulation of CA1 pyramidal cells could therefore play the role of a spatial map.

These findings have recently been confirmed with multi-unit electrodes that allow one to record from 70 to 150 CA1 pyramidal cells at once. Astonishingly, it is possible to *predict* the path taken by the rat on the basis of these recordings (Wilson and McNaughton 1993). This is a remarkable top-down stimulatory finding.

Top-down and bottom-up experiments of both the stimulatory and inhibitory variety are quite common in neurobiology. They are common because the findings of such experiments, taken together, reveal aspects of and thereby place constraints on the hierarchical organization of a mechanism. More specifically, when experiments of this sort go well, they place constraints on the possibilities for integrating the different levels. Top-down and bottom-up experiments help to situate an item, like the hippocampus or the NMDA receptor, within the context of a higher-level mechanism. They also identify components in the mechanisms that produce higher-level activities and properties. These experiments tell us what the relevant entities and activities are, how they are nested in component–subcomponent relations, and how the activities of the component entities fit into their mechanistic context. Persistent failure to situate an item within a hierarchical mechanism, or persistent failure to uncover a lower-level mechanism for that item, prunes mechanism schemata involving that item from the hypothesis space.

This role of these experiments in placing constraints on the integration of the different levels of a hierarchy is even more apparent in multilevel experiments, from which our last example is drawn.

## 2.5.3 Multilevel Experiments

In late 1996, researchers at MIT, Columbia, and Cal Tech published a series of papers describing the effects of highly specific genetic deletions (or "gene knockouts") on entities and activities at multiple levels in the spatial-memory hierarchy (McHugh et al. 1996; Rottenberg et al. 1996; Tsien et al. 1996a, 1996b).

The experiments that are our focus are bottom-up and inhibitory. Specifically, the researchers invented a molecular scalpel for deleting the *NMDAR1* gene, a gene encoding an essential subunit of the NMDA receptor (Tsien et al. 1996a), and for deleting it only in the pyramidal cells of the CA1 region of the mouse hippocampus. The deletion was also timed to occur only after normal hippocampal development is thought to be complete. The trick was to couple the deletion to a promoter of a gene that is expressed selectively in CA1 pyramidal cells and that is expressed only in the later developmental stages of the hippocampus. This intervention technique gives researchers finer-grained spatial and temporal resolution in their manipulation of the brain's activities than has ever been possible. This, in turn, provides higher spatio-temporal resolution on the organization of the mechanism of spatial memory.

These experiments have been praised as the first to investigate the mechanism of spatial memory, "at all levels in a single set of experiments, from molecular changes through altered patterns of neuronal firing to impaired learning" (Roush 1997, p. 32), and for taking an important first step toward the "dream of neurobiology ... to understand all aspects of interesting and important cognitive phenomena – like memory – from the underlying molecular mechanisms through behavior" (Stevens 1996, p. 1147). More specifically, we claim that these experiments advance the goal of integrating the different levels in this multilevel mechanism.

Knockout mice, those without functional NMDA receptors, had difficulty escaping the Morris water maze. They performed far worse than controls in learning to escape. When placed in a maze *without* a platform, control mice concentrated their swimming in the platform's previous location. Knockout mice swam about randomly (Tsien et al. 1996b). Multi-unit recordings from CA1 pyramidal cells in the knockout mice revealed significant impairments in spatial map formation. The researchers found, to their surprise, that CA1 cells in the knockout mice *did* exhibit place-related firing. But the place fields were much larger and much less sharply defined. These deficits in spatial map formation can reasonably be attributed to the absence of LTP at synapses lacking functional NMDA receptors. The researchers found that knocking out the NMDA receptor eliminated LTP induction in CA1 and not in any other region of the brain (Tsien et al. 1996b).

This complicated experiment is a bottom-up inhibitory experiment with detection at multiple levels. The intervention technique intervenes to perturb the NMDA receptor by deleting the *NMDAR1* gene. The detection techniques register the results of this intervention on LTP, spatial map formation, and spatial memory. There is a lot to be said about the strength of these experimental findings, but this is not our focus here. Instead, we are interested in how these multilevel experiments constrain hypotheses about the integration of multilevel mechanisms.

This set of experiments is designed to test a popular sketch of the multilevel mechanism of spatial memory. It is a sketch because we are not remotely in a position to trace out all of the mechanisms at all of the different levels. Instead, this sketch is a hypothesis of how the components at different levels are integrated with one another. In particular, it is the hypothesis that the NMDA receptor is a necessary component of the mechanism of LTP, which is a necessary component of the mechanism of spatial map formation, which is a necessary component of the mechanism of spatial memory. If these nesting relationships do hold, then knocking out an essential gene for the NMDA receptor would be expected to eliminate the induction of LTP, to eliminate spatial map formation, and to leave the mice hopelessly lost in the Morris water maze. The rough conformity of the findings to these expectations is heartening in this respect.

It is important to note, however, that the genetic deletion did not eliminate spatial map formation in the CA1 region of the hippocampus. Instead, knocking out this essential gene made the map less precise. This lack of precision still had behavioral implications, and so it is consistent with some role for LTP within the context of the spatial-memory mechanism as a whole. However, this anomalous finding forced the researchers to rethink the role of CA1 LTP in the context of this mechanism. The persistence of place-related firing in CA1 suggests that place fields must be established in an earlier stage of the mechanism. So, the role of CA1 within the context of the spatial-memory mechanism is not the formation of spatial maps. Instead, the researchers suggest – with characteristic caution – that CA1 has the role of "learn[ing] associations between entorhinal inputs and place [information projecting from the CA3 region of the hippocampus]" (McHugh et al. 1996, p. 1347). This finding, in other words, constrains our understanding of the role of CA1 LTP in the spatial-memory mechanism; as this case of revision suggests, it also helps to establish the productive organization of the components of the mechanism as a whole. (For a more systematic discussion of

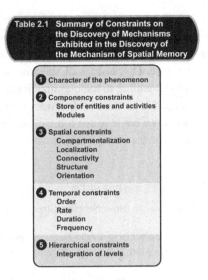

Table 2.1  Summary of Constraints on the Discovery of Mechanisms Exhibited in the Discovery of the Mechanism of Spatial Memory

❶ Character of the phenomenon

❷ Componency constraints
   Store of entities and activities
   Modules

❸ Spatial constraints
   Compartmentalization
   Localization
   Connectivity
   Structure
   Orientation

❹ Temporal constraints
   Order
   Rate
   Duration
   Frequency

❺ Hierarchical constraints
   Integration of levels

anomaly resolution, see Darden and Cook 1994; Darden 1991, 1992; also see Chapter 9, this book.)

It is in this way that multilevel experiments furnish constraints on the integration of components at multiple levels. These hierarchical constraints, in conjunction with the character of the phenomenon and the spatial, temporal, and componency constraints, guide and delimit the discovery of mechanisms.

## 2.6  CONCLUSION

It is now possible to step back and look at this discovery process as a whole. The continuing discovery of the mechanism of spatial memory has proceeded gradually through the piecemeal elaboration and revision of mechanisms at multiple levels and through the gradual integration of the components at each of these levels. This discovery process has been guided by a sketch, replete with black boxes, of how this mechanism is organized. Certain of these black boxes are starting to open with the accumulation of constraints on how they are to be filled in. We are beginning to understand the details of LTP and, perhaps more important, we are beginning to evaluate more precisely the role of LTP within the context of the spatial-memory mechanism. The same is true for spatial map formation in the hippocampus and the opening and closing of the NMDA receptor.

This discovery process is guided by constraints on the organization of the mechanism that are summarized in Table 2.1. These constraints come from many different specialties within neurobiology: neuroanatomists, clinical psychologists, electrophysiologists, and molecular neurobiologists are all contributing different constraints on the emerging organization of this hierarchical mechanistic structure. Discussions of scientific discovery in neurobiology should proceed by attending to the organizational structure of mechanisms. The product shapes the process of discovery. In understanding both the product and the process, we come to see more clearly what is involved in understanding phenomena by describing mechanisms.

#### REFERENCES

Barinaga, M. (1999), "New Clues to How Neurons Strengthen Their Connections," *Science* 284: 1755–1757.

Bechtel, William and Robert C. Richardson (1993), *Discovering Complexity: Decomposition and Localization as Strategies in Scientific Research*. Princeton, NJ: Princeton University Press.

Bogen, James and J. Woodward (1988), "Saving the Phenomena," *Philosophical Review* 97: 303–352.

Chapuis, N., M. Durup, and C. Thinus-Blanc (1987), "The Role of Exploratory Experience in a Shortcut in Golden Hamsters *(Mesocricetus auratus),*" *Animal Learning and Behavior* 15: 174–178.

Craver, Carl F. (1998), *Neural Mechanisms: On the Structure, Function, and Development of Theories in Neurobiology*. Doctoral dissertation, Department of History and Philosophy of Science, University of Pittsburgh, Pittsburgh, PA.

Craver, Carl F. (2001), "Role Functions, Mechanisms, and Hierarchy," *Philosophy of Science* 68: 53–74.

Darden, Lindley (1991), *Theory Change in Science: Strategies from Mendelian Genetics*. New York: Oxford University Press.

Darden, Lindley (1992), "Strategies for Anomaly Resolution," in Ronald Giere (ed.), *Cognitive Models of Science*. Minnesota Studies in the Philosophy of Science, v. 15. Minneapolis, MN: University of Minnesota Press, pp. 251–273.

Darden, Lindley (1996), "Generalizations in Biology: Essay Review of K. Schaffner's *Discovery and Explanation in Biology and Medicine,*" *Studies in History and Philosophy of Science* 27: 409–419.

Darden, Lindley and Michael Cook (1994), "Reasoning Strategies in Molecular Biology: Abstractions, Scans, and Anomalies," in David Hull, Mickey Forbes, and Richard M. Burian (eds.), *PSA 1994*, v. 2. East Lansing, MI: Philosophy of Science Association, pp. 179–191.

Engert, E. and T. Bonhoeffer (1999), "Dendritic Spine Changes Associated with Hippocampal Long-Term Synaptic Plasticity," *Nature* 399: 66–70.

Frey, U., and R. G. Morris (1998), "Synaptic Tagging: Implications for Late Maintenance of Hippocampal Long-Term Potentiation," *Trends in Neuroscience* 21: 181–188.

Hacking, Ian (1988), "On the Stability of the Laboratory Sciences," *Journal of Philosophy* 85: 507–514.

Hacking, Ian (1992), "The Self-Vindication of the Laboratory Sciences," in A. Pickering (ed.), *Science as Practice and Culture*. Chicago, IL: University of Chicago Press, pp. 29–64.

Hebb, D. O. (1949), *The Organization of Behavior*. New York: Wiley.

Kuno, M. (1995), *The Synapse: Function, Plasticity, and Neurotrophism*. Oxford: Oxford University Press.

Lederberg, Joshua S. (1995), "Notes on Systematic Hypothesis Generation and Application to Disciplined Brainstorming," in *Working Notes: Symposium: Systematic Methods of Scientific Discovery*. AAAI Spring Symposium Series. American Association for Artificial Intelligence, Stanford, CA: Stanford University, pp. 97–98.

Machamer, Peter, Lindley Darden, and Carl F. Craver (2000), "Thinking about Mechanisms," *Philosophy of Science* 67: 1–25.

Maletic-Savatic, M., R. Malinow, and K. Svoboda (1999), "Rapid Dendritic Morphogenesis in CA1 Hippocampal Dendrites Induced by Synaptic Activity," *Science* 283: 1923–1926.

Malinow, R. (1998), "Silencing the Controversy in LTP?" *Neuron* 21: 1226–1227.

McHugh, T. J., K. I. Blum, J. Z. Tsien, S. Tonegawa, and M. A. Wilson (1996), "Impaired Hippocampal Representation of Space in CA1-Specific NMDARI Knockout Mice," *Cell* 87: 1339–1349.

Morris, R. G. M., P. Garrud, J. N. P. Rawlins, and J. O'Keefe (1982), "Place Navigation Impaired in Rats with Hippocampal Lesions," *Nature* 297: 681–683.

O'Keefe, J. and J. Dostrovsky (1971), "The Hippocampus as a Spatial Map: Preliminary Evidence from Unit Activity in the Freely Moving Rat," *Brain Research* 34: 171–175.

Olton, D. S. and R. J. Samuelson (1976), "Remembrances of Places Passed: Spatial Memory in Rats," *Journal of Experimental Psychology: Animal Behavior Processes* 2: 97–116.

Rottenberg, A., M. Mayford, R. D. Hawkins, E. R. Kandel, and R. U. Muller (1996), "Mice Expressing Activated CaMKII Lack Low Frequency LTP and Do Not Form Stable Place Cells in the CA1 Region of the Hippocampus," *Cell* 87: 1351–1361.

Roush, W. (1997), "New Knockout Mice Point to Molecular Basis of Memory," *Science* 275: 32–33.

Schaffner, Kenneth (1993), *Discovery and Explanation in Biology and Medicine*. Chicago, IL: University of Chicago Press.

Scoville, W. B. and B. Millner (1957), "Loss of Recent Memory after Bilateral Hippocampal Lesions," *Journal of Neurology, Neurosurgery, and Psychiatry* 20: 11–20.

Sherry, D. and S. Healy (1998), "Neural Mechanisms of Spatial Representation," in S. Healy (ed.), *Spatial Representation in Animals*. Oxford: Oxford University Press, pp. 133–157.

Shi, S., Y. Hayashi, R. S. Petralia, S. H. Zaman, R. J. Wenthold, K. Svoboda, and R. Malinow (1999), "Rapid Spine Delivery and Redistribution of AMPA Receptors after Synaptic NMDA Receptor Activation," *Science* 284: 1811–1816.

Skipper, Robert (1999), "Selection and the Extent of Explanatory Unification," *Philosophy of Science (Supplement)* 66: S196–S209.

Stevens, C. F. (1996), "Spatial Learning and Memory: The Beginning of a Dream," *Cell* 87: 1147–1148.

Tolman, E. (1948), "Cognitive Maps in Rats and Men," *Psychological Review* 55: 189–208.

Tolman, E. and C. Honzick (1930), "Introduction and Removal of Reward and Maze Performance in Rats," *University of California Publications in Psychology* 4: 257–275.

Tsien, J. Z., D. E. Chen, D. Gerber, C. Tom, E. Mercer, D. Anderson, M. Mayford, and E. R. Kandel (1996a), "Subregion- and Cell Type-Restricted Gene Knockout in Mouse Brain," *Cell* 87: 1317–1326.

Tsien, J. Z., P. T. Huerta, and S. Tonegawa (1996b), "The Essential Role of Hippocampal CA1 NMDA Receptor-Dependent Synaptic Plasticity in Spatial Memory," *Cell* 87: 1327–1338.

Wilson, M. A. and B. McNaughton (1993), "Dynamics of the Hippocampal Ensemble Code for Space," *Science* 261: 1055–1058.

Zola-Morgan, S. and L. Squire (1993), "Neuroanatomy of Memory," *Annual Review of Neuroscience* 16: 547–563.

# 3

## Strategies in the Interfield Discovery
## of the Mechanism of Protein Synthesis[1]

### 3.1 INTRODUCTION

Crucial pieces of the mechanism of protein synthesis were discovered by biochemists and molecular biologists in the 1950s and 1960s. At the outset, the approaches of these fields were very different, focusing on different components and finding different aspects of the mechanism from different ends. By about 1965, the results from the different approaches were integrated. The scientific work leading to this integration reveals general strategies for discovering mechanisms.

This instance of interfield integration, like many other discovery episodes in the biological sciences, crucially involves discovering a mechanism. Focusing centrally on *mechanisms* provides new ways of thinking about discovery, interfield integration, and reasoning strategies for scientific change. Philosophers of science have separately analyzed scientific discovery, interfield relations, mechanisms, and reasoning strategies. This chapter brings together these disparate topics. A unified approach yields reasoning strategies in discovering mechanisms that integrate results from different fields.

[1] This chapter was originally published as Darden, Lindley and Carl F. Craver (2002), "Strategies in the Interfield Discovery of the Mechanism of Protein Synthesis," *Studies in History and Philosophy of Biological and Biomedical Sciences* 33: 1–28. This work was supported by National Science Foundation grant SBR-9817942. Lindley Darden's work was also supported by a General Research Board Award from the Graduate School of the University of Maryland, College Park. The characterization of mechanism is due to our work with Peter Machamer, with whom we've had many fruitful discussions. Discussions with Jeff Ramsey about chemical bonding were helpful. Paul Zamecnik kindly supplied reprints for several of his papers. We thank Payman Farsaii and Scott James for research assistance and the following for comments: Chris Cosans, Nancy Hall, Sandra Herbert, Marcia Kraft, Jeff Lewandowski, and Eric Saidel. Sotiris Kotsanis suggested considering how entities and their properties are changed as the mechanism operates. Brad Rives challenged fuzzy metaphysical categories, at least some of which we have sharpened.

Many philosophers of science (e.g., Popper 1965) have been skeptical about finding methods for reasoning in discovery. Even those who have had much to say about scientific change (e.g., Kuhn 1962; Laudan 1977; Kitcher 1993) have not even discussed reasoning strategies for discovering new paradigms, traditions, or practices. Nonetheless, a few philosophers have worked on discovery (e.g., Nickles, ed. 1980a, 1980b; Meheus and Nickles, eds., 1999). Reasoning in discovery is a more tractable problem if discovery is viewed as an extended process of construction, evaluation, and revision (Darden 1991). One fruitful reasoning strategy is to search for a solution to a problem in one field by relating it to items in another field (Darden and Maull 1977; see Chapter 5, this book; Bechtel 1984, 1986, 1988; Darden 1991). A new perspective is emerging by focusing on interfield relations in the discovery of mechanisms.

Some philosophers have argued for the importance of mechanisms in science (Wimsatt 1972; Brandon 1985; Glennan 1996; Machamer, Darden, Craver 2000; see Chapter 1, this book) and in molecular biology in particular (Burian 1996a; Crick 1988). Wimsatt, for example, says that, "At least in biology, most scientists see their work as explaining types of phenomena by discovering mechanisms..." (Wimsatt 1972, p. 67). In their pioneering work, Bechtel and Richardson (1993) elucidate decomposition and localization strategies for discovering mechanisms in simple and complex systems.

This chapter discusses two additional strategies for interfield, mechanism discovery: schema instantiation and forward/backward chaining. Schema instantiation is the application of an abstract mechanism framework and the search for components to fill in its details. The strategy of forward/backward chaining involves reasoning about known (or hypothesized) mechanism components to fill gaps in the understanding of the productive continuity of the mechanism, either forward or backward.

The protein synthesis case is a rich case for investigating reasoning in an interfield discovery episode. Molecular biologists and biochemists brought different ideas and techniques to the problem of how proteins are synthesized. At the outset, they worked on different ends of the mechanism. Eventually, their results were integrated to produce a single description of the mechanism. Protein synthesis comprises one of the core mechanisms in the two fields: the mechanism for gene expression for molecular biology and the mechanism for the synthesis of enzymes and structural proteins important in the study of metabolism in biochemistry. Consequently, understanding this interfield discovery episode illuminates reasoning in a significant achievement of twentieth-century biology that integrated results from two fields: molecular biology and biochemistry.

Although they are in need of reinterpretation from the perspective of reasoning in interfield mechanism discovery, valuable historical accounts of parts
of the protein synthesis story exist. The history of this episode has been
told both by the scientists involved, in autobiographical accounts (e.g., Crick
1988; Gros 1979; Hoagland 1990, 1996; Jacob 1988; Watson 1962, 1968,
2000; Zamecnik 1962a, 1962b, 1969, 1976, 1979, 1984), and by historians
of molecular biology and biochemistry (e.g., Burian 1996a; Chadarevian and
Gaudilliere 1996; Gaudilliere 1993, 1996; Judson 1996; Kay 2000; Morange
1998; Olby 1970; Thieffry and Burian 1996). There are aspects of this story
(mostly its biochemical side) that we will not discuss. However, we will weave
together some of Rheinberger's (1997) new historical work concerning Paul
Zamecnik's biochemical research with more familiar accounts of molecular
biological work. Rheinberger's emphasis on Zamecnik's "experimental system" highlights a previously neglected aspect of experimental scientific work,
as well as the changes in the problem-contexts in which that system was used.
However, Rheinberger neglects Zamecnik's (e.g., 1969) explicit discussion of
the use of that experimental system in the search for the *mechanism of protein
synthesis*. Unlike previous work, we stress interfield integration through and
strategies for mechanism discovery.

We begin in Section 3.2 with a brief characterization of mechanisms, the
abstract schemata that are used in their description and their discovery, and two
reasoning strategies that make use of this characterization: namely, schema
instantiation and forward/backward chaining. In Section 3.3, we interpret the
discovery of protein synthesis as a case of interfield mechanism discovery,
paying particular attention to the strategies used in this scientific episode.
Finally, Section 3.4 abstracts from examples in the case study, and discusses
the two general strategies for mechanism discovery in more detail.

### 3.2   MECHANISMS, SCHEMATA, STRATEGIES

An abstract characterization of mechanisms aids in analyzing this historical
case and in finding strategies for mechanism discovery:

> Mechanisms are entities and activities organized such that they are productive
> of regular changes from start or set-up to finish or termination conditions.
> (Machamer, Darden, and Craver 2000; see Chapter 1, this book)

Types of entities include ions, macromolecules (e.g., proteins and the nucleic
acids, DNA and RNA), and cellular structures, such as ribosomal particles,
which are composed of both RNA and proteins. Types of activities include
geometrico-mechanical activities, such as lock and key docking of an enzyme

and its substrate, and electro-chemical activities, such as strong covalent bonding and weak hydrogen bonding.

Entities having certain kinds of properties are necessary for the possibility of acting in certain ways, and certain kinds of activities are only possible when there are entities having certain kinds of properties. Entities and activities are interdependent (Machamer, Darden, and Craver 2000; see Chapter 1, this book). For example, appropriate chemical valences are necessary for covalent bonding, polar charges are necessary for hydrogen bonding, and appropriate shapes are necessary for lock and key docking.

Mechanisms are made of components that work together to do something. The entities and activities are organized in productive continuity from beginning to end. One goal in discovering a mechanism is to reveal the mechanism's productive continuity. Determining the temporal boundaries of the mechanism – that is, the set-up and the termination conditions – allows work to proceed from both ends to find the intermediate stages. Looking forward, each stage must give rise to, allow, drive, or make the next. Conversely, looking back, each stage must have been produced, driven, or allowed by the previous stage(s). Consequently, the reasoning strategy of forward chaining from the (perhaps hypothesized) set-up conditions and backward chaining from the termination conditions is a fruitful research strategy for finding the productive continuity of a mechanism.

In addition to using components in the forward/backward strategy, another strategy for discovering a mechanism is schema instantiation. Mechanism schemata are abstract frameworks for mechanisms. They contain placeholders for the components of the mechanism (both entities and activities) and indicate, with variable degrees of abstraction, how the components are organized. Often, these placeholders characterize a component's role in the mechanism. Discovering a mechanism involves specifying and filling in the details of a schema, that is, instantiating it by moving to a lower degree of abstraction. As we will see, diagrams and equations are often employed to depict graphically the schematic organization of mechanisms.

3.3   DISCOVERING THE MECHANISM OF PROTEIN SYNTHESIS
1953–1965: BIOCHEMISTRY AND MOLECULAR BIOLOGY

The discovery of the mechanism of protein synthesis was an interfield discovery. Both biochemists and molecular biologists contributed to it.

A group of biochemists were working to understand a mechanism for assembling polypeptides. They took the end of the mechanism to be a protein, consisting of amino acids held together by strong covalent bonds. By the 1940s, when MD-turned-biochemist Paul Zamecnik began his work, biochemists had discovered more than twenty amino acids and had elucidated the nature of the linkages between them in peptide bonds. Zamecnik and his colleagues, especially Mahlon Hoagland, sought to understand energetic intermediates between free amino acids and their linkage in polypeptides (recalled in Zamecnik 1962a, 1969, 1979, 1984; Hoagland 1990, 1996). They were thus working backward from peptide bonds to the mechanisms of polypeptide assembly, focusing on chemical reactions and energy requirements for such strong covalent bonds to form. Biochemists often used in vitro experimental systems, such as Zamecnik's cell-free rat-liver preparation. As Zamecnik put it graphically: "The biochemist traditionally studies living cells by smashing them to bits and trying to analyze the function of their parts." (Zamecnik 1958, p. 118).

The other players in the episode were molecular biologists, such as James Watson and Francis Crick, who took the beginning of the mechanism to be DNA. They sought to understand the "genetic code," as it came to be called, by which the order of the bases in DNA is related to the order of amino acids in proteins. They were thus reasoning forward from DNA to ordered amino acid sequences in proteins. Molecular biologists focused on weak hydrogen bonds and on determining macromolecular structure. Their experimental techniques were often grounded in x-ray crystallography and the building of scale models, which had earned a good reputation in Watson and Crick's work on the structure of DNA (Watson and Crick 1953a) and Linus Pauling's work on protein structure (Pauling and Corey 1950). Molecular biologists also used genetic techniques, such as crossbreeding, to investigate the role of DNA in genetic mechanisms.

Zamecnik contrasted the approaches of the two fields:

> As in the building of a tunnel, digging is going on from two sides of this mound of uncertainty, in the hope of meeting in the middle. Investigators primarily interested in protein synthesis are moving back to a study of ribonucleic acid metabolism, while those interested primarily in the gene and DNA metabolism are studying interrelationships with ribonucleic acid from the other side. It has become quite clear that ribonucleic acid is the connecting link between the hereditary message of the gene and its enzymic expression.
>
> (Zamecnik 1962a, p. 47)

These molecular biologists and biochemists differed from each other in the techniques they used, in the parts of the mechanism they investigated, and in their attention to different aspects of the productive continuity in the mechanism. While biochemists, such as Zamecnik, were homogenizing rat livers and tracing centrifuge fractions, some molecular biologists were crystallizing macromolecules and subjecting them to x-ray analysis, and yet others were doing genetic crosses with bacteria and the bacterial virus, bacteriophage. While biochemists were subjecting the protein end of the mechanism to chemical analysis, molecular biologists began with the genetic material of DNA. While molecular biologists traced the "flow of information" (Crick's phrase, 1958, p. 144), the biochemists studied the flow of matter or energy in the mechanism. While molecular biologists questioned how the genetic information contained in the order of bases along the DNA double helix served to order the amino acids in proteins, biochemists investigated the energy requirements for binding free amino acids in the strong, covalent, peptide bonds of proteins.

Despite these numerous differences, fruitful interfield interactions between molecular biology and biochemistry served to integrate their findings. As work proceeded from both ends of the mechanism, it converged in the middle, as we will see, with the discovery of new entities (three types of RNA and activated amino acids) and their activities. The result was an understanding of the productive continuity of the protein synthesis mechanism from beginning to end.[2]

### 3.3.1  Diagrams of 1953–1954

The differences between these fields are nicely illustrated by comparing Zamecnik's and Watson's diagrams in Figure 3.1.

Zamecnik's diagram of 1953 shows the components of an in vitro, cell-free protein synthesis system. As Rheinberger (1997) ably recounts, Zamecnik's experimental system was constructed from homogenized and centrifuged components of rat-liver cells. Work with this system provided Zamecnik with a mechanism sketch illustrated in the diagram. It includes several components (some of which are numbered in the diagram): continued energy production

---

[2] We take this to be a successful case of discovery in science. It is difficult to imagine that proteins are not usually synthesized by the mechanism involving DNA and RNA that we discuss here, even though, since the 1960s, much more has been discovered about this mechanism. Furthermore, some special-case anomalies are now known, such as in RNA viruses and in cases where amino acids are added or removed post-translationally. The discovery of introns, RNA splicing, and RNA editing added additional stages to the mechanism in eucaryotes. We have neglected the history of work on the enzymes that catalyze various stages in this mechanism; an exciting new chapter begins with the recent hypothesis that the ribosome itself functions as a ribozyme in the formation of the peptide bond (Cech 2000). However, even if the proposed mechanism schema of protein synthesis is revised in light of subsequent evidence, the reasoning strategies discussed here are still general ones for discovering plausible mechanisms.

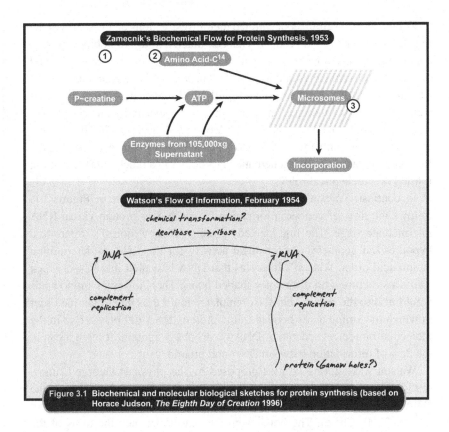

Figure 3.1 Biochemical and molecular biological sketches for protein synthesis (based on Horace Judson, *The Eighth Day of Creation* 1996)

by means of the formation of ATP (a molecule which supplies energy) from precursors (1); radioactive amino acids that were added to the system by the researchers (2); and a centrifuge fraction (unnumbered) coming in from below, which was presumed to contain enzymes. These all come together on the microsomes (3). Zamecnik later showed that the microsome had as a component a ribonucleoprotein particle, even later called a "ribosome," which was the functional unit (Zamecnik 1958). The microsomes were hypothesized to be the location of protein synthesis, as the arrow pointing to "incorporation" indicates. "Incorporation" traces the flow of radioactive counts from the labeled amino acids to what was presumed to be (and later shown to be) a polypeptide chain. This chain consisted of amino acid subcomponents that were covalently bonded with peptide bonds. For Zamecnik and his colleagues, the termination condition to be understood was peptide bond formation, and understanding this required finding its energy requirements, as well as the possible intermediates between free amino acids and polypeptide chains (Zamecnik 1969).

Zamecnik's lab group concentrated their investigations on the stages of the mechanism near the end product. This continued a tradition in biochemistry, begun earlier in the century, in which the nature of the peptide bond and the chemical formulae of amino acids had been elucidated (Zamecnik 1969; Rheinberger 1997). Biochemical and cytological studies had shown that RNA was part of the microsomes and was somehow associated with protein synthesis. (Brachet's and Caspersson's work on this is discussed in Thieffry and Burian 1996; Rheinberger 1997.) However, it was unclear in 1954 what the RNA did. It was not even a schematic placeholder in a biochemical equation; it filled no biochemical role.

In contrast, Watson's 1954 diagram (see the lower part of Figure 3.1) follows the flow of genetic information from DNA to protein via an RNA intermediate stage. The line labeled "protein (Gamow holes?)" refers to a hypothesized geometrico-mechanical activity for the RNA in determining amino acid order. Watson's idea was that DNA is copied into RNA, which forms a structure with differently shaped holes. Then, different amino acids would fit into the holes. The RNA template would thus determine the order in which the amino acids bonded in the protein. RNA did play a role in the molecular biological schema of DNA –> RNA –> protein. It filled a gap in the flow of information between DNA and protein.

Watson was altering an idea proposed by the physicist George Gamow (1954). Gamow had proposed that proteins were synthesized directly on the DNA double helix, by fitting into "holes" in the helix (discussed in detail in Kay 2000, Ch. 4). The holes were the spaces between the turns of the helix. Gamow thought that different groups of three or four bases in the DNA might produce differently shaped holes. This was his way of filling in Watson and Crick's brief suggestion that it "seems likely that the precise sequence of the bases is the code which carries the genetical information" (1953b, p. 966).

Gamow speculated about the nature of the code – that is, the relation between a particular group of bases (presumed to be 3 or 4) and a particular amino acid (discussed in Kay 2000). However, Gamow was unaware of the biochemical and cytological evidence that proteins are not synthesized directly on DNA but are associated with RNA in the cytoplasm. Consequently, the entity whose structure was relevant was at a subsequent stage of the mechanism; it was not the structure of DNA but of RNA. Carrying Gamow's idea forward to the next stage, Watson proposed that the cytoplasmic RNA had "holes," whose shapes were determined by the RNA bases surrounding the holes. Different amino acids were assumed to fit into different holes, bringing them into close spatial proximity so that peptide bonds could form. Finally, the protein would leave the RNA template. In 1954, Watson speculated that RNA might have a helical shape (Rich and Watson 1954a, 1954b; discussed in Watson 1962, 2000).

The contrasts between Zamecnik's and Watson's schematic mechanism diagrams graphically depict the differences between the approaches of biochemists and molecular biologists mentioned briefly above. They had different starting points. Watson began with DNA, Zamecnik with the energy requirements of peptide bond formation. The molecular biologists wished to show how the order of the bases in the DNA (the genetic material) determined the order of the amino acids in the protein. Zamecnik's lab investigated high-energy intermediates back along the path to free amino acids.

The differences between the fields were clearly recognized at the time. Hoagland, a colleague in Zamecnik's lab, drew this contrast in recounting his 1955 announcement of amino acid activation at a meeting of molecular biologists:

> The palpable indifference with which the audience received the news showed how tightly closed the door between the biochemistry and molecular biology compartments was. The focus of the meeting, of course, was on how the polymerizing system ordered its units, not on how it energized them.
>
> (Hoagland 1996, p. 78)

Zamecnik also reflected back on the differences:

> If molecular biology is the domain of large molecules, the study of protein synthesis was rooted in the simple biochemical desire to understand how the energy barrier from free amino acid to peptide was overcome. The numerous connections of protein synthesis with the nucleotide-nucleic acid family were unanticipated in the early days. (Zamecnik 1984, p. 466)

Despite these differences, researchers in both fields recognized the importance of understanding the mechanism of protein synthesis. Both were speculating about some role for RNA. They followed each other's work, even though the molecular biologists had little interest in the energy requirements of the mechanism. Zamecnik recounted a visit by Watson in 1954 during which they discussed the newly discovered double helix model of DNA. Zamecnik expressed disappointment that the bases are inside the helix; it seemed as if they were less available to play some role in protein synthesis (Zamecnik 1969, p. 5). Researchers in both fields knew of Frederick Sanger's work in the late 1940s and early 1950s in which he determined the sequence of amino acids in insulin and thereby showed that various prior hypotheses about protein structures were incorrect (recalled in Crick 1988; Zamecnik 1962a;

discussed in Chadarevian 1996). They approached the role of RNA from both ends, with their own techniques and perspectives on the mechanism, moving forward from DNA and backward from peptide bond formation.

### 3.3.2 Molecular Biological Work: Three-Dimensional Structure

In 1954, Watson and Rich attempted to find the structure of RNA by using the same techniques that had led to the discovery of the double helical structure of DNA. They produced fibers and took x-ray photographs, but the results were inconclusive and did not provide sufficient evidence for the hypothesized RNA helix (Rich and Watson 1954a, 1954b).

RNA viruses, such as tobacco mosaic virus (TMV), produce their own proteins in the host. In 1952, Watson had learned x-ray crystallography techniques by investigating the structure of TMV (Watson 1968). TMV is rod-shaped, but other small RNA viruses are spherical. Crick and Watson (1957, p. 12) noted that spherical RNA viruses have a similar RNA content, as well as a similar shape, to the microsomes of cells. In their work on viral structure, Crick and Watson speculated that the amino acid sequence of the viral protein "is determined by the molecular structure of the RNA of the infecting virus . . . " and "the 'coding' implied . . . is relatively simple" (Crick and Watson 1957, p. 7). However, their hypotheses of the structure of viruses (Crick and Watson 1956) did not illuminate the structure of the microsomes or the code for relating RNA bases to amino acids. RNA viruses were not sufficiently like microsomes to provide a simpler system for studying the RNA intermediate in protein synthesis more generally.

In sum, the x-ray techniques, the studies of viral structure, and the search for a presumed three-dimensional structure of RNA did not bring the same success for the structures and activities of RNA in protein synthesis as it had for the structure of DNA and the mechanism of DNA replication. Extending the research program from the success with DNA structure was a plausible approach, but these early molecular biological efforts did not elucidate the mechanism of protein synthesis.

### 3.3.3 Draining the Biochemical Bog

While Watson and Crick were investigating RNA structure, Zamecnik's lab was busy "draining the biochemical bog" (Rheinberger 1997) of the homogenized rat-liver system for cell-free protein synthesis. They were grinding up rat livers and subjecting them to centrifugation in order to separate components and thereby to investigate their relative roles in the mechanism. Such work

74

allowed the discovery of previously unknown entities in centrifuge fractions that were required for the mechanism to operate. By 1954, the cell-free system for the incorporation of $^{14}$C-labeled amino acids into a polypeptide chain was working. In addition to the amino acids, the system required microsomes, an undifferentiated centrifuge fraction (the 105,000g supernatant), and ATP (a molecule that provides energy), but no DNA. When Hoagland joined Zamecnik's lab, his research focused on the question of whether an intermediate step occurred in the reaction between free amino acids and the formation of peptide bonds (Zamecnik 1969; Hoagland 1990). This question was within the traditional biochemical approach of finding intermediates in chemical reactions and in investigating the energy requirements for the formation of covalent bonds – in this case, the peptide bond. Hoagland did indeed find a high-energy intermediate in the reaction, as he backward chained along the pathway to free amino acids. In Zamecnik's (1984) biochemical notation, this reaction is characterized as follows:

$$aa_1 + pppA + E_1 <-> aa.pA_1.E_1 + pp$$

The amino acid (aa) combines with ATP (adenosine triphosphate, here pppA), and the reaction is catalyzed by an enzyme (E). Two phosphates (pp) are released and the other product of the reaction is an activated amino acid, called "aminoacyl-AMP" (aa.pA $_1$). This reaction provided "a mechanism for activation of amino acids" (Hoagland 1955, p. 288). Thus, as Hoagland backward chained from bound amino acids in proteins to free amino acids, he found the first of two intermediates. This aminoacyl-AMP was the expected high-energy intermediate of an activated amino acid. The other was a bit of RNA that changed from being "junk" to having an important role in the mechanism, a role not anticipated in the biochemical reaction schemata (Hoagland 1990, Ch. 5; Rheinberger 1997). Transfer RNA was one of several RNAs that would be found to play roles in the mechanism and whose discovery would fill gaps in the understanding of its productive continuity.

### 3.3.4   *Adaptor RNA and Soluble RNA to Transfer RNA*

After the biochemical work had shown that the ribonucleoprotein particles, not the lipoprotein membranes of microsomes, are associated with protein synthesis (Zamecnik 1958), Watson began work on three-dimensional ribosomal structure. He thus continued to pursue his view that the three-dimensional structure was the crucial property of RNA for enabling its activity (discussed in Watson 1962; 2000).

In contrast to the reaction equations of the biochemists, the molecular biological notation depicts information flow. As of about 1957, the sequential stages in the mechanism schema might have been depicted in the following way:

one DNA sequence –>

one nucleoprotein particle in the microsome –>

one protein chain

Crick began to think about the requirements to build a three-dimensional structural model of RNA with twenty differently shaped holes. As he recounted: "Well, since we can't find the structure of RNA from the data, let's do it the other way around by assuming that there are twenty different cavities and trying to build a structure that *had* twenty different cavities. And as soon as you put it that way, you saw that it was almost impossible to *do*" (quoted in Judson 1996, p. 283). Some of the amino acids have very similar shapes. Unable to construct any possible scale model, Crick became skeptical of Watson's "holes in RNA" hypothesis.

Crick hypothesized using the weak electro-chemical activity of hydrogen bonding, not the geometrico-mechanical activity of Watson's schema, for the role of ordering amino acids. Crick recounts his reasoning:

> The main idea was that it was very difficult to consider how DNA or RNA, in any conceivable form, could provide a direct template for the side-chains of the twenty standard amino acids. What any structure *was* likely to have was a specific pattern of atomic groups that could form hydrogen bonds. I therefore proposed a theory in which there were twenty adaptors (one for each amino acid), together with twenty special enzymes. Each enzyme would join one particular amino acid to its own special adaptor. This combination would then diffuse to the RNA template. An adaptor molecule could fit in only those places on the nucleic acid template where it could form the necessary hydrogen bonds to hold it in place. Sitting there, it would have carried its amino acid to just the right place it was needed.                                    (Crick 1988, pp. 95–96)

This idea was first suggested in 1955 in a widely circulated but unpublished paper (Crick 1988, p. 95). In its first published form, Crick speculated about the nature of the molecules that played the role of adaptors: "These might be any sort of small molecule – amino sugars, for example – but an obvious class would consist of molecules based on di-, tri-, or tetranucleotides" (Crick 1957, p. 26). These nucleotides would be particularly suited to play the role of the complementarily charged items needed for hydrogen bonding. Nucleic acid bases have polar structures with slight charges. Such charges were already known to hold the double helix of DNA together; maybe they also played a role in the protein synthesis mechanism.

Meanwhile, Zamecnik and Hoagland had been investigating RNA synthesis as an adjunct research project to their investigation of protein synthesis. In their centrifuged preparations, they found a soluble RNA fraction that differed from the heavier microsomal RNA. To their surprise, this soluble RNA was covalently bound to the $^{14}$C-labeled amino acids (Zamecnik 1969, p. 6; Hoagland 1990, p. 94; Judson 1996, p. 324; Rheinberger 1997, p. 155).

In 1956, the biochemical work in Zamecnik's lab was integrated with the molecular biology work after Watson visited Hoagland in Zamecnik's lab. Anything serving the role of an adaptor in the protein synthesis mechanism would have to bind to amino acids. When Watson saw Hoagland's results, he told Hoagland of Crick's as-then-unpublished idea of an adaptor RNA that attached to the amino acids (Hoagland 1990, p. 93).

Hoagland recalls:

> I was bowled over by the ingenuity and beauty of Francis's idea and sensed that it had to be the explanation of our experimental findings. An image arose in my mind: we biochemical explorers were hacking our way through a dense jungle to discover a beautiful long-lost temple, while Francis Crick, flying gracefully overhead on gossamer wings of theory, waited for us to see the goal he already was gazing down upon. (Hoagland 1990, p. 94)

He also told less happy stories about this episode (quoted in Rheinberger 1997, p. 157). Naturally, scientists prefer to figure out for themselves what role in a mechanism is played by an entity they discover. Of this episode, Rheinberger, who stresses the differences between the two approaches, said: "Biochemical reasoning in terms of metabolic intermediates came to be confronted with reasoning in terms of genetic information transfer" (Rheinberger 1997, p. 158).

In 1959, Hoagland, Zamecnik, and their colleague Mary Stephenson published a figure (Figure 3.2) that incorporated this idea of a transfer RNA, the small saw-toothed pieces, which were still referred to as soluble RNA. They are shown binding to amino acids (aa) and bringing them to the supposedly helix-shaped ribosomal RNA comprising the ribosome. The saw-toothed patterns, although depicted geometrico-mechanically, were said to represent hydrogen bonding between complementary charges on bases of the soluble RNAs and the ribosomal RNA (Hoagland et al. 1959, p. 111). The solid lines along the string of amino acids represent the usual peptide bonds. This diagram integrates the perspectives of both fields. Molecular biology provided the hypothesized helical structure of the template RNA and the activity

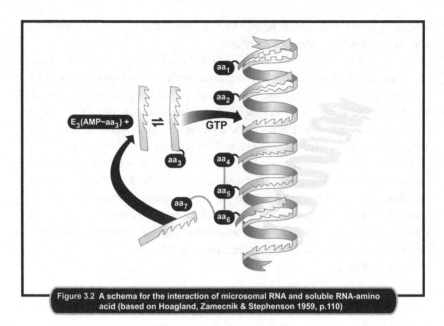

Figure 3.2 A schema for the interaction of microsomal RNA and soluble RNA-amino acid (based on Hoagland, Zamecnik & Stephenson 1959, p.110)

of hydrogen bonding. Biochemistry provided the activated amino acids that become covalently bonded, first to transfer RNA and then to each other in the protein.

Furthermore, the biochemical results served to revise details of Crick's adaptor hypothesis. Transfer RNA, as soluble adaptor RNA came to be called, turned out to be larger than Crick proposed. Hoagland (1996, p. 79) said that Crick should have expected the adaptor to be larger than the coding trinucleotide because it also needed a specific active site to attach to its specific amino acid. Crick had not sufficiently considered the side reaction in the mechanism that Hoagland had extensively investigated. Prior to the coding stage (on which Crick focused), the amino acid attached to the adaptor RNA. Thus, the RNA had to be large enough to have two active sites, one to bind the amino acid and another to hydrogen bond to template RNA. Consideration of all its roles in the mechanism would have shown that the adaptor RNA had to be larger than a di- or tri-nucleotide.

Again, the molecular biologists and the biochemists had concentrated on different subcomponents of the mechanism, even different activities of a given subcomponent. They reached there by different routes. Crick was reasoning forward about the activities of nucleic acids in information transfer, while Hoagland was chaining backward along the path to the activated amino acids and empirically isolating them in an in vitro system where he, surprisingly, found an attached RNA.

Crick gave the term "information" a precise characterization after he stated what he dubbed the "Central Dogma," which "states that once 'information' has passed into protein *it cannot get out again.* In more detail, the transfer of information from nucleic acid to nucleic acid or from nucleic acid to protein may be possible, but transfer from protein to protein, or from protein to nucleic acid is impossible. Information means here the *precise* determination of sequence, either of bases in the nucleic acid or of amino acids in residues in the protein" (Crick 1958, p. 153).

### 3.3.5 *Anomalies for the Ribosome as Template*

In the late 1950s, anomalies began to emerge for the view of ribosomes as carrying the information for ordering the amino acids. In 1958, two Russians, Belozerskii and Spirin, showed that "the DNA of different microorganisms had greatly different base ratios. . . . The base composition of the total RNA of these same organisms hardly varied at all . . . " (Crick 1959, pp. 35–36). Most of the RNA in the cell is found in the ribosomes. If DNA is transcribed into ribosomal RNA, one would expect the base ratios of the DNA and the riboso- mal RNA in a given species to be similar. They were not. This anomalous data, confirmed by two groups, challenged the role proposed for the ribosome as a template in the mechanism of protein synthesis. The ribosomal base anomaly indicated a problem about the DNA to RNA step in the proposed mech- anism schema because the expected relationship between the two was not found.

Crick (1959) generated a set of alternative hypotheses to resolve this anomaly, localized in various stages of the mechanism schema. We will not take the space here to consider each of them; notably, he did not include the postulation of an as-yet-undiscovered type of RNA having a base composition like DNA. This is the idea of a separate messenger RNA, different from either transfer RNA or ribosomal RNA. The discovery of such a messenger RNA is how the ribosomal base anomaly was soon resolved in 1961. Reflecting back on this reasoning later (Crick 1988), Crick says that there were plenty of RNAs available; that is, the ribosome had been shown to have two RNA components of different sizes. So, there seemed no need to postulate an as-yet-undetected type of RNA to serve as the expected template.

It is instructive to note the difference here between adaptor RNA and messenger RNA from the perspective of the mechanism's activities. When only one type of RNA was known – namely, the ribosome – Crick needed a second type, the adaptor, to postulate hydrogen bonding. Hydrogen bonding occurs between complementary charges on different molecules, so the use of the activity of hydrogen bonding demanded a second type of molecule, probably RNA, to bond to ribosomal RNA. Hence, Crick hypothesized a second type, adaptor RNA, to play this role. No such demand was present that required messenger RNA; Crick was not forced to postulate it by reasoning forward about the activities of the mechanism or their correlative entities. Instead, it was discovered during the resolution of an empirical anomaly about the rate of protein synthesis in in vivo genetic experiments. (For a general discussion of anomaly driven theory change, see Darden 1991, 1995, see Chapter 10, this book, 1998; Darden and Cook 1994.)

### 3.3.6   Rate Anomaly and Messenger RNA

In addition to the ribosomal base anomaly, another empirical anomaly arose for the idea of a stable ribosome as the template for protein synthesis in bacteria. Work by Arthur Pardee, Francois Jacob, and Jacques Monod produced the puzzling experimental result. Jacob and Monod were working on the control of protein synthesis using genetic techniques and observing the effects of mutants (Judson 1996; Morange 1998). In the process of studying its control mechanism, they discovered a new component of the primary mechanism of protein synthesis. As Crick said in reflecting on reasoning in mechanism discovery: One must sort out effects due to the nature of a mechanism itself from effects due to its control when trying to unscramble a complex biological system (Crick 1988, p. 111). Forward and backward chaining require sorting out the components of a primary mechanism from the components of mechanisms that control it.

In the famous, so called PaJaMo bacterial mating experiment, protein synthesis began very quickly after a functional gene entered recipient *E. coli* bacteria lacking that gene (Pardee, Jacob, and Monod 1959; discussed in Olby 1970; Schaffner 1974a; Judson 1996; Morange 1998). Ribosomes were assumed to be stable particles, requiring some time for synthesis. If the ribosomal RNA had to be synthesized on the incoming DNA (the functional gene), and then the ribosomal particle had to be assembled, one would not expect the very rapid initiation of protein synthesis. A greater time lag would be expected. Again, the molecular biologists were chaining forward from the gene to the next stage in protein synthesis; they were considering the time that the assembly of the entities (presumed to be ribosomal templates) in the next stage would require.

Three alternative hypotheses were devised to account for the surprising rapidity of the initiation of protein synthesis (Olby 1970). Each stage in the proposed molecular biological mechanism schema served as a site of hypothesis formation. Reasoning forward from the DNA, the first hypothesis proposed that the DNA of the gene itself could serve as the site of protein synthesis. But this hypothesis was problematic because protein synthesis was already known to be associated with the ribosomes, not with DNA. Maybe protein synthesis in bacteria differed from the systems previously studied by biochemists, such as Zamecnik's cell-free, rat-liver system. Bacterial DNA is not in a separate nucleus but is in the cytoplasm itself. Monica Riley, Pardee's student, thought the hypothesis of synthesis on the bacterial DNA might be the correct hypothesis (Riley, personal communication).

A second hypothesis was that ribosomes were rapidly synthesized after the DNA entered the recipient cell. But ribosomes were known to be stable particles with at least two RNA subunits, so there seemed to be insufficient time to synthesize new ones before protein synthesis started. The third hypothesis was that a new RNA, with a base sequence like that of the DNA, was synthesized quickly. This DNA-like RNA (also called X or "tape" RNA) would then use the existing stable ribosomes as the sites for protein synthesis. (For further discussion of these three alternatives, see Olby 1970; Jacob 1988.)

Thus, we see that reasoning forward about the mechanism's stages and the time needed for each stage introduces temporal constraints on the description of the mechanism (Craver and Darden 2001; see Chapter 2, this book). If a hypothesized stage would be expected to take longer than a time-course experiment shows that it does, then the resulting anomaly indicates the need for a change in the hypothesized mechanism. Something else was needed to instantiate the role of the template, something that can form more rapidly than ribosomes, something with a base sequence like that of DNA.

Some empirical evidence already existed for a DNA-like RNA, but it had been misinterpreted (discussed in Watson 1962; Judson 1996, pp. 414–415, 418–422). New experiments were done to try to provide more direct evidence for this "tape RNA" (Jacob 1988) or "messenger RNA," as Jacob and Monod named it in 1961:

> The property attributed to the structural messenger of being an unstable intermediate is one of the most specific and novel implications of this scheme . . . This leads to *a new concept of the mechanism of information transfer*, where the protein synthesizing centers (ribosomes) play the role of non-specific constituents which can synthesize different proteins, according to specific instructions which they receive from the genes through M-RNA.
>
> (Jacob and Monod 1961, p. 353; emphasis added)

This unstable intermediate in the mechanism was a new type of RNA. Messenger RNA was transcribed from DNA and thus carried the genetic code for ordering the amino acids during protein synthesis. This new type of RNA would be formed quickly, as was required by the rate at which synthesis began after the DNA entered the recipient bacterium. Flow of information – namely, the order of bases and the order of amino acids – was again the focus of the molecular biologists. (Compare Crick's statement that he would not discuss the flow of energy or matter but the flow of information in protein synthesis in Crick 1958, p. 144.)

New experimental work provided good evidence for the existence of messenger RNA. Sydney Brenner and Francois Jacob succeeded in differentially labeling old ribosomal RNA and newly synthesized messenger RNA (Brenner et al. 1961). Also, in Watson's lab, Francois Gros added radioactively labeled uracil to *E. coli* and detected an unstable RNA that was neither ribosomal nor transfer RNA (Gros et al. 1961). All the different RNA components of the mechanism had been found. (For more on the discovery of messenger RNA, see Gaudilliere 1996; Gros 1979; Jacob 1988; Judson 1996; Morange 1998; Olby 1970; Rheinberger 1997; Watson 1962.)

Subsequent work, utilizing both biochemical and, to a lesser extent, genetic techniques led to cracking the genetic code (Nirenberg and Matthaei 1961; Crick et al. 1961). The work on the code is one with more interfield competition between biochemists and molecular biologists than the cooperation discussed thus far. That story has been told elsewhere (Crick 1988; Judson 1996; Kay 2000).

From that competition emerged an integrated account of the triple, non-overlapping code operating within the protein synthesis mechanism. The base sequence in DNA is transcribed into messenger RNA, which moves to the ribosomes, the site for the subsequent stages. A specific triplet codon on the messenger RNA hydrogen bonds to its complementary anticodon on a transfer RNA, which is attached to its specific activated amino acid. As the transfer RNAs bond sequentially to the messenger RNA, the amino acids are brought into appropriate proximity so that peptide bonds form. Incorporation of amino acids occurs in a specific linear order, based on the order of the codons in the messenger RNA. The ribosomes are merely the site where the messenger RNA attaches and is held in an appropriate orientation for these stages in the mechanism to occur. The ribosome no longer played the role of template in the information flow schema; that role was now filled by the tape-like messenger RNA. The type of mechanism had changed from a mold-like template to a tape along which the ribosome moved, sequentially incorporating amino acids (Crick 1988).

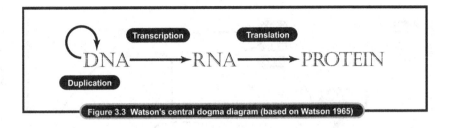

Figure 3.3 Watson's central dogma diagram (based on Watson 1965)

### 3.3.7 Summary of Protein Synthesis Case

Diagrams by Watson represent the integrated understanding of the mechanism of protein synthesis as of the 1960s. The diagrams come from Watson's (1965) *The Molecular Biology of the Gene*, the first molecular biology textbook. Just as the Zamecnik group began incorporating the molecular biological views of structure and hydrogen bonding in their 1959 diagram, Watson's more detailed 1965 diagrams integrate the biochemical findings of the activation of amino acids and their covalent bonding to soluble (later "transfer") RNA.

Watson's abstract schema for the mechanism of protein synthesis is "often called the central dogma" (Watson 1965, p. 297). Figure 3.3 illustrates the components of the mechanism in a very schematic way. DNA is transcribed into messenger RNA, which is translated into protein. Parts of this mechanism are depicted in much more detail in two other figures. Figure 3.4 shows the roles of all three types of RNAs and the other entities in the mechanism, including the DNA, studied by molecular biologists, and the amino acids, studied by biochemists. Activities include *transcription* of DNA to messenger RNA and *translation* of messenger RNA, utilizing transfer RNAs and ribosomes, into protein. Translation is carried out, most importantly, by the activity of *weak hydrogen bonding* between slightly charged complementary bases in the transfer RNA and the messenger RNA. The biochemical findings are shown in the lower right: amino acids (AA) combine with ATP to form the high-energy intermediate (AA~AMP), which then covalently bonds to soluble RNA (sRNA), which then moves to the ribosome. Further biochemical entities are depicted: "Enzymes involved in protein synthesis" and another energy molecule (GTP), whose role in the mechanism was then unknown. Another of Watson's figures (Figure 3.5) graphically depicts the hydrogen bonding. Hydrogen bonding between three bases and their complements is how the genetic code is "read." The ribosome, with its two circular parts, moves along the messenger RNA in the successive stages of the mechanism. The growing peptide chain of amino acids extends from the top of the diagram. Note the structural details and the way the activity of hydrogen bonding

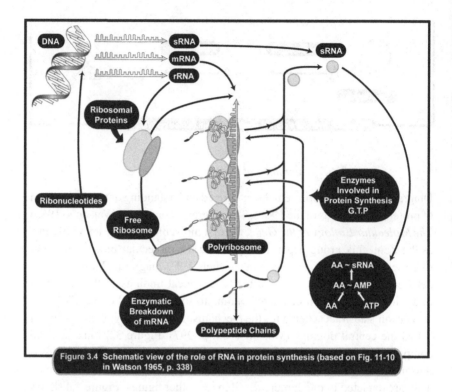

Figure 3.4 Schematic view of the role of RNA in protein synthesis (based on Fig. 11-10 in Watson 1965, p. 338)

is depicted extending between the lines representing the charged bases of the transfer RNA and messenger RNA. Structures of macromolecules and weak hydrogen bonding are important features of the molecular biologist's schema of protein synthesis.

The discovery of all the key components in the mechanism of protein synthesis required the integration of aspects of the mechanism studied by both fields. Molecular biologists worked forward from the DNA. Biochemists worked backward from peptide bonds to activated amino acids. They met in the middle and both contributed to the elucidation of the roles of the three types of RNA and the activities of hydrogen bonding, covalent bonding, and energetic intermediates.

### 3.4 STRATEGIES FOR DISCOVERING MECHANISMS

The discovery of the mechanism of protein synthesis illustrates a number of strategies for discovering mechanisms that move beyond those discussed in previous work (Darden 1991; Bechtel and Richardson 1993). This section

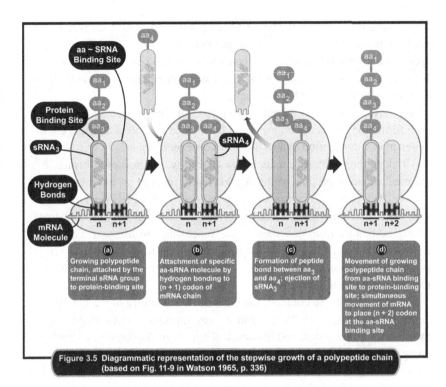

Figure 3.5 **Diagrammatic representation of the stepwise growth of a polypeptide chain** (based on Fig. 11-9 in Watson 1965, p. 336)

discusses those strategies in a more general way. Discovery cannot be relegated to wooly flights of imagination, unconstrained conjecture, or single "a-ha" moments. Rather, there are more or less reliable yet inherently fallible strategies for discovery. Discovery is construed broadly to include construction, evaluation, and anomaly resolution.

### 3.4.1 Productive Continuity

Mechanisms – in particular, the mechanism of protein synthesis – are entities and activities organized such that they are productive of regular changes (Machamer, Darden, and Craver 2000; see Chapter 1, this book). The entities and activities are organized spatially and temporally in a productive continuity in which one stage produces or gives rise to the next (Craver and Darden 2001; see Chapter 2, this book). This productive continuity may be traced by following what flows or what is passed, transmitted, preserved, transformed, prevented, or allowed from one stage of the mechanism to the next.

In reasoning in the discovery of mechanisms, one proceeds with the goal of eliminating gaps in the description of the mechanism's productive continuity.

These gaps may appear as spatial or temporal gaps, where no role can as yet be schematized. Alternatively, a gap may simply involve an inability to specify how a given role in the mechanism schema is to be filled. In searching for the productive continuity of the mechanism, one must find an activity for each entity and an entity for each activity.

We have seen two different conceptions of this productive continuity: an informational conception proposed by molecular biologists and a chemical/energetic conception investigated by some biochemists.

Crick delineated flow of matter, flow of energy, and flow of information in the mechanism of protein synthesis and clearly focused on information flow (Crick 1958, p. 144). What mattered to molecular biologists such as Watson and Crick was tracing information flow, that is, the preservation of linear order and pattern from one stage of the mechanism to the next. Productive continuity in this case is measured in terms of constancy (or at least partial preservation) of linear order from DNA bases through messenger RNA bases to amino acids in the protein.

For biochemists, in contrast, the productive continuity of the mechanism depends on the flow of matter and energy through various stages of the mechanism, usually depicted in chemical equations. Zamecnik sought high-energy intermediates (e.g., the activated amino acids) to fill the productive gap between free amino acids and strong covalent bonds holding them together in the polypeptide chains, that is, to understand how the energy barrier from free amino acid to peptide was overcome.

The chemical/energetic and the informational senses of productive continuity were integrated with the discovery of the continuous mechanism involving the flow of matter, energy, and information, all together.

This case reveals two general types of strategies for reasoning to eliminate gaps in the understanding of the productive continuity of a mechanism or, more positively, for discovering the organization that sustains that productive continuity. These strategies are, first, schema instantiation and, second, forward/backward chaining. We consider these in turn.

### 3.4.2  Schema Instantiation

One strategy for constructing a hypothetical mechanism is to instantiate an abstract schema (Darden and Cain 1989; see Chapter 8, this book; Machamer, Darden, Craver 2000; see Chapter 1, this book). Such a schema provides the framework of the mechanism. Schemata typically specify roles, black boxes, at varying degrees of abstraction with more or less detail specified. The schema terms can then be filled with the mechanism's entities and activities as they are hypothesized and discovered.

In the discovery of the mechanism of protein synthesis, biochemists and molecular biologists were using different schemata. Biochemists tried to instantiate chemical reaction schemata in chemical equations to show how an energetically unfavorable reaction could occur. These chemical schemata guided researchers to seek high-energy intermediates and led Hoagland and Zamecnik to see the importance of activated amino acids. The schemata represented in biochemists' equations exhibit the productive continuity explicitly as balance of matter, reshuffling molecules on either side of the equations. Such chemical-reaction schemata suggest finding balanced equations, high-energy intermediates, enzymes, and conditions leading the reaction to go in one direction rather than the other.

Molecular biologists instead used a schema depicting the flow of information, which Crick characterized as the precise determination of sequence of either bases or amino acids (Crick 1958). That flow was further specified in Watson's version of the central dogma. (For the discussion of the contrast between Watson's and Crick's versions of the central dogma, see Keyes 1999a, 1999b.) Watson worked to find the pieces to instantiate the following schema:

DNA –> RNA –> protein

As of 1952, what Watson knew about the RNA portion in the middle was very sketchy, a mere black box (role), specified only as a template. Subsequent work explored different ways of filling that black box, of finding the RNA machinery, and of thereby eliminating this gap in the mechanism's productive continuity and information flow. In contrast, the biochemists' schemata had no role for RNA. Thus, when it was empirically found to be associated with protein synthesis, that produced a puzzle as to what role it played.

As discussed in Section 3.3, the mechanism schema of protein synthesis underwent several changes. Initially, there were two schemata for the different fields. The molecular biologists' schema:

(1m) DNA –>helical RNA –>protein

ordered amino acids via a geometrico-mechanical activity.

The biochemist's schema:

(1b) amino acids + ATP + other centrifuge fractions –> activated aa complex –> protein focused instead on the energy requirements of the reaction. These schemata were integrated in an interim schema:

(2) DNA –>ribosomal RNA + activated aa-tRNA complex –>protein

with ordering via hydrogen bonding activity. In the face of empirical anomalies for the idea of the ribosomal template and the subsequent discovery of messenger RNA, the integrated schema was then transformed to:

(3) DNA –>messenger RNA + activated aa-tRNA complex + ribosomal RNA –> protein, with ordering via hydrogen bonding activity between triplet codons.

These changes show that the schema became more complicated, with one type of RNA being changed to two, then to three types. Also, the activity of hydrogen bonding came to play a crucial role in the linear ordering of the amino acids. The biochemical work on the role of ATP in the activation of amino acids was integrated into the schema as a source of energy for peptide bond formation. These interfield relations lead to the integration of previously separate schemata into a single productively continuous description of the mechanism.

Schema instantiation, however, is but one of the strategies for discovering mechanisms. Gaps in the understanding of a mechanism's productive continuity can also be filled by relying on what we know about entities, activities, and their interdependencies. This knowledge is deployed in the strategies of forward/backward chaining.

### 3.4.3 Forward and Backward Chaining to Find Productive Continuity

When reasoning about productive continuity in mechanisms, one can reason from either end of the mechanism, that is, by forward or backward chaining. Because the choice of the beginning and the end of the mechanism is somewhat relative and context-dependent, this forward-backward strategy applies equally well to any convenient starting or ending point in a mechanism. Furthermore, this reasoning strategy can be applied to branching mechanisms, which can be followed forward or backward along side branches, and to feedback and other cyclic mechanisms, which can be arbitrarily viewed as beginning somewhere in the cycle.

Forward and backward chaining are made possible by the intelligible relationships between interdependent entities and activities. Entities and a specific subset of their properties enable the activities in which they engage (given appropriate conditions). Furthermore, activities require distinct types of entities and properties of those entities as the basis for such acts.

Scientists come to mechanism construction with prior knowledge about types of entities and activities that facilitates forward/backward chaining. Typically, a field will have a store of knowledge about types of entities and activities that figure in proposed mechanisms in that field at a given time (Craver and Darden 2001; see Chapter 2, this book). As we have seen, during the 1950s and 1960s in molecular biology, that store included three-dimensional macromolecular nucleic acids and proteins (often having helical shapes), as well as weak types of chemical bonding, such a hydrogen bonding. Biochemists of that period (as exemplified in Zamecnik's work) typically hypothesized chemical reactions by considering smaller molecules, such as amino acids, with particular bonding properties, such as valence. Bonding activities were typically strong ones, such as covalent and ionic bonding. High-energy molecules, such as ATP, and catalysts, such as enzymes, were in the biochemists' store. (In our discussion of the historical case, we omitted the discovery of the enzymes involved in the protein synthesis mechanism; that's another part of the story.)

Forward and backward chaining each have two subtypes, one for entities and one for activities. Attending to entities during forward chaining, one may use what is known or conjectured about the *activity-enabling properties of entities*. Such attention allows one to speculate as to the kinds of activities that a given entity can engage in to produce the next stage of the mechanism. Alternatively, one may use knowledge of the occurrence of an activity in the mechanism to conjecture as to the consequences of that activity for entities and properties in the next stage. This is conjecturing about *activity consequences*.

Conversely, when backward chaining, the properties of an entity can provide clues as to the activities that produced it. That is, one or more of an entity's properties may serve as an *activity signature*, a property that signals to the researcher the prior occurrence of some activity. Alternatively, during backward chaining, one may find *entity signatures* of activities, that is, properties of activities that provide clues as to what entities in a prior stage may have lead to the occurrence of those activities.

We consider each of these in turn.

### 3.4.4 Activity-Enabling Properties of Entities

Quite general properties of entities enable the activities in which they can engage. Such properties include three-dimensional structure and size, as well as their orientation and location *in situ*. Structures can promote or prohibit geometrico-mechanical activities. Three-dimensional shapes can be open or closed, narrow or wide, exposing or concealing. This taxonomy of shapes, notice, is closely tied to the activities in which entities with such shapes can engage. An open entity permits movement through it more or less as it is narrow or wide. The same can be said of structural relations between two or more entities: their shapes may be complementary; one entity may be inside, behind, or above another; or they may be touching or distant. Again, these relational spatial properties can permit or prohibit activities. So, discovering the structural properties of an entity can often give clues as to the kinds of activities in which it is likely to engage in the next stage of the mechanism. An example is the hint about the nature of replication provided by the double helix structure of DNA (Watson and Crick 1953a, 1953b).

Activity-enabling properties of entities are not limited to such spatial and structural properties. Entities may also have different kinds of charges and molecules have valences, both of which affect the kinds of bonding activities in which they can engage. Charges have different strengths, different spheres of influence, and different arrangements within the three-dimensional structure of the entity.

In forward chaining, one may ask, what could these entities with these properties in these set-up conditions be expected to do? For example, molecular biologists often attend to the activity-enabling properties of macromolecules, such as their charges, their three-dimensional structures, and the orders of their components. The bases in the double helix of DNA, once the helix is opened, can be expected to guide complementary bonding with other bases. The slight charges on the polar bases are the properties that enable the DNA bases to engage in hydrogen bonding if charged molecules, complementary to them, are available. So, it was reasonable for Crick to conjecture hydrogen bonding between DNA bases and complementary RNA bases within the protein synthesis mechanism. RNA bases were chosen as the complements because empirical work had shown that the mechanism involved RNA at a subsequent stage. Another possible way to forward chain makes use of another property of the DNA double helix. Gamow used the structure of DNA and possible combinations of the four bases to conjecture the occurrence of a geometrico-mechanical activity (fitting into Gamow holes) that could order the amino acids in proteins.

These examples illustrate the ways that the activity-enabling properties of entities can be fruitfully used in forward chaining to conjecture the activities of the next stage of the mechanism.

90

### 3.4.5   Activity Consequences

Activities have such properties as rate, duration, strength, and sphere of influence. Forward chaining by reasoning about the next stage of the mechanism, one may ask: What is expected of the entities in the subsequent stage, given the prior occurrence of some activity? In a case where hydrogen bonding has occurred between complementarily charged polar molecules, for example, one expects that their charges have been neutralized and are not available to be activity-enabling properties for the next stage. Also, one expects to find a loosely associated (via hydrogen bonding) complex of molecules in that subsequent stage. Because such bonds are weak and easily broken, an even more subsequent stage may involve dissolving them. The hydrogen bonding of a triplet codon on messenger RNA to the anticodon on its corresponding transfer RNA is a transient phenomenon, maintained long enough for the attached amino acid to join the growing polypeptide chain. The hydrogen bond soon dissolves, as the next amino acid is brought into place.

On the other hand, suppose there is evidence for strong covalent bonding, such as finding the larger amount of energy that is required to break the bond. Then one knows that the molecule that is formed by such bonding is stable and, in any subsequent stage in which the molecule is broken apart, a given amount of energy will be required. Furthermore, in the covalent bonding activity, the valences of the bonding atoms have been used and are not available for additional bonding. The changes are a consequence of the activity of covalent bonding having occurred.

In contrast to these electro-chemical activities, geometrico-mechanical activities, such as lock and key docking, also have characteristic activity consequences. As the docking activity occurs, various stresses and strains may be transmitted through the new, larger structure, possibly changing other active sites, and permitting new geometrico-mechanical activities in the next stage. This sort of change occurs in what is called "allosteric" regulation, when one molecule docks at one site, and changes the shape of the molecule. (Electro-chemical charges are often also active in allosteric changes.) Such changes may then permit or stop some other activity. In further work on the mechanisms that control protein synthesis, such allosteric changes were found to play an important role (on Monod's discovery of allosteric regulation as the "second secret of life," see Judson 1996, Part III).

Knowledge of these general entity-enabling and entity-changing properties of activities allows one to reason forward about the next stage(s) of a mechanism, given knowledge that an activity occurred in a previous stage.

### 3.4.6   Activity Signatures of Entities

Now, consider reasoning backward rather than forward. Backward chaining may use properties of the entities at a later stage of the mechanism to conjecture the nature of the preceding stages. One asks how such entities could have been produced or what activities could have given rise to, driven, made, or allowed this later stage. Decomposing an end product may show its ingredients and provide hints as to what activities could have assembled those ingredients into the product.

As we have seen, beginning with the synthesized protein, biochemists decomposed it into amino acids and found that amino acids are joined by covalent peptide bonds. Once the formation of peptide bonds was shown to be energetically unfavorable, then an energy source was known to be necessary in a preceding step. A search began for high-energy intermediates. These end products carried activity signatures that aided backward chaining.

### 3.4.7   Entity Signatures of Activities

The characteristic features of an activity may provide clues as to the entities that engaged in it. Distinct kinds of activities require distinct kinds of entities with distinct kinds of properties to produce them. Suppose experimental evidence indicates that weak hydrogen bonding occurs in the mechanism, then one knows that slightly polar molecules must be present. Or, if one conjectures that an activity carries out some role in a mechanism, then that will demand certain properties of entities that can engage in it. Once Crick conjectured that amino acids were ordered via a hydrogen bonding activity, then this stage in the hypothesized mechanism required two types of complementarily, polarly charged molecules. Hence, not only a template RNA was needed but also an adaptor. When covalent bonding occurs, then entities with appropriate valences must be present. Furthermore, it is likely that enzymes and an energy source are present to carry out this energetically unfavorable reaction.

If lock and key docking has occurred, then two complementarily shaped entities were available. Locks and keys must have complementary structures, of appropriate sizes, appropriately oriented to each other, and they must come into contact to allow for such geometrico-mechanical activities.

In sum, the discovery of the mechanism of protein synthesis exemplifies the strategy of forward/backward chaining, going from entities to activities and back again.

### 3.5 CONCLUSION

Many (perhaps most) of the important discoveries in the biological sciences have been discoveries of mechanisms. So, our conclusions about the discovery of the mechanism of protein synthesis will likely generalize to other cases. Centering historical and philosophical analyses on mechanisms reveals previously neglected aspects of scientific discovery, reasoning, and theory change. Attending to mechanisms illuminates patterns in the organization of biological knowledge, which have various implications for constructing, evaluating, and revising that knowledge over time.

Here we have discussed two discovery strategies (i.e., schema instantiation and forward/backward chaining) that immediately suggest themselves once one thinks carefully about mechanisms. No doubt, this is just the beginning.

### REFERENCES

Bechtel, William (1984), "Reconceptualizations and Interfield Connections: The Discovery of the Link Between Vitamins and Coenzymes," *Philosophy of Science* 51: 265–292.

Bechtel, William (1986), "Introduction: The Nature of Scientific Integration," in W. Bechtel (ed.), *Integrating Scientific Disciplines*. Dordrecht: Nijhoff, pp. 3–52.

Bechtel, William (1988), *Philosophy of Science: An Overview for Cognitive Science*. Hillsdale, NJ: Lawrence Erlbaum.

Bechtel, William and Robert C. Richardson (1993), *Discovering Complexity: Decomposition and Localization as Strategies in Scientific Research*. Princeton, NJ: Princeton University Press.

Belozersky, Andrei N. and Alexander S. Spirin (1958), "A Correlation between the Compositions of Deoxyribonucleic and Ribonucleic Acids," *Nature* 182: 111–112.

Brandon, Robert (1985), "Grene on Mechanism and Reductionism: More Than Just a Side Issue," in Peter Asquith and Philip Kitcher (eds.), *PSA 1984*, v. 2. East Lansing, MI: Philosophy of Science Association, pp. 345–353.

Brenner, Sydney, Francois Jacob, and M. Meselson (1961), "An Unstable Intermediate Carrying Information From Genes to Ribosomes for Protein Synthesis," *Nature* 190: 576–581.

Burian, Richard M. (1996a), "Underappreciated Pathways Toward Molecular Genetics as Illustrated by Jean Brachet's Cytochemical Embryology," in S. Sarkar (ed.), *The Philosophy and History of Molecular Biology: New Perspectives*. Dordrecht: Kluwer, pp. 67–85.

Cech, T. R. (2000), "The Ribosome is a Ribozyme," *Science* 289: 878.

Chadarevian, Soraya de (1996), "Sequences, Conformation, Information: Biochemists and Molecular Biologists in the 1950s," *Journal of the History of Biology* 29: 361–386.

Chadarevian, Soraya de and Jean-Paul Gaudilliere (1996), "The Tools of the Discipline: Biochemists and Molecular Biologists," Introduction to Special Issue, *Journal of the History of Biology* 29: 327–330.

Craver, Carl F. and Lindley Darden (2001), "Discovering Mechanisms in Neurobiology: The Case of Spatial Memory," in Peter Machamer, R. Grush, and P. McLaughlin (eds.), *Theory and Method in the Neurosciences*. Pittsburgh, PA: University of Pittsburgh Press, pp. 112–137.

Crick, Francis (unpublished MS of 1955), "On Degenerate Templates and the Adaptor Hypothesis: A Note for the RNA Tie Club."

Crick, Francis (1957), "Discussion Note," in E. M. Crook (ed.), *The Structure of Nucleic Acids and Their Role in Protein Synthesis: Biochemical Society Symposium* 14 (February 18, 1956). London: Cambridge University Press, pp. 25–26.

Crick, Francis (1958), "On Protein Synthesis," *Symposium of the Society of Experimental Biology* 12: 138–167.

Crick, Francis (1959), "The Present Position of the Coding Problem," *Structure and Function of Genetic Elements: Brookhaven Symposia in Biology* 12: 35–39.

Crick, Francis (1988), *What Mad Pursuit: A Personal View of Scientific Discovery*. New York: Basic Books.

Crick, Francis, Leslie Barnett, Sydney Brenner, and R. J. Watts-Tobin (1961), "General Nature of the Genetic Code for Proteins," *Nature* 192: 1227–1232.

Crick, Francis and James D. Watson (1956), "Structure of Small Viruses," *Nature* 177: 473–475.

Crick, Francis and James D. Watson (1957), "Virus Structure: General Principles," in *Ciba Foundation Symposium on The Nature of Viruses*. London: J. & A. Churchill, pp. 5–13.

Darden, Lindley (1991), *Theory Change in Science: Strategies from Mendelian Genetics*. New York: Oxford University Press.

Darden, Lindley (1995), "Exemplars, Abstractions, and Anomalies: Representations and Theory Change in Mendelian and Molecular Genetics," in James G. Lennox and Gereon Wolters (eds.), *Concepts, Theories, and Rationality in the Biological Sciences*. Pittsburgh, PA: University of Pittsburgh Press, pp. 137–158.

Darden, Lindley (1998), "Anomaly-Driven Theory Redesign: Computational Philosophy of Science Experiments," in Terrell W. Bynum and James Moor (eds.), *The Digital Phoenix: How Computers are Changing Philosophy*. Oxford: Blackwell, pp. 62–78.

Darden, Lindley and Joseph A. Cain (1989), "Selection Type Theories," *Philosophy of Science* 56: 106–129.

Darden, Lindley and Michael Cook (1994), "Reasoning Strategies in Molecular Biology: Abstractions, Scans, and Anomalies," in David Hull, Micky Forbes, and Richard M. Burian (eds.), *PSA 1994*, v. 2. East Lansing, MI: Philosophy of Science Association, pp. 179–191.

Darden, Lindley and Nancy Maull (1977), "Interfield Theories," *Philosophy of Science* 44: 43–64.

Gamow, George (1954), "Possible Relation between Deoxyribonucleic Acid and Protein Structures," *Nature* 173: 318.

Gaudilliere, Jean-Paul (1993), "Molecular Biology in the French Tradition? Redefining Local Traditions and Disciplinary Patterns," *Journal of the History of Biology* 26: 473–498.

Gaudilliere, Jean-Paul (1996), "Molecular Biologists, Biochemists, and Messenger RNA: The Birth of a Scientific Network," *Journal of the History of Biology* 29: 417–445.

Glennan, Stuart S. (1996), "Mechanisms and The Nature of Causation," *Erkenntnis* 44: 49–71.

Gros, Francois (1979), "The Messenger," in Andre Lwoff and Agnes Ullmann (eds.), *Origins of Molecular Biology: A Tribute to Jacque Monod.* New York: Academic Press, pp. 117–124.

Gros, Francois, Howard Hiatt, Walter Gilbert, Chuck G. Kurland, R. W. Risebrough, and James D. Watson (1961), "Unstable Ribonucleic Acid Revealed By Pulse Labeling of *E. coli*," *Nature* 190: 581–585.

Hoagland, Mahlon B. (1955), "An Enzymic Mechanism for Amino Acid Activation in Animal Tissues," *Biochimica et Biophysica Acta* 16: 288–289.

Hoagland, Mahlon B. (1990), *Toward the Habit of Truth.* New York: Norton.

Hoagland, Mahlon B. (1996), "Biochemistry or Molecular Biology? The Discovery of 'Soluble RNA'," *Trends in Biological Sciences Letters (TIBS)* 21: 77–80.

Hoagland, Mahlon B., Paul Zamecnik, and Mary L. Stephenson (1959), "A Hypothesis Concerning the Roles of Particulate and Soluble Ribonucleic Acids in Protein Synthesis," in R. E. Zirkle (ed.), *A Symposium on Molecular Biology.* Chicago, IL: Chicago University Press, pp. 105–114.

Jacob, Francois (1988), *The Statue Within: An Autobiography.* New York: Basic Books.

Jacob, Francois and Jacques Monod (1961), "Genetic Regulatory Mechanisms in the Synthesis of Proteins," *Journal of Molecular Biology* 3: 318–356.

Judson, Horace F. (1996), *The Eighth Day of Creation: The Makers of the Revolution in Biology.* Expanded Edition. Cold Spring Harbor, NY: Cold Spring Harbor Laboratory Press.

Kay, Lily E. (2000), *Who Wrote the Book of Life? A History of the Genetic Code.* Stanford, CA: Stanford University Press.

Keyes, Martha (1999a) "The Prion Challenge to the "Central Dogma" of Molecular Biology, 1965–1991, Part I: Prelude to Prions," *Studies in the History and Philosophy of Biological and Biomedical Sciences* 30: 1–19.

Keyes, Martha (1999b) "The Prion Challenge to the "Central Dogma" of Molecular Biology, 1965–1991, Part II: The Problem with Prions," *Studies in the History and Philosophy of Biological and Biomedical Sciences* 30: 181–218.

Kitcher, Philip (1993), *The Advancement of Science: Science without Legend, Objectivity without Illusions.* New York: Oxford University Press.

Kuhn, Thomas (1962), *The Structure of Scientific Revolutions.* Chicago, IL: University of Chicago Press.

Laudan, Larry (1977), *Progress and Its Problems.* Berkeley, CA: University of California Press.

Machamer, Peter, Lindley Darden, and Carl F. Craver (2000), "Thinking About Mechanisms," *Philosophy of Science* 67: 1–25.

Meheus, Joke, and Thomas Nickles (eds.) (1999), *Scientific Discovery and Creativity: Case Studies and Computational Approaches*. Special Issue of *Foundations of Science* 4 (4).

Morange, Michel (1998), *A History of Molecular Biology*. Trans. from French by Matthew Cobb. Cambridge, MA: Harvard University Press.

Nickles, Thomas (ed.) (1980a), *Scientific Discovery, Logic, and Rationality*. Dordrecht: Reidel.

Nickles, Thomas (ed.) (1980b), *Scientific Discovery: Case Studies*. Dordrecht: Reidel.

Nirenberg, Marshall W. and J. H. Matthaei (1961), "The Dependence of Cell-Free Protein Synthesis in *E. coli* Upon Naturally Occurring or Synthetic Polyribonucleotides," *Proceedings of the National Academy of Sciences* (USA) 47: 1588–1602.

Olby, Robert (1970), "Francis Crick, DNA, and the Central Dogma," in Gerald Holton (ed.), *The Twentieth Century Sciences*. New York: W. W. Norton, pp. 227–280.

Pardee, Arthur B., Francois Jacob, and Jacques Monod (1959), "The Genetic Control and Cytoplasmic Expression of 'Inducibility' in the Synthesis of beta-galatosidase," *Journal of Molecular Biology* 1: 165–178.

Pauling, Linus and Robert B. Corey (1950), "Two Hydrogen-Bonded Spiral Configurations of the Polypeptide Chain," *Journal of the American Chemical Society* 72: 5349.

Popper, Karl R. (1965), *The Logic of Scientific Discovery*. New York: Harper Torchbooks.

Rheinberger, Hans-Jörg (1997), *Toward a History of Epistemic Things: Synthesizing Proteins in the Test Tube*. Stanford, CA: Stanford University Press.

Rich, Alexander and James D. Watson (1954a), "Physical Studies on Ribonucleic Acid," *Nature* 173: 995–996.

Rich, Alexander and James D. Watson (1954b), "Some Relations Between DNA and RNA," *Proceedings of the National Academy of Sciences* 40: 759–764.

Schaffner, Kenneth (1974a), "Logic of Discovery and Justification in Regulatory Genetics," *Studies in History and Philosophy of Science* 4: 349–385.

Thieffry, Denis and Richard M. Burian (1996), "Jean Brachet's Alternative Scheme for Protein Synthesis," *Trends in the Biochemical Sciences* 21 (3): 114–117.

Watson, James D. ([1962] 1977), "The Involvement of RNA in the Synthesis of Proteins," in *Nobel Lectures in Molecular Biology 1933–1975*. New York: Elsevier, pp. 179–203.

Watson, James D. (1965), *Molecular Biology of the Gene*. New York: W. A. Benjamin.

Watson, James D. (1968), *The Double Helix*. New York: New American Library.

Watson, James D. (2000), *A Passion for DNA: Genes, Genomes, and Society*. Cold Spring Harbor, NY: Cold Spring Harbor Laboratory Press.

Watson, James D. and Francis Crick (1953a), "A Structure for Deoxyribose Nucleic Acid," *Nature* 171: 737–738.

Watson, James D. and Francis Crick (1953b), "Genetical Implications of the Structure of Deoxyribonucleic Acid," *Nature* 171: 964–967.

Wimsatt, William (1972), "Complexity and Organization," in Kenneth F. Schaffner and Robert S. Cohen (eds.), *PSA 1972, Proceedings of the Philosophy of Science Association*. Dordrecht: Reidel, pp. 67–86.

Zamecnik, Paul C. (1953), "Incorporation of Radioactivity from DL-Leucine-1-C[14] into Proteins of Rat Liver Homogenate," *Federation Proceedings* 12: 295.

Zamecnik, Paul C. (1958), "The Microsome," *Scientific American* 198 (March): 118–124.

Zamecnik, Paul C. (1962a), "History and Speculation on Protein Synthesis," *Proceedings of the Symposia on Mathematical Problems in the Biological Sciences* 14: 47–53.

Zamecnik, Paul C. (1962b), "Unsettled Questions In the Field of Protein Synthesis," *Biochemical Journal* 85: 257–264.

Zamecnik, Paul C. (1969), "An Historical Account of Protein Synthesis, With Current Overtones – A Personalized View," *Cold Spring Harbor Symposia on Quantitative Biology* 34: 1–16.

Zamecnik, Paul C. (1976), "Protein Synthesis – Early Waves and Recent Ripples," in A. Kornberg, B. L. Horecker, L. Cornudella, and J. Oro (eds.), *Reflections in Biochemistry*. New York: Pergamon Press, pp. 303–308.

Zamecnik, Paul C. (1979), "Historical Aspects of Protein Synthesis," *Annals of the New York Academy of Sciences* 325: 269–301.

Zamecnik, Paul C. (1984), "The Machinery of Protein Synthesis," *Trends in Biochemical Sciences (TIBS)* 9: 464–466.

# 4

## Relations Among Fields

### Mendelian, Cytological, and Molecular Mechanisms[1]

#### 4.1   INTRODUCTION

Philosophers of biology have debated the nature of the relations between Mendelian genetics and molecular biology for some fifty years. They have proposed a variety of relations between the fields, including reduction, replacement, and explanatory extension. This chapter proposes a new analysis: the two fields discovered separate but serially connected mechanisms. These hereditary mechanisms have different *working entities* and the mechanisms operate at different times in an *integrated temporal series of hereditary mechanisms*. This analysis better characterizes the practice of biologists than previous accounts, as evidenced both by the historical development of the two fields and by presentations of the results of the two fields in contemporary textbooks.

Accounts of formal reduction played many roles in philosophical analyses of science in the second half of the twentieth century. Reduction was seen both as the relation among theories at different levels of organization at a given time (sometimes called "microreduction") and as the relation between

[1] This chapter was originally published as Darden, Lindley (2005), "Relations Among Fields: Mendelian, Cytological, and Molecular Mechanisms," in Carl F. Craver and Lindley Darden (eds.), Special Issue: "Mechanisms in Biology" of *Studies in History and Philosophy of Biological and Biomedical Sciences* 36: 357–371. This work was supported by the U.S. National Science Foundation grant SBR-9817942 and by an award from the General Research Board of the Graduate School at the University of Maryland. Thanks to David Hull and Ken Schaffner, my teachers, who introduced me to this problem many years ago; to David Didion, who pointed out that independent assortment occurs prior to segregation; to Joan Straumanis for the metaphor, genes on chromosomes like passengers in a train; and to the following for their help: Carl Craver, Nathaniel Goldberg, Nancy Hall, Peter Machamer, Greg Morgan, Jonathan Roy, Rob Skipper, Jim Tabery, Ken Waters, the DC History and Philosophy of Biology discussion group, and an anonymous referee.

predecessor and successor theories. Furthermore, reduction was tied closely to explanation. The connection between what was to be explained (the explanandum) and what did the explaining (the explanans, usually general laws) was claimed to be (usually) deduction (Hempel 1965). Hence, the deduction of the reduced theory (or the observations that it explained) from the reducing theory in formal reduction permitted the claim that the reducing theory explained the reduced theory (or its observation statements). The status of the reduced theory after a reduction was different in different accounts of formal reduction. In some accounts of reduction (e.g., Kemeny and Oppenheim 1956; Oppenheim and Putnam 1958), the reduced theory was eliminated; the reducing theory explained all the observations previously explained by the eliminated theory. In other accounts (e.g., Nagel 1961), one theory was derived from another, but the reduced theory might still be useful in some way. Scientific progress was viewed as carrying out more and more reductions to lower levels of organization (Oppenheim and Putnam 1958).

In analyses of the relations between Mendelian genetics and molecular biology, these roles for reduction have been criticized. Progress seemed to have occurred with the development of molecular biology, but how was this progress to be characterized? Did molecular biology microreduce Mendelian genetics or was it a successor that replaced Mendelian genetics? Could all the findings of Mendelian genetics be best explained at the molecular level? Did such molecular explanations entail the deduction of Mendel's laws from general laws of molecular biology? What were these general laws of molecular biology? Were the theories in the fields best characterized as consisting of laws? After the development of molecular biology, what was the status of Mendelian genetics?

Attempts to answer these questions about the relations between Mendelian genetics and molecular biology have been fraught with difficulties. It is argued here that integration of a temporal series of mechanisms with different working entities is the appropriate way to characterize the relations between Mendelian genetics and molecular biology. Cytology furnished the mechanisms of Mendelian heredity. The discovery of new mechanisms by molecular biology might be considered explanatory extension of Mendelian genetics, but this analysis differs from a previous account, as we will see. This extension occurred as a result of the discovery of mechanisms that illuminated *black boxes* noted by Mendelian geneticists but not accessible by Mendelian/cytological techniques. These molecular mechanisms occur before and after the chromosomal mechanisms, including the mechanisms of DNA replication, mutation, and protein synthesis.

Furthermore, this chapter argues that this analysis in terms of mechanisms provides alternative accounts of the following issues. The structure of biological theories in these fields is best analyzed by appeal to mechanism schemas and not by appeal to sets of laws or argument schemata. Further, explanations of various phenomena consisted of describing the mechanisms that produced them, not by logically deducing anything from anything. Progress occurred in this case, not by reduction or replacement, not by adding premises to argument schemata, not by appeal to the smallest size components, but by discovering mechanisms with working entities of different sizes.

The following sections elaborate and support this analysis of the relations between Mendelian genetics and molecular biology in terms of serially integrated mechanisms. Section 4.2 characterizes the fields to be discussed. Section 4.3 criticizes a selection of previous philosophical analyses and critiques. Section 4.4 characterizes mechanisms, mechanism schemas, sketches, and working entities. Section 4.5 is a historical account tracing the discovery of mechanisms in Mendelian genetics, cytology, and molecular biology. Section 4.6 provides evidence for the contemporary integration of hereditary mechanisms by examining an exemplary contemporary molecular biology textbook.

## 4.2  THE FIELDS OF MENDELIAN GENETICS, MOLECULAR BIOLOGY, AND THEIR NEIGHBORS

Because the topic here is the relation between the fields of Mendelian genetics and molecular biology, a few words about the identification of these fields and their neighbors are in order. Although the institutional and professional aspects of scientific disciplines are frequently complex, the conceptual components of scientific fields can often be delineated. These include the central problem, a domain of phenomena related to the problem, techniques and methods, and general knowledge encapsulated in concepts, laws, theories, or mechanism schemas that aim to provide solutions to the central problem (cf. Darden and Maull 1977; see Chapter 5, this book).

Several fields are at issue here. The field of *classical Mendelian genetics* emerged in 1900 with its central problem being the explanation of patterns of inheritance of characteristics. The technique used was artificial breeding of organisms with variant characteristics. Empirical regularities were explained by appeal to the two Mendelian laws of segregation and independent assortment, which state regularities in the behavior of hypothetical genes. The theory of the gene also included claims about linkage of genes arranged linearly in groups and crossing over between alleles in the same linkage group (e.g., Morgan 1926, p. 25; Darden 1991).

In the nineteenth century, the central problem of the field of *cytology* was to find the basic units of organisms. Cell theory solved that problem; the field moved on to the microscopic study of stained cells and their components. The study of chromosomes (darkly staining string-like structures in the nucleus) began in the late nineteenth century, and details about chromosomal behavior in mitosis (normal cell division) and meiosis (formation of gametes in which the chromosome number is halved) were available in the early twentieth century (see, e.g., Wilson 1900; Hughes 1959). Chromosomal behavior during meiosis, it will be argued, provided the mechanisms producing the regularities noted in Mendel's laws. *Cytology* became *cell biology* in the 1950s and 1960s when new techniques provided ways of studying the functions of newly observed ultrastructure of cells (Davis 1980, p. 209). The field underwent further change after the rise of molecular biology.

The field that came to be called *molecular biology* emerged in 1953, I argue, with the discovery of the double helix structure of DNA. The term "molecular biology" was coined in 1938[2] and was used by some x-ray crystallographers to label their work on the three-dimensional structure of macromolecules. However, in the 1950s and 1960s, the field of molecular biology drew not only on x-ray crystallography and structural chemistry to study the structure of macromolecules but also on theoretical work on the genetic code to find relations between nucleic acids and proteins, as well as genetic breeding techniques applied to microorganisms. The central problem of this post-1953 early molecular biology was the nature of the gene – how it replicates, mutates, and produces proteins. Molecular biologists solved these problems about the gene by finding mechanism schemas for DNA replication, point mutation, protein synthesis, and gene regulation, emphasizing weak forms of bonding, such as hydrogen bonding, and by using microorganisms as model organisms (see, e.g., Watson 1965; Morange 1998).

The field of *biochemistry* emerged in the early twentieth century with a focus on proteins and enzymes, as well as other chemical components in the metabolism of living things. In the 1930s, when enzymes were found to be proteins and proteins were found to be macromolecules, one of biochemistry's many problems was the discovery of the different amino acids composing

---

[2] In 1938, Warren Weaver of the Rockefeller Foundation coined the term and some of the early x-ray crystallographers used it to refer to their work on the structure of biological macromolecules, such as hair (Olby 1994). However, the field, as codified in Watson's (1965) *Molecular Biology of the Gene*, began in 1953. The term "molecular biology" came into widespread usage after the founding of the *Journal of Molecular Biology* in 1959.

proteins. An associated problem for some biochemists was to understand the energetics of the chemical reaction that produces the strong covalent bonds between amino acids to form peptide bonds. Thus, although the two fields of early molecular biology and biochemistry investigated molecules within a similar size range, their problems, techniques, and hypothesized mechanisms were, and to some extent still are, different (see, e.g., White et al. 1954; Kohler 1982).

By about 1970, early molecular biology had solved its central problem about the nature of the gene, at least in procaryotes (i.e., microorganisms, such as bacteria, without an organized nucleus or chromosome). In its next phase, molecular biology added the study of eucaryotes (i.e., organisms with an organized nucleus and chromosomes). This later phase merged with *cell biology* (see, e.g., Alberts et al. 1983). Its central problem was and continues to be to elucidate the molecular structures and mechanisms within and between cells. Especially relevant for our purposes here, for example, is the mechanism of crossing over between homologous chromosomes.

Molecular biology developed a repertoire of techniques for manipulating the genetic material that enabled other fields to "go molecular." Such expansion of molecular biology into cell biology and other areas has led some to distinguish *molecular genetics* from molecular biology. Philosophers (e.g., Kitcher 1984) often use "molecular genetics" synonymously with early molecular biology. Alternatively, "molecular genetics" may refer only to results produced by cross-breeding variants to produce hybrid organisms (a usage extended from techniques of Mendelian genetics to genetic manipulations in bacteria). Another usage of "molecular genetics" is to refer to any study of genetics at the molecular level, that is, any study of the molecular biology of the gene. Whether biologists' usage of "molecular genetics" has any identifiable historical trajectory is unclear.

"Molecular biology" is the name of the historical field with relations to Mendelian (sometimes called "classical") genetics. Historians have chronicled its history (e.g., Morange 1998); scientists in the period we will discuss often called themselves "molecular biologists." Francis Crick, for example, said: "I myself was forced to call myself a molecular biologist because when inquiring clergymen asked me what I did, I got tired of explaining that I was a mixture of crystallographer, biophysicist,[3] biochemist and geneticist, an explanation which in any case they found too hard to grasp" (quoted in Stent 1969, p. 36).

---

[3] On the usage of the term "biophysics" at Cambridge University in England, see Chadarevian (2002).

Not only is it important to identify the fields at issue here, we also need to delineate the theories, theoretical entities, and size levels. Sometimes philosophers assumed that one could identify a theoretical entity, which was used in a theory appropriately located at a level of organization for any given scientific field, or even a larger branch of science. For example, the units – atom, molecule, cell – figured in atomic theory, chemical theory, and cell theory, which corresponded to the branches of science – physics, chemistry, and biology (see, e.g., Oppenheim and Putnam 1958). The view that theories at lower levels would explain all the observations at higher levels constituted a reductive account of the goals of and progress in science.

However, this oversimplified identification of units, theories, fields, and levels breaks down for the fields of interest here. It is difficult to see at what level in a hierarchy Mendelian genetics, with its study of phenotypic characters and hypothetical genes, is to be placed when compared to cytology and the study of chromosomes. Genes were claimed to be parts of chromosomes, but no one suggested reducing cytology to genetics (or vice versa). Furthermore, molecular biologists and biochemists investigated many of the same units and reactions at roughly the same size level but used different techniques and perspectives. Molecular biologists and biochemists sometimes found their work to be complementary, such as work on the role of RNAs in the mechanism of protein synthesis (see Darden and Craver 2002; see Chapter 3, this book), and sometimes used their differing techniques to compete, such as the biochemical (e.g., Nirenberg and Matthaei 1961) verses genetic (e.g., Crick et al. 1961) work on the genetic code (Kay 2000). Although Mendelian genetics could be analyzed as having two laws as part of the theory of the gene, molecular biology did not fit this account of theory structure. Nothing was found to fill the role of general laws of molecular biology. As we will see, problems of identifying levels, fields, theories, laws, and units plagued the attempts to analyze the relations between fields in terms of reduction. These problems spawned new accounts of theory structure and explanation.

### 4.3 PREVIOUS WORK ON THE RELATIONS BETWEEN MENDELIAN GENETICS AND MOLECULAR BIOLOGY

Kenneth Schaffner used and developed Ernst Nagel's (1961, Ch. 11) analysis of derivational theory reduction to argue for the reduction of classical Mendelian genetics to molecular biology and refined it over many years (summarized in Schaffner 1993). The goal of formal reduction was to logically deduce the laws of classical genetics from the laws of molecular biology. Such a derivation required that all the terms of Mendelian genetics not in molecular biology be connected via "correspondence rules." Hence, Schaffner endeavored to find molecular equivalents of such terms as "gene" as well as "predicate terms," such as "is dominant." (One allele of a gene is said to be "dominant" over another if the character associated with that gene appears in the hybrid offspring of a cross between pure breeding parents. For example, in a cross between tall and short pea plants, tall is dominant over short.)

David Hull (1974) criticized formal reduction, argued against Schaffner's claims, and suggested instead that perhaps molecular biology replaced classical genetics. Hull's critiques focused on the problem of connectibility of terms. A close look at their dispute showed that it hinged on debates about mechanisms of gene expression. Hull said:

> Even if all gross phenotypic traits are translated into molecularly characterized traits, the relations between Mendelian and molecular predicate terms express prohibitively complex, many-many relations. Phenomena characterized by a single Mendelian predicate term can be produced by several different types of molecular mechanisms. Hence, any reduction will be complex. Conversely, the same types of molecular mechanisms can produce phenomena that must be characterized by different Mendelian predicate terms. Hence, reduction is impossible. (Hull 1974, p. 39)

Schaffner criticized Hull for claiming that the same molecular mechanism could give rise to phenomena labeled with different Mendelian terms. "Different molecular mechanisms can appropriately be invoked in order to account for the same genetically characterized relation, as the genetics is less sensitive. The same molecular mechanisms can also be appealed to in order to account for different genetic relations, but only if there are further differences at the molecular level," such as different initial conditions (Schaffner 1993, p. 444). Empirical investigation was required to determine the molecular mechanisms of gene expression (Schaffner 1993, p. 439).

One artifact of the connectibility of terms required in formal reduction was that philosophers focused so much attention on dominance. The dominance of one allele over another during gene expression was found to have many exceptions. In 1926, when T. H. Morgan stated the theory of the gene, dominance was not included as a component of the theory (Morgan 1926, p. 25). Finding molecular mechanisms for dominance, which was later called "dosage effect" (Darden 1991, p. 72), was not a central problem in early molecular biology, as we will see. Such connectibility of terms and other items requiring attention in a formal reduction analysis were peripheral to the concerns of scientists, as both Schaffner (1974b) and Hull (1974) realized. The idealized

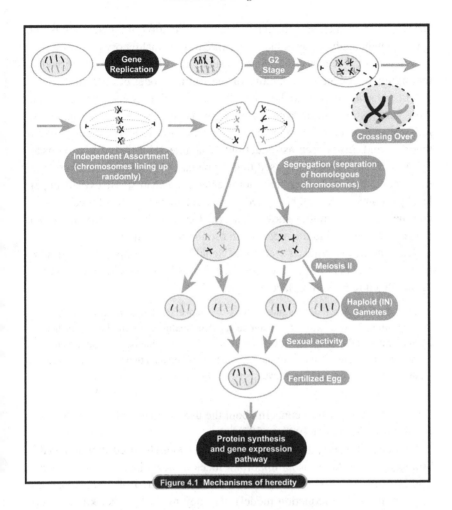

Figure 4.1 Mechanisms of heredity

formal reduction relation, even if it could have been imposed on some version of the historically developing fields (or some logical reconstruction of their theories), did not serve to capture the practice of scientists. When scientists use the term "reduction," they mean something different from that captured in the formal philosophical account.

Note that the mechanisms at issue in the Schaffner–Hull debate were located in the black box in Figure 4.1 labeled "gene expression." Mendelian genetics had no account of the mechanisms producing gene expression. It is not surprising that studying molecular mechanisms of gene expression produced refinements in the understanding of the network of relations among gene products during the production of the phenotypic traits that were accessible to Mendelian geneticists. Such illumination of a black box is not appropriately analyzed as either replacement or formal theory reduction.

Wimsatt (1976) analyzed Schaffner and Hull's dispute. He strengthened the argument for the peripherality of the formal model of reduction to the practice of science. Instead, he noted: "At least in biology, most scientists see their work as explaining types of phenomena by discovering mechanisms, rather than explaining theories by deriving them or reducing them to other theories, and *this* is seen as reduction, or as integrally tied to it" (Wimsatt 1976, p. 671). Wimsatt proposed that "explanatory reduction" (microreduction) should be recast as a search for lower-level mechanisms to explain upper level phenomena. As will be argued below, a characterization of mechanisms and their working entities shows that one does not always move down to a lower size level to find the mechanism producing a phenomenon.

Responding to Wimsatt's critique, Schaffner (1993, pp. 490–499) analyzed relations between the "causal/mechanical" view of reduction and the "generalized reduction/replacement" model:

> ... reductions in science frequently have two aspects: (1) ongoing advances that occur in piecemeal ways, for instance, as some features of a model are further elaborated at the molecular level or perhaps a new "mechanism" is added to the model, and (2) assessments of the explanatory fit between two theories (viewed as a collection of models) or even between two branches of science.
>
> (Schaffner 1993, p. 495)

But Schaffner expressed concern about the use of "mechanism," an "unanalyzed term" (Schaffner 1993, p. 287).[4]

Darden and Maull (1977; see Chapter 5, this book) focused attention on the bridges between fields as an important locus of new discoveries in science. The bridges, we claimed, might be identities (required of correspondence rules in the formal reduction model), but they might be other kinds. Interfield relations included part-whole relations (e.g., genes are parts of chromosomes), structure-function relations (e.g., an identified molecule functions as the repressor in gene regulation), and cause and effect relations. Sometimes the relations were elaborated in an "interfield theory," such as the chromosome theory of Mendelian heredity (i.e., genes as parts of chromosomes). What was important, we argued, was to find the relations, not to formally derive anything from anything. However, we missed the importance of mechanisms in our analysis.

---

[4] I thank Jim Tabery for the plausible interpretation of Schaffner's (1993) Ch. 6 ("Explanation and Causation") as Schaffner's own attempt to analyze causal mechanisms, using Salmon's and Mackie's work, and connect causal mechanistic explanation with explanation via generalizations.

Moving beyond debates about formal theory reduction, Kitcher (1984, 1989) and Waters (1990) advanced the discussion about the relations between the fields of Mendelian genetics and (what Kitcher called) molecular genetics. Kitcher criticized a reductive approach and Waters defended "informal" aspects of reduction.

Developing an analysis of theory structure in terms of argument schemata, Kitcher argued that the relation between Mendelian and molecular genetics was "explanatory extension" (Kitcher 1984, p. 371). The theory of molecular genetics provided a refined and expanded set of premises when compared to the argument schemata of classical genetics (Kitcher 1989, pp. 440–442). However, classical genetics retained its own schema. For example, the independent assortment of genes (Mendel's second law) was explained, according to Kitcher, by instantiating a pairing and separation schema, thereby showing that chromosomal pairing and separation was a unifying natural kind. Such unification would be lost if attention was focused on the gory details at the molecular level. The cytological level thus constituted an "autonomous level of biological explanation" (Kitcher 1984, p. 371).

On the other hand, in order to solve problems of gene replication, mutation, and action, Kitcher claimed that the "gory" molecular details were required and were part of the expanded premise set of the schema labeled "Watson–Crick" (Kitcher 1989, p. 441). Among the premises of the "Watson–Crick" schema, for example, were "transcription, post-transcriptional modification and translation for the alleles in question," along with details of "cell biology and embryology" for the organisms in question (Kitcher 1989, pp. 440–442). An explanation of a particular pattern of distribution of progeny phenotypes in a genetic cross resulted from instantiating the appropriate schema: the variables were filled with the details from the particular case and the conclusion derived from the premises.

Waters (1990) criticized Kitcher's arguments. First, Waters noted that other instances of pairing and separation processes, other than the separation of paired chromosomes, were only hypothetical. It was difficult to see what was achieved by unifying chromosomal processes with other imagined pair separations, carried out by other kinds of hypothetical forces, in a pair-separation schema. Second, Waters criticized the uninformativeness of the "gory" molecular details. Waters claimed that our understanding was enhanced by the molecular models for the mechanism of crossing over. Such increased understanding constituted "informal reduction" as opposed to formal, derivational reduction.

In the remainder of the chapter, I argue for a different view in which mechanisms play a prominent role. I argue that the two fields of Mendelian genetics and molecular biology investigated separate but serially connected mechanisms with different working entities that operate at different times in (what is now known to be) an integrated temporal series of hereditary mechanisms. One explains, for example, a particular distribution of phenotypic traits from a genetic cross as resulting from a series of mechanisms connecting parent(s) to offspring. Along with Kitcher, I argue that the phenomena summarized in Mendel's laws of segregation and independent assortment were (and are) explained by the behavior of chromosomes. However, the reason has nothing to do with unification into a natural kind of pair-separation processes but because the chromosomes are the working entities of the mechanisms of meiosis. The analysis of working entities in a mechanism provides an alternative to Kitcher's analysis, but also serves to block the reductive move that one always gains understanding by investigating the gory details of the smallest entities present.

Waters correctly noted that the molecular level is the appropriate one for finding the mechanism(s?) of crossing over. However, the molecular mechanism of crossing over operates at a stage in a temporal sequence of hereditary mechanisms between the pairing and then the separation processes that Kitcher discussed. Kitcher and Waters were talking past each other because they were not arguing about the same mechanisms. Different genetic mechanisms, as we will see, have different working entities and operate at different times; tasks in discovering mechanisms are to find the working entities at whatever level of organization and the temporal sequence in which they act.

Also with Kitcher, I argue for abstract schemas of varying scope as a way of representing general knowledge in these fields. However, the schemas are not argument schemata but mechanism schemas. (Here, I use "schemas" rather than "schemata" as the plural of "schema.") Rather than premises and conclusions, mechanism schemas describe the mechanism's entities and activities and their productively continuous organization from beginning to end. Mechanism schemas are often depicted in diagrams. Diagrams perspicuously show the structures of the entities, as well as spatial arrangements and temporal stages. Furthermore, mechanism schemas have various degrees of abstraction (Darden 1996), not merely the two-place relation of a variable and its value.

This mechanistic analysis, I claim, better captures the practice of biologists – with their frequent talk and diagrams of mechanisms – than do the analyses of the relations between the fields in terms of formal reduction, informal reduction, replacement, and explanatory extension via expanded argument schemata.

With Wimsatt, I argue that biologists often explain phenomena by describing mechanisms. "Mechanism" is no longer an unanalyzed term. The analysis to be discussed below differs in some respects from the decompositional view in Wimsatt (1976) and developed by Glennan (1996, 2002). For them, the behavior of the system was to be explained by decomposing a system into its parts and explaining its behavior by the interactions of its parts. Instead, I argue, finding the mechanism that produces a phenomenon may require not further decomposition of a system but instead going "up" in size level. For example, finding the mechanism of segregation of genes did not require decomposing genes into their parts but required finding the wholes, the chromosomes, on which the parts, the genes, ride. (For comparisons of the decompositional and this alternative view of mechanisms, see Machamer, Darden, Craver 2000; see Chapter 1, this book; Tabery 2004). To this analysis we now turn.

## 4.4  MECHANISMS, MECHANISM SCHEMAS, MECHANISM SKETCHES

This analysis of the relations between Mendelian genetics and molecular biology makes use of several concepts, some analyzed in previous work, such as *mechanism, mechanism schema, mechanism sketch*, and others introduced here, such as *working entities*. This section discusses these concepts and then uses them to explicate the relations among these fields.

Mechanisms have components that work together to do something. One identifies some phenomenon of interest (in the sense of Bogen and Woodward 1988) or some task that is carried out (Bechtel and Richardson 1993). One then seeks the mechanism that produces the phenomenon or carries out the task. Previous work on the concept of mechanism provides this characterization:

> Mechanisms are entities and activities organized such that they are productive
> of regular changes from start or set-up to finish or termination conditions.
> (Machamer, Darden, Craver 2000; see Chapter 1, this book)

Types of entities include macromolecules (e.g., proteins and the nucleic acids, DNA and RNA); subcellular structures, such as ribosomal particles (composed of RNA and proteins); chromosomes (composed of DNA and proteins); and cells. Types of activities include geometrico-mechanical activities, such as lock and key docking of an enzyme and its substrate; and chemical bonding activities, such as the forming of strong covalent bonds and weak hydrogen bonds. The entities and activities are organized in *productive continuity* from beginning to end; that is, each stage gives rise to the next. Entities having certain kinds of activity-enabling properties allow the possibility of acting in certain ways, and certain kinds of activities are only possible when there are entities having certain activity-enabling properties (Darden and Craver 2002; see Chapter 3, this book; Darden 2002).

*Working entities* engage in activities within a mechanism. Various general features of entities (whether working or not) aid in identifying the working entities of a mechanism. An entity may have a spatio-temporal location. An entity may have a clear boundary, such as a membrane bounding it. An entity may be composed of chemically bonded subparts that are not similarly bonded to the parts of other entities. It may be composed of specific chemicals that differ from chemicals in the surroundings. It may be robustly detectable, that is, accessed by using different techniques (Wimsatt 1981). It may be stable over some period of time, as are chromosomes, or it may be rapidly synthesized and degraded, as are some messenger RNAs. It may have a developmental history; that is, it may be formed during embryological development. It may have an evolutionary history; that is, it may be a descendant in a lineage.

In addition to these general features of entities, working entities in mechanisms have additional features. A working entity *acts* in a mechanism. It may move from one place to another. It has activity-enabling properties. It may have one or more *localized* active sites. For example, the centromeres of chromosomes are active sites that attach to the spindles during the mechanisms of meiosis. Similarly, enzymes have localized active sites that bind to substrates. Alternatively, the active sites may be *distributed* throughout the entity, as are the slightly charged bases along the entire double helix, which serve as the active sites in DNA replication.

Working entities in a given mechanism may be and often are different sizes. For example, ions, macromolecules, and cell organelles may all be working entities in the same mechanism, such as the mechanism of protein synthesis. Because working entities in a given mechanism may be of different sizes, mechanism levels may not correspond tidily to size levels. Of course, all biological entities are composed of smaller parts; however, most subcomponents do not change during the activities of the working entities of which they are parts. For example, atomic nuclei are parts of working entities but merely stable subcomponents. Atomic nuclei are not *working* entities or active sites in the DNA replication mechanism. They are parts of the structure, buried away behind electrons from active sites. In other conditions, nuclei of atoms can become working entities (e.g., in nuclear fission mechanisms when atoms are split). But during DNA replication, atomic nuclei are not working entities. Similarly, genes outside the centromere are just along for the ride on the chromosomes during meiosis; they only become working entities during the operation of developmental mechanisms occurring later.

Perhaps surprisingly, genes are not the working entities in any of the hereditary mechanisms except gene expression. Genes have no "role function" (Craver 2001) in the mechanisms of chromosomal pairing and separation. Like atomic nuclei in the mechanism of DNA replication, genes are buried within chromosomal packaging during chromosomal pairing and separation. Genes are along for the ride, like passengers on a train; they are not working entities, as those mechanisms operate. The entire chromosomes are the working entities, and their centromeres are the active sites.

Similarly, in DNA replication, the entire DNA molecule is the working entity, with the polar charges of individual bases as the active sites. The genes are not working entities or active sites during DNA replication.

Only during the mechanisms of gene expression do genes become working entities; they are working segments of DNA molecules (except in RNA viruses) that are active in mechanisms for the transcription of DNA segments into RNA. Individuating genes can be problematic. Bacterial genes are easier to individuate because they are usually continuous segments of a DNA molecule. Because eucaryotic genes have introns (as well as other complicating factors), identifying the DNA segments making up the genes has proved a challenging task. What one needs to know in order to identify a gene is the mechanism in which it participates. Sometimes one reasons backward from a gene's product, such as a protein, to locate the DNA segment(s) that produced it. (For numerous kinds of mechanisms involved in gene action, see Fogle 2000.)

One form of a structural gene is a linear sequence of bases that is transcribed into messenger RNA, whose linear sequence of bases is translated into the linear sequence of amino acids in a protein (see Figure 4.3). Other genes are transcribed into RNAs that play other roles (e.g., transfer RNA and ribosomal RNA). Debate has occurred as to whether regulatory elements should be called genes (Waters 1994). This is just a terminological dispute; what is important is what the DNA segment does in the transcription mechanism. Some segments of DNA work as part of the control mechanism, while others are transcribed. Subsequently, in organisms with introns, some regions of the pre-messenger RNA are spliced out to produce messenger RNA. One need not argue about whether the gene includes regulatory and intronic regions, as long as the particular roles of these DNA segments in the transcription mechanism are understood. The role in a mechanism is what is important.

Scientists rarely depict all the particular details when describing a mechanism; representations are usually schematic, often in diagrams. A *mechanism schema* is a truncated abstract description of a mechanism that can be instantiated by filling it with more specific descriptions of component entities and

111

activities. An example is Watson's (1965) diagram of his version of the central dogma of molecular biology:[5]

DNA –> RNA –> protein

This is a schematic representation (with a high degree of abstraction) of the mechanism of protein synthesis, which can be instantiated with details of DNA base sequence, complementary RNA sequence, and the corresponding order of amino acids in the protein produced by the mechanism (see Figure 4.3).

In contrast, a mechanism *sketch* cannot (yet) be instantiated. Components are (as yet) unknown. Sketches may have *black boxes* for missing components whose function is not yet known. A more developed sketch may have boxes whose functional role (Craver 2001) is known or conjectured, but what specific entities and activities carry out that function in the mechanism are (as yet) unknown. Sketches guide further work to fill the black boxes. (Biologists may use the term "model" to refer to a schema, a sketch, or an instantiation of a schema.)

Mendel's laws sketched regularities found to be produced by chromosomal mechanisms during meiosis. The general knowledge in molecular biology is best characterized not in terms of laws or a theory but as a set of mechanism schemas (Machamer, Darden, Craver 2000; see Chapter 1, this book; Craver 2002a). These are schemas for such mechanisms as DNA replication (and repair), protein synthesis, and gene regulation. They have domains of applicability of varying scope, from the widely found mechanism of protein synthesis to the myriad different mechanisms of gene regulation.

An adequate description of hereditary mechanisms shows the wider context into which any given mechanism fits. Hereditary mechanisms operate in a temporal series stretching from parent(s) to offspring. A goal in understanding heredity is to find well-supported schemas for the mechanisms of heredity, as depicted in Figure 4.1.

## 4.5   HISTORICAL DEVELOPMENTS: DISCOVERING HEREDITARY MECHANISMS

This section discusses one way to identify the relations between fields: by tracing the historical discoveries of hereditary mechanisms in the twentieth

---

[5] For differences between Watson's and Crick's versions of the central dogma, see Darden (1995; see Chapter 10, this book); and Keyes (1999a, 1999b).

century in the fields of Mendelian genetics, cytology, and molecular biology. This history provides evidence for the claim that the relations among these fields are best understood from the perspective of the relations among the mechanisms they discovered.

Seminal publications are good sources for tracing the development of fields. The field of classical Mendelian genetics began in 1900 with the rediscovery and reinterpretation (de Vries 1900; Correns 1900) of Gregor Mendel's 1865 paper. The field developed significantly in the hands of T. H. Morgan and his colleagues, whose early work was presented in their 1915 book, *The Mechanism of Mendelian Heredity* (Morgan et al. 1915). The culmination of that work can conveniently be marked with Morgan's publication of a 1926 book, *The Theory of the Gene* (Morgan 1926).

The primary technique of classical Mendelian genetics was cross-breeding of variants of plants and animals, noting the distributions of the variant phenotypic characters through several generations, and, finally, making inferences about hypothetical genes associated with those characters. Geneticists made two inferences about genes (and their alleles): namely, that they exhibited segregation and independent assortment, called "Mendel's first and second laws." In order to state these laws, it is useful to consider a typical breeding experiment. Suppose pure breeding tall and short pea plants are crossbred.[6] The hybrid offspring are called the "$F_1$ generation," for the "first filial generation." All the $F_1$ plants are tall. Consequently, tall is said to be "dominant" over short. When the plants in the $F_1$ generation are allowed to self-fertilize, then the $F_2$ plants occur in the ratio of three tall to one short. The short ones thereafter breed true, but the tall split in the next generation into a 1:2 ratio of pure breeding tall to those that again behave as hybrids. Such 3:1 ratios in the $F_2$ generation are an empirical regularity that was observed by early geneticists in many species of plants and animals. One variant dominating over the other in the $F_1$ generation was found only in some cases; sometimes the character of the hybrid might be a blend form that is intermediate between the two parents or look quite different from the parents. Thus, dominance was not a general empirical regularity (Darden 1991), nor was it listed as a component of the theory of the gene by Morgan (Morgan 1926, p. 25).

To account for the regular 3:1 ratios, geneticists sketched aspects of the mechanism operating during the formation of gametes (i.e., sperm and eggs in animals) of the hybrid. A relationship between a phenotypic character and

---

[6] For the contrast between Gregor Mendel's historical work and this account (in, e.g., Morgan 1919), see Darden (1991, Ch. 4).

an allele of a gene was assumed (e.g., one allele is associated with the tall character in peas and its corresponding allele with the character for short). Usually in a sexually breeding organism there are two alleles of a given gene. During the formation of gametes, the two alleles separate (i.e., segregate) so that each gamete receives one but not the other. This regularity is "Mendel's first law." This law thus sketches the behavior of some sort of mechanism operating during the formation of gametes to separate the paired alleles of a gene.

When two traits in peas are followed through two generations, during an experiment such as the one described previously, the regularity in the $F_2$ generation is found to be 9:3:3:1. For example, if a tall plant that produces yellow peas is crossed with a short plant that produces green peas, the two traits behave independently. At the beginning of Mendelism, no separate law was formulated to express this independence. Segregation just seemed to be operating in each trait separately. Only after exceptions were found was Mendel's second law explicitly formulated.

In 1906, Bateson and his associates found a case in which two traits exhibited segregation but the traits did not produce the expected 9:3:3:1 ratios (Bateson et al. 1906). Bateson referred to the phenomenon as "coupling" because the mechanism that he proposed consisted of some "allelomorphs" (as he called them) being attracted to others. However, Morgan renamed the phenomenon "linkage" (e.g., Morgan and Lynch 1912) when he and his colleagues proposed an alternative mechanism. They explained the lack of independent assortment as due to the anomalous genes being linked on chromosomes. In 1919, Morgan explicitly separated Mendel's two laws and formulated "Mendel's second law" as the claim that genes in different linkage groups assort independently during the formation of gametes (Morgan 1919; Monaghan and Corcos 1984; Darden 1991). The regular behavior of chromosomal mechanisms produced the phenomena encapsulated in Mendel's two laws of segregation and independent assortment of different linkage groups. The Morgan group was able to make use of work on chromosomes already carried out by cytologists (Wilson 1900).

By 1900, cytologists had shown that microscopically visible chromosomes occurred in pairs that separated during the formation of gametes so that each gamete had one half the usual parental number. Independently, Walter Sutton (1903) and Theodore Boveri (1904) proposed that the hypothetical hereditary factors exhibiting Mendelian segregation were in or on the chromosomes. The chromosome theory of Mendelian heredity was thus an interfield theory, postulating a part-whole relation between visible chromosomes and hypothetical genes, and thereby integrating findings from the fields of cytology and genetics[7] (Darden and Maull 1977; see Chapter 5, this book).

Morgan (1909, 1910a) had originally opposed the chromosome theory, arguing that the amount of correlation of characters expected if they were linked on chromosomes had not been found. He changed his view in 1910 as a result of his work on the fruit fly *Drosophila* (Morgan 1910b, 1911a). Patterns of inheritance allowed Morgan to infer that sometimes pieces were switched between homologous chromosomes. This mechanism of crossing-over occurred after pairing but prior to separation (see Figure 4.1); it served to produce less correlation of characters because alleles were reshuffled. As a result, Morgan abandoned his objection to the chromosome theory.

Morgan, with his students Calvin Bridges, A. H. Sturtevant, and H. J. Muller, actively pursued the relations between chromosomes and "factors" (in 1917, Morgan adopted Johannsen's 1909 term "gene"). In 1915, in *The Mechanism of Mendelian Heredity*, they stated: "the chromosomes furnish exactly the kind of mechanism that the Mendelian laws call for" (Morgan et al. 1915, p. viii).

Key hereditary problems unsolved by Mendelian geneticists and unilluminated by drawing on cytology were the chemical nature of genes (speculated to be proteins), how genes replicated, how genes mutated and then faithfully reproduced those mutations, and how genes related to phenotypic characters. The black boxes of Figure 4.1 show the mechanisms not illuminated by Mendelian/cytological techniques.

The modern field of molecular biology began, I argue, in 1953 with the discovery of the DNA double helix by James Watson and Francis Crick (Watson and Crick 1953a, 1953b; Olby 1994; Judson 1996; Morange 1998). The new field drew on work in several other fields to solve the questions about genes unsolved by Mendelian genetics. These fields included x-ray crystallography, for determining the structures of macromolecules; structural chemistry, especially Linus Pauling's work on weak forms of chemical bonding, such as hydrogen bonding (e.g., Pauling 1939; Pauling and Corey 1950); and, to a much lesser extent, biochemistry, for its study of the chemical analysis of proteins and nucleic acids, as well as energy requirements of strong covalent bonding.

---

[7] Although our concern here is with the way geneticists used findings from cytology, inferences went both ways, as is often a mark of interfield theories, as discussed in Darden and Maull (1977; see Chapter 5, this book). New claims were made about chromosomes: namely, their random assortment, based on the connection to genetics; also, new use was made of the mechanisms of meiosis to explain Mendel's laws.

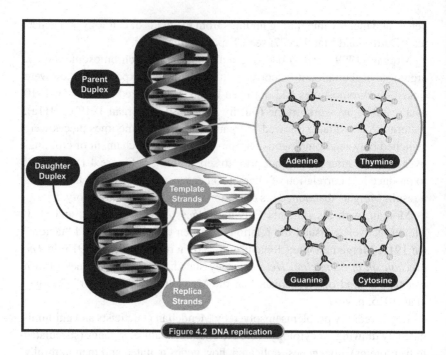

Figure 4.2 DNA replication

The model organisms of early molecular biology were bacteria (i.e., procaryotes) and their viruses. Choice of such model organisms, which lacked a nucleus and organized chromosomes, indicated that the central problems addressed by the field in its early days were not problems about the molecular details of chromosome pairing, crossing-over, and separation, as one would expect on a reduction analysis. Instead, the molecular biologists tackled the problems of finding the nature of the genes (i.e., nucleic acids, not proteins) and the mechanisms for gene replication, mutation, and expression.

As Watson and Crick (1953a) noted, the structure of DNA immediately suggested a copying mechanism. Their 1953 sketch served as the framework for the discovery of the DNA replication mechanism. The DNA helix opened, they suggested, to serve as two templates for the assembly of their complements (Figure 4.2). The structure also immediately suggested one way that mutations might form, what we now call "point mutation." An error in copying occurs when a base substitution was made that departs from the usual A-T, G-C base pairing during DNA replication.

Interestingly, genes proved not to be the working entities of either gene replication or gene mutation. The working entity of the mechanism of gene replication was found to be an entire DNA double helix molecule. As in the chromosomal mechanisms of meiosis, the genes were found to be just along

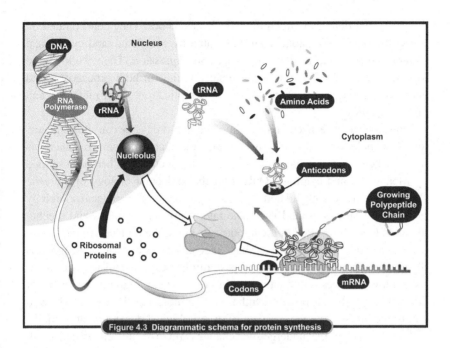

Figure 4.3 Diagrammatic schema for protein synthesis

for the ride during the replication of helices. Moving down in size, the working entity for the formation of a point mutation was found to be the single base of a DNA double helix, smaller than a DNA segment constituting a functional gene. The genes became the working entities only in the mechanisms of gene expression.

The efforts of early molecular biologists in investigating the mechanisms of gene expression began with work on the mechanism of protein synthesis and its regulation. One part of this work resulted in the discovery of the mechanism sketch DNA→RNA→protein. This sketch was elaborated into a detailed understanding (Figure 4.3) of the mechanism of protein synthesis (Darden and Craver 2002; see Chapter 3, this book) and the reading of the genetic code (Kay 2000). Another part of this work resulted in the understanding of some mechanisms of gene regulation, especially the mechanism of derepression in the *lac* operon of *E. coli*, that served to turn a group of genes (i.e., an operon) on and off, depending on environmental conditions (Jacob and Monod 1961; Morange 1998).

This brief account of the development of the field of molecular biology from 1953 to 1970 showed that early molecular biologists discovered different mechanisms than the Mendelian/cytological ones operating during meiosis in eucaryotes. They focused on the nature of the gene rather than seeking

117

the molecular details about chromosomal mechanics. They concentrated on filling the black boxes totally unilluminated by Mendelian and cytological techniques: gene replication, mutation, and expression. Their work served to elucidate mechanisms more universally found (in both procaryotes and eucaryotes): namely, mechanisms of DNA replication, point mutation, and protein synthesis.

Now that we have discussed early molecular biology, we can situate historically the molecular work on some of the details of chromosomal mechanisms found in hereditary mechanisms of sexual reproduction that had been studied by Mendelism and cytology. Only after the molecular biology of the *gene* (Watson 1965) was elucidated did attention turn to the molecular biology of the *cell* (Alberts et al. 1983). The later work on structures and mechanisms of the cell resulted in the hypotheses about the molecular mechanism of crossing-over in eucaryotes. These mechanisms were found in domains of smaller scope – namely, only in sexually breeding organisms with organized chromosomes – in contrast to the more widely found mechanisms of DNA replication and protein synthesis. The molecular biology of the cell contributed to solving the problem of embryological development in multicellular eucaryotes that undergo such changes. Yet to be fully elucidated are all the steps in mechanisms of gene expression between gene(s) and phenotypic character(s).

Looking back at the history of Mendelian genetics, cytology, and molecular biology, it is amazing that the classical geneticists were able to infer as much as they did about genes simply by following the transmission of phenotypic characters through several generations. That they were able to detect crossing-over, segregation, linkage, and independent assortment of different linkage groups was perhaps attributable to these phenomena resulting from chromosomal mechanisms. In the mechanisms of gene replication, mutation, and expression, the molecules and their parts become the working entities. These molecular mechanisms were unilluminated by the breeding techniques of Mendelian genetics or the tracing of chromosomal mechanics by cytologists. These molecular mechanisms served to fill some of the gaps between the inferred genes and their phenotypic characters studied by Mendelian genetics.

### 4.6   CONTEMPORARY ACCOUNT OF THE RELATIONS AMONG HEREDITARY MECHANISMS

The historical development of the two fields discussed in the previous section provides one form of evidence for the claim that the two fields studied different mechanisms in an integrated, temporal series of hereditary mechanisms. Examination of an exemplary contemporary textbook of molecular biology provides another form of evidence for this claim and, furthermore, shows that the meiotic chromosomal mechanisms of Mendelian heredity have not been eliminated but have been integrated with the prior and subsequent molecular ones (see Figure 4.1).

The first 1965 edition of James Watson's *The Molecular Biology of the Gene* codified the newly emerged field of molecular biology. This seminal text, after two more editions (Watson 1970, 1977), culminated in a 1988 multi-authored work (Watson et al. 1988). Chapter 1, "The Mendelian View of the World," opened with a discussion of meiosis. Mendel's crosses and Mendel's laws were diagrammed, and they were explained, the authors claimed, by the "chromosome theory of heredity." Recounting Sutton's work, the authors noted: "He postulated that the yellow- and green-seeded genes are carried on a certain pair of chromosomes and that the round- and wrinkled-seeded genes are carried on a different pair. This hypothesis immediately explains the experimentally observed 9:3:3:1 ratios" (Watson et al. 1988, p. 12). Again, in the summary of the chapter: "Mendel proposed that a hereditary factor (now known to be a gene) for each hereditary trait is given by each parent to each of its offspring. The physical basis for this behavior is the distribution of homologous chromosomes during meiosis: One (randomly chosen) of each pair of homologous chromosomes is distributed to each haploid cell. . . . For many years, the structure of genes and the chemical way in which they control cellular characteristics were a mystery" (Watson et al. 1988, p. 23).

Later chapters detailed the mechanisms of DNA replication (including mutation and repair) and the associated molecular mechanisms of recombination in crossing-over. For example, after discussing competing hypotheses about crossing-over: "In retrospect it is obvious what mechanism most precisely aligns DNA molecules in crossing over, because we can hardly imagine any other: Complementary base-pairing between strands unwound from two different chromosomes puts the chromosomes in exact register" (Watson et al. 1988, p. 316). An entire part of the book was devoted to the "Steps in Protein Synthesis." This was followed by discussion of gene expression mechanisms, first in the regulation of protein synthesis in bacteria and then a final part indicating the black boxes associated with "Facing Up to Eucaryotic Cells."

This textbook account supports the analysis of contemporary understanding of the mechanisms of heredity. The black boxes of Figure 4.1 were filled, in part, by the mechanisms depicted in Figures 4.2 and 4.3. The mechanisms

119

for sexually breeding organisms occur in a temporal sequence of DNA replication, chromosomal duplication, crossing-over, chromosomal random assortment, chromosomal segregation, germ cell formation, organismal mating, gametic fertilization, and, finally, gene expression during development, thereby producing phenotypic characters.

#### 4.7 CONCLUSION

To return to the philosophical analyses, we see that Mendelian genetics has not been reduced to molecular biology nor replaced by it. As both Schaffner and Hull realized, the formal model of reduction does not capture the practice of biologists, either in the way the two fields developed historically or as depicted in an influential contemporary textbook. As Kitcher argued, the appropriate explanatory level for pairing and separation was and is that of the chromosomes. As Waters argued, the molecular level was the appropriate one for finding the mechanism of crossing-over. That mechanism operates after chromosomes pair but before they separate. As Kitcher argued, molecular biology was an explanatory extension of Mendelian genetics and cytology.

What now needs to be added to this analysis is the view for which this chapter has argued. The fields of Mendelian genetics and molecular biology are best characterized as investigating different, serially integrated hereditary mechanisms. The mechanisms operate at different times and are composed of different working entities of different sizes. One does not always make progress by moving to lower size levels. The important interfield bridge between Mendelian genetics and cytology was neglected by most previous philosophical accounts. The working entities of the mechanisms of Mendelian heredity are chromosomes, whose movements serve to segregate alleles and independently assort genes in different linkage groups. The regularities captured in Mendel's laws of segregation and independent assortment were and still are explained by the chromosomal mechanisms of meiosis, as the Morgan group's work showed and as depicted in recent textbooks. The behaviors of chromosomes in meiosis provide "the mechanisms of Mendelian heredity." The working entities of numerous mechanisms of the molecular biology of the gene are larger and smaller segments of DNA plus related molecules. Molecular DNA mechanisms filled black boxes unilluminated by Mendelian/cytological techniques. Progress in genetics occurred not by reduction or replacement but by discovering new mechanisms and integrating them into the temporal series of hereditary mechanisms.

120

REFERENCES

Alberts, Bruce, Dennis Bray, Julian Lewis, Martin Raff, Keith Roberts, and James D. Watson (1983), *Molecular Biology of the Cell*. New York: Garland.

Bateson, William, E. R. Saunders, and R. C. Punnett (1906), "Experimental Studies in the Physiology of Heredity," *Reports to the Evolution Committee of the Royal Society III*. Reprinted in R. C. Punnett (1928), *Scientific Papers of William Bateson*, v.2. Cambridge: Cambridge University Press, pp. 152–161.

Bechtel, William and Robert C. Richardson (1993), *Discovering Complexity: Decomposition and Localization as Strategies in Scientific Research*. Princeton, NJ: Princeton University Press.

Bogen, James and James Woodward (1988), "Saving the Phenomena," *Philosophical Review* 97: 303–352.

Boveri, Theodor (1904), *Ergebnisse über die Konstitution der chromatischen Substanz des Zellkerns*. Jena: G. Fischer.

Chadarevian, Soraya de (2002), *Designs for Life: Molecular Biology after World War II*. New York: Cambridge University Press.

Correns, Carl ([1900] 1966), "G. Mendel's Law Concerning the Behavior of Progeny of Varietal Hybrids," translated from German and reprinted in C. Stern and E. Sherwood (eds.), *The Origin of Genetics, A Mendel Source Book*. San Francisco, CA: W. H. Freeman, pp. 119–132.

Craver, Carl F. (2001), "Role Functions, Mechanisms, and Hierarchy," *Philosophy of Science* 68: 53–74.

Craver, Carl F. (2002a), "Structures of Scientific Theories," in Peter K. Machamer and M. Silberstein (eds.), *Blackwell Guide to the Philosophy of Science*. Oxford: Blackwell, pp. 55–79.

Crick, Francis, Leslie Barnett, Sydney Brenner, and R. J. Watts-Tobin (1961), "General Nature of the Genetic Code for Proteins," *Nature* 192: 1227–1232.

Darden, Lindley (1991), *Theory Change in Science: Strategies from Mendelian Genetics*. New York: Oxford University Press.

Darden, Lindley (1995), "Exemplars, Abstractions, and Anomalies: Representations and Theory Change in Mendelian and Molecular Genetics," in James G. Lennox and Gereon Wolters (eds.), *Concepts, Theories, and Rationality in the Biological Sciences*. Pittsburgh, PA: University of Pittsburgh Press, pp. 137–158.

Darden, Lindley (1996), "Generalizations in Biology," Essay Review of K. Schaffner's *Discovery and Explanation in Biology and Medicine*. *Studies in History and Philosophy of Science* 27: 409–419.

Darden, Lindley (2002), "Strategies for Discovering Mechanisms: Schema Instantiation, Modular Subassembly, Forward/Backward Chaining," *Philosophy of Science (Supplement)* 69: S354–S365.

Darden, Lindley and Carl Craver (2002), "Strategies in the Interfield Discovery of the Mechanism of Protein Synthesis," *Studies in History and Philosophy of Biological and Biomedical Sciences* 33: 1–28.

Darden, Lindley and Nancy Maull (1977), "Interfield Theories," *Philosophy of Science* 44: 43–64.

Davis, Bernard D. (1980), "Frontiers of the Biological Sciences," *Science* 209: 78–89.

Reasoning in Biological Discoveries

Fogle, T. (2000), "The Dissolution of Protein Coding Genes in Molecular Biology," in Peter Beurton, Raphael Falk, and Hans-Jörg Rheinberger (eds.), *The Concept of the Gene in Development and Evolution*. New York: Cambridge University Press, pp. 3–25.

Glennan, Stuart S. (1996), "Mechanisms and The Nature of Causation," *Erkenntnis* 44: 49–71.

Glennan, Stuart S. (2002), "Rethinking Mechanistic Explanation," *Philosophy of Science (Supplement)* 69: S342–S353.

Hempel, Carl G. (1965), *Aspects of Scientific Explanation*. New York: The Free Press, Macmillan.

Hughes, Arthur (1959), *A History of Cytology*. New York: Abelard-Schuman.

Hull, David (1974), *Philosophy of Biological Science*. Englewood Cliffs, NJ: Prentice-Hall.

Jacob, Francois and Jacques Monod (1961), "Genetic Regulatory Mechanisms in the Synthesis of Proteins," *Journal of Molecular Biology* 3: 318–356.

Johannsen, Wilhelm (1909), *Elemente der Exakten Erblichkeitslehre*. Jena: G. Fischer.

Judson, Horace F. (1996), *The Eighth Day of Creation: The Makers of the Revolution in Biology*. Expanded Edition. Cold Spring Harbor, NY: Cold Spring Harbor Laboratory Press.

Kay, Lily E. (2000), *Who Wrote the Book of Life? A History of the Genetic Code*. Stanford, CA: Stanford University Press.

Kemeny, J. and P. Oppenheim (1956), "On Reduction," *Philosophical Studies* 7: 6–17.

Keyes, Martha (1999a), "The Prion Challenge to the "Central Dogma" of Molecular Biology, 1965–1991, Part I: Prelude to Prions," *Studies in the History and Philosophy of Biological and Biomedical Sciences* 30: 1–19.

Keyes, Martha (1999b), "The Prion Challenge to the "Central Dogma" of Molecular Biology, 1965–1991, Part II: The Problem with Prions," *Studies in the History and Philosophy of Biological and Biomedical Sciences* 30: 181–218.

Kitcher, Philip (1984), "1953 and All That: A Tale of Two Sciences," *The Philosophical Review* 93: 335–373.

Kitcher, Philip (1989), "Explanatory Unification and the Causal Structure of the World," in Philip Kitcher and Wesley Salmon (eds.), *Scientific Explanation*. Minnesota Studies in the Philosophy of Science, v. 13. Minneapolis, MN: University of Minnesota Press, pp. 410–505.

Kitcher, Philip (1999), "The Hegemony of Molecular Biology," *Biology & Philosophy* 14: 195–210.

Kohler, Robert E. (1982), *From Medical Chemistry to Biochemistry: The Making of a Biomedical Discipline*. New York: Cambridge University Press.

Machamer, Peter, Lindley Darden, and Carl Carver (2000), "Thinking About Mechanisms," *Philosophy of Science* 67: 1–25.

Mendel, Gregor ([1865] 1966), "Experiments on Plant Hybrids," translated from German and reprinted in Curt Stern and Eva Sherwood (eds.), *The Origin of Genetics, A Mendel Source Book*. San Francisco, CA: W. H. Freeman, pp. 1–48.

Monaghan, Floyd and A. Corcos (1984), "On the Origins of the Mendelian Laws," *The Journal of Heredity* 75: 67–69.

Morange, Michel (1998), *A History of Molecular Biology*. Trans. from French by Matthew Cobb. Cambridge, MA: Harvard University Press.

Morgan, Thomas Hunt (1909), "What Are 'Factors' in Mendelian Explanations?" *American Breeder's Association Report* 5: 365–368.

Morgan, Thomas Hunt (1910a), "Chromosomes and Heredity," *American Naturalist* 44: 449–496.

Morgan, Thomas Hunt (1910b), "Sex-Limited Inheritance in *Drosophila*," *Science* 32: 120–122.

Morgan, Thomas Hunt (1911a), "An Attempt to Analyze the Constitution of the Chromosomes on the Basis of Sex-Limited Inheritance in *Drosophila*," *Journal of Experimental Zoology* 11: 365–413.

Morgan, Thomas Hunt (1917), "The Theory of the Gene," *American Naturalist* 51: 513–544.

Morgan, Thomas Hunt (1919), *The Physical Basis of Heredity*. Philadelphia, PA: J. B. Lippincott Co.

Morgan, Thomas Hunt (1926), *The Theory of the Gene*. New Haven, CT: Yale University Press.

Morgan, Thomas Hunt and Clara J. Lynch (1912), "The Linkage of Two Factors in *Drosophila* that Are Not Sex-Linked," *Biological Bulletin* 23: 174–182.

Morgan, Thomas Hunt, A. H. Sturtevant, H. J. Muller, and Calvin B. Bridges (1915), *The Mechanism of Mendelian Heredity*. New York: Henry Holt and Company.

Nagel, Ernest (1961), *The Structure of Science*. New York: Harcourt, Brace, and World.

Nirenberg, M. W. and J. H. Matthaei (1961), "The Dependence of Cell-Free Protein Synthesis in *E. coli* Upon Naturally Occurring or Synthetic Polyribonucleotides," *Proceedings of the National Academy of Sciences* 47: 1588–1602.

Olby, Robert (1994), *The Path to the Double Helix: The Discovery of DNA*. Revised Edition. Mineola, NY: Dover.

Oppenheim, Paul and Putnam, Hilary (1958), "Unity of Science as a Working Hypothesis," in H. Feigl, M. Scriven, and G. Maxwell (eds.), *Concepts, Theories, and the Mind-Body Problem*, Minnesota Studies in the Philosophy of Science, v. 2. Minneapolis, MN: University of Minnesota Press, pp. 3–36.

Pauling, Linus (1939), *The Nature of the Chemical Bond*. Ithaca, NY: Cornell University Press.

Pauling, Linus and Robert B. Corey (1950), "Two Hydrogen-Bonded Spiral Configurations of the Polypeptide Chain," *Journal of the American Chemical Society* 72: 5349.

Schaffner, Kenneth (1974b), "The Peripherality of Reductionism in the Development of Molecular Genetics," *Journal of the History of Biology* 7: 111–139.

Schaffner, Kenneth (1993), *Discovery and Explanation in Biology and Medicine*. Chicago, IL: University of Chicago Press.

Stent, Gunther (1969), *The Coming of the Golden Age: A View of the End of Progress*. Garden City, NY: American Museum of Natural History Press.

Sutton, Walter (1903), "The Chromosomes in Heredity," *Biological Bulletin* 4: 231–251.

Tabery, James G. (2004), "Synthesizing Activities and Interactions in the Concept of a Mechanism," *Philosophy of Science* 71: 1–15.

Vries, Hugo de ([1900] 1966), "The Law of Segregation of Hybrids." Translated from German and reprinted in C. Stern and E. Sherwood (eds.), *The Origin of Genetics, A Mendel Source Book*. San Francisco, CA: W. H. Freeman, pp. 107–117.

Waters, C. Kenneth (1990), "Why the Anti-Reductionist Consensus Won't Survive the Case of Classical Mendelian Genetics," in Arthur Fine, Micky Forbes, and Linda Wessels (eds.), *PSA 1990*, v. 1, East Lansing, MI: Philosophy of Science Association, pp. 125–139.

Waters, C. Kenneth (1994), "Genes Made Molecular," *Philosophy of Science* 61: 163–185.

Watson, James D. (1965), *Molecular Biology of the Gene*. New York: W. A. Benjamin.

Watson, James D. (1970), *Molecular Biology of the Gene*. 2nd ed. New York: W. A. Benjamin.

Watson, James D. (1977), *Molecular Biology of the Gene*. 3rd ed. New York: W. A. Benjamin.

Watson, James D., Nancy H. Hopkins, Jeffrey W. Roberts, Joan Argetsinger Steitz, and Alan M. Weiner (1988), *Molecular Biology of the Gene*. 4th ed. Menlo Park, CA: Benjamin/Cummings.

Watson, James D. and Francis Crick (1953a), "A Structure for Deoxyribose Nucleic Acid," *Nature* 171: 737–738.

Watson, James D. and Francis Crick (1953b), "Genetical Implications of the Structure of Deoxyribonucleic Acid," *Nature* 171: 964–967.

White, Abraham, Philip Handler, Emil L. Smith, and DeWitt Stetten, Jr. (1954), *Principles of Biochemistry*. New York: McGraw-Hill.

Wilson, Edmund B. (1900), *The Cell in Development and Inheritance*. 2nd ed. New York: Macmillan.

Wimsatt, William (1976), "Reductive Explanation: A Functional Account," in A. C. Michalos, C. A. Hooker, G. Pearce, and R. S. Cohen (eds.), *PSA 1974*. Dordrecht: Reidel, pp. 671–710.

Wimsatt, William C. (1981), "Robustness, Reliability, and Overdetermination," in M. Brewer and B. Collins (eds.), *Scientific Inquiry and the Social Sciences*. San Francisco, CA: Jossey-Bass, pp. 124–163.

# II

## Reasoning Strategies

### Relating Fields, Resolving Anomalies

# 5

# Interfield Theories[1]

## 5.1 INTRODUCTION

Interactions between different areas or branches or fields of science have often been obscured by current emphasis on the relations between different scientific theories. Although some philosophers have indicated that different branches may be related, the actual focus has been on the relations between theories within the branches. For example, Ernest Nagel has discussed the reduction of one branch of science to another (1961, Ch. 11). But the relation that Nagel describes is really nothing more than the derivational reduction of the *theory* or *experimental* law of one branch of science to the theory of another branch.

We, in contrast to Nagel, are interested in the interrelations between the areas of science that we call *fields*. For example, cytology, genetics, and biochemistry are more naturally called fields than theories. Fields may have theories within them, such as the classical theory of the gene in genetics; such theories we call *intrafield* theories. In addition, and more important for our purposes here, interrelations between fields may be established via *interfield* theories. For example, the fields of genetics and cytology are related via the chromosome theory of Mendelian heredity. The existence of such interfield theories has been obscured by analyses such as Nagel's that erroneously conflate theories and fields and see interrelations as derivational reductions.

The purpose of this chapter is, first, to draw the distinction between field and *intra*field theory, and, then, more importantly, to discuss the generation of heretofore unrecognized *inter*field theories and their functions in relating

---

[1] This chapter was originally published as Darden, Lindley and Nancy Maull (1977), "Interfield Theories," *Philosophy of Science* 44: 43–64.

two fields.[2] Finally, we wish to mention the implications of this analysis for reduction accounts and for unity and progress in science.

By analysis of a number of examples, we will show that a field is an area of science consisting of the following elements: a central problem, a domain consisting of items taken to be facts related to that problem,[3] general explanatory factors and goals providing expectations as to how the problem is to be solved,[4] techniques and methods, and, sometimes but not always, concepts, laws, and theories which are related to the problem and which attempt to realize the explanatory goals. A special vocabulary is often associated with the characteristic elements of a field.[5] Of course, we could attempt to associate institutional and sociological factors with the elements of a field, but such an attempt would fail to serve the purpose of our discussion. We are interested in conceptual, not sociological or institutional, change. Thus, the elements of a field are conceptual, not sociological, of primary interest to the philosopher, not the sociologist.

The elements are also historical. Fields in science emerge, evolve, sometimes even cease to be. (We have not yet explored the latter phenomenon of decline.) Although any or all of the elements of the field may have existed separately in science, they must be brought together in a fruitful way for the field to emerge. Such an emergence is marked by the recognition of a promising way to solve an important problem and the initiation of a line of research in that direction. For instance, what comes to be the central problem of a field may have been a long-unsolved puzzle and the techniques may have been used elsewhere, but the field emerges when someone sees that those techniques yield information relevant to the problem. Or, perhaps, a new concept

---

[2] We are not using "theory" in the sense of "deductive system." In spite of the difficulties of providing a general analysis of "theory," we have retained the term because the developments with which we are concerned are called "theories" by their originators. Furthermore, the interfield developments are solutions to theoretical problems, as we shall see.

[3] "Domain" is used here in the sense analyzed by Dudley Shapere (1974a).

[4] Stephen Toulmin emphasizes the importance of explanatory problems and goals in Toulmin (1972, Chs. 2 and 3).

[5] A special vocabulary is not a formal language, but a specialized part of the natural language. Nor is a special vocabulary a theoretical vocabulary, since the terms in it may be associated with the domain or techniques of the field as well as its problem solutions. A term may become part of the special vocabulary by specialization within the field (e.g., "mutation" became specialized in genetics) or may be introduced as a new term (e.g., "epistasis" was introduced in genetics). Further examples of such terms are the following: in genetics, "test cross," "cis," "trans," "locus"; in cytology, "meiosis," "mitosis," 'karyotype," "chromosome"; in biochemistry, "respiratory quotient," "ligase," "citric acid cycle"; in physical chemistry, "bond angle," "secondary structure," and "optical rotation." For further discussion of special vocabularies, see Maull Roth (1974) and Maull (1977).

is proposed, giving new insight into an old puzzling problem and generating a line of research.

Because the convergence of the elements of a field can be identified historically, the emergence of a field can often be dated. Then, scientists who were part of the new field can be identified: they used the techniques of the field to solve its central problem. Others who had worked on the central problem in other ways or who had used the techniques for other purposes were not members of the field. The lone precursor who worked on the problem with the techniques but did not found an ongoing line of research, may, with hindsight, be called a geneticist, a biochemist, or whatever, even though the field (and perhaps even the term designating the field) did not exist at the time.

Of the terms current in philosophy of science which refer to categories broader than theory, the one which has most similarities to "field" is Stephen Toulmin's "discipline."[6] What Toulmin classes as a discipline, we would probably also class as a field. Included in his examples of disciplines are genetics, along with physics, atomic physics, chemistry, biochemistry, biology, and evolutionary biology (1972, pp. 141, 145, 146, 180). From these examples we see that Toulmin encounters a difficulty which is also a problem for an analysis in terms of fields: criteria seem to be needed to distinguish between disciplines, subdisciplines, and supradisciplines. For instance, is atomic physics a subdiscipline of physics or is atomic physics the discipline and physics the supradiscipline? Toulmin gives no way of distinguishing disciplines from smaller or larger units; as a result, his examples are somewhat confusing.

In this chapter, we discuss fields that are within the broader scientific areas of biology and chemistry, but we do not give criteria for distinguishing more inclusive from less inclusive categories. We suspect that the level at which an analysis is carried out may depend on the questions being asked and the historical period being examined. For example, historical examinations of science in the nineteenth century might well ask the question – when did biology emerge as a field in science? But twentieth-century historians of

---

[6] Other current broader categories include Imre Lakatos's "research programme" (Lakatos 1970) and Thomas Kuhn's "paradigm" or "disciplinary matrix" (Kuhn 1970). These have fewer similarities to fields and are fraught with more difficulties than Toulmin's analysis. For further discussion, see Darden (1974). Two further comments about fields are in order here. This analysis is not intended as a demarcation between science and nonscience. There may well be fields of nonscience with some of the same elements indicated here for scientific fields. Criteria other than those we have provided would be necessary to distinguish nonscientific fields from scientific ones. Secondly, this analysis does not presuppose that all of science can be neatly divided into mutually exclusive fields. Such division would not be expected of things which evolve. Further investigation is necessary to determine the limits of applicability of the term 'field' in other cases.

biology are more likely to treat fields within what has become the broader area of biology. Thus, the formulation of time-independent criteria for the delineating of fields will be difficult and might even serve to obscure important aspects of the historical development of science.

Toulmin's lists of the components of a discipline are numerous: "body of concepts, methods, and fundamental aims" (Toulmin 1972, p. 139); "a communal tradition of procedures and techniques for dealing with theoretical or practical problems" (p. 142); "(i) the current explanatory goals of the science, (ii) its current repertory of concepts and explanatory procedures, and (iii) the accumulated experience of the scientists working in this particular discipline" (p. 175). Taken collectively, these components are similar to the elements of a field.[7]

Although similarities exist between field and discipline, there are several reasons why we have not adopted Toulmin's term. First, even though the components of disciplines are similar to those of fields, they are not identical. The central problem, domain, and techniques will play important roles in our analysis; they are not found in Toulmin's lists. Furthermore, we find it difficult to use or to analyze such components as "accumulated experience of scientists." But most important, Toulmin's notion of a discipline is embedded in an epistemology: Knowledge is the result of a selection process much like the selection processes proposed by evolutionary theory for biological organisms. Although Toulmin's analysis is provocative, we would rather not commit our analysis to his "evolutionary epistemology." The legitimate use of an evolutionary analogy, we believe, can only be discovered by a detailed investigation of science, for example, by examination of interactions among the elements of fields.

Examples of fields will now be examined in more detail. Cytology in its early days had the central problem – what are the basic units of organisms? This problem was solved by the postulation of the cell theory and its subsequent elaboration and confirmation in the nineteenth century. Afterwards, the problem for cytologists (or "cell biologists," as they have come to be called) became the characterization of different types of cells, of organelles within cells, and of their various functions. The problem is tackled primarily with the technique of microscopic analysis.

The field of genetics, on the other hand, has as its central problem the explanation of patterns of inheritance of characteristics. The characteristics

---

[7] In comparing fields and disciplines, Toulmin said that the field is what the discipline is concerned with. "A discipline is an activity." (Private conversation with LD, March 12, 1974.)

may be either gross phenotypic differences, such as eye color in the fly *Drosophila*, as investigated in classical genetics, or molecular differences, such as loss of enzyme activity, as investigated in modern transmission genetics. The patterns of inheritance are investigated with the technique of artificial breeding. The laws of segregation and independent assortment (Mendel's laws), once their scope was known and they were well confirmed, became part of the domain to be explained. For many of the early geneticists, though not all, the goal was to solve the central problem by the formulation of a theory involving material units of heredity (genes) as explanatory factors. In attempting to realize the goal, T. H. Morgan and his associates formulated the theory of the gene of classical genetics. Extension of the theory and techniques from *Drosophila*, Morgan's model organism, to microorganisms marked the modern phase of the field of genetics, a phase which may be called "modern transmission genetics."

The central problem of biochemistry is the determination of a network of interactions between the molecules of cellular systems and their molecular environment; these molecules and their interrelations are the items of the domain. As was the case with genetics in which laws became part of the domain, here too the solution to a problem may contribute new domain items. For example, the Krebs cycle was part of the solution to the problem of determining the interactions between molecules and became, in turn, part of the domain of biochemistry; its relation to other complex pathways then posed a new problem. Many techniques of biochemistry are aimed at the reproduction of in vivo systems in vitro, that is, the "test tube" simulation of the chemical reactions that occur in living things.

The determination of the structure and three-dimensional configuration of molecules has become the concern of physical chemistry.[8] Thus, the central problem of physical chemistry is the determination of the interactions of all parts of a molecule relative to one another, under varying conditions. The domain of physical chemistry is the parts of molecules and their interactions. Physical chemistry has evolved complex techniques for the determination of

---

[8] Determination of the structure and three-dimensional configuration of molecules was not, at the turn of the century, the concern of physical chemists, like Wilhelm Ostwald, who were interested only in energy relations in biological systems. However, organic chemists (e.g., Emil Fischer) were interested in the structural analysis of molecules. But, by 1910, even Ostwald, who had been influenced by Ernst Mach's positivistic view of science, admitted that molecules exist, thus removing skepticism about the application of the kinetic techniques of physical chemistry to problems about the structure of molecules. For an excellent account of the interaction between physical chemistry and organic chemistry, see Fruton (1972).

the structure and conformation of molecules: x-ray diffraction, mass spectrometry, electron microscopy, and the measurement of optical rotation.

With these examples of fields in mind, we may contrast fields and *intra*field theories. A field at one point in time may not contain a theory, or may consist of several competing theories, or may have one rather successful theory. Well-confirmed laws and theories may become part of the domain and a more encompassing theory may be sought to explain them. Although theories within a field may compete with one another, in general, fields do not compete, nor do theories in different fields compete. Furthermore, one field does not reduce another field; reduction in the sense of derivation would be impossible between such elements of a field as techniques and explanatory goals.[9]

Even though fields do not bear the relations formerly thought to exist between theories, fields may be related to one another. Indeed, our main concern here is with the relations between fields which serve to generate a different type of theory, the *interfield theory*, which sets out and explains the relations between fields. Our task now is to discuss the conditions which lead to the generation of interfield theories. The discussion of general features of generation will be followed by examples of interfield theories: the chromosome theory of Mendelian heredity bridging the fields of cytology and genetics; the operon theory relating the fields of genetics and biochemistry; and the theory of allosteric regulation connecting the fields of biochemistry and physical chemistry. The examples will then serve as a basis for characterizing the general functions of interfield theories.

## 5.2 THE GENERATION OF INTERFIELD THEORIES

An interfield theory functions to make explicit and explain relations between fields. Relations between fields may be of several types; among them are the following:

(1) A field may provide a *specification of the physical location* of an entity or process postulated in another field. For example, in its earliest formulation, the chromosome theory of Mendelian heredity postulated that the Mendelian genes were *in* or *on* the chromosomes; cytology provided the physical location of the genes. With more specific knowledge, the theory explained the

[9] We are not taking a stand as to whether, in some possible cases, a theory in one field may be derived (in the sense of reduction) from a theory in another field. However, our examples here do not indicate that any such reduction has occurred; on the contrary, a main point of this chapter is that an analysis in terms of interfield theories, not reductions, is the appropriate analysis in these important cases in biology.

relation in more detail: the genes are part of (in) the chromosomes. Thus, the relation became more specific, a *part-whole* relation.

(2) A field may provide the *physical nature* of an entity or process postulated in another field. Thus, for example, biochemistry provided the physical nature of the repressor, an entity postulated in the operon theory.

(3) A field may investigate the *structure* of entities or processes, the *function* of which is investigated in another field. Physical chemistry provides the structure of molecules whose function is described biochemically.

(4) Fields may be linked *causally*, the entities postulated in one field providing the causes of effects investigated in the other. For example, the theory of allosteric regulation provides a causal explanation of the interaction between the physicochemical structure of certain enzymes and a characteristic biochemical pattern of their activity.

These types of relations are not necessarily mutually exclusive; as the examples indicate, structure-function relations may also be causal.

Several different types of reasons may exist for generating an interfield theory to make explicit such relations between fields. First, relationships between two fields may already be known to exist prior to the formulation of the interfield theory. We refer to such pre-established relationships as *background knowledge*. For example, prior to the proposal of the operon theory, the fields of genetics and biochemistry were known to be related; to cite one of many instances, the physical nature of the gene was specified biochemically as DNA. Thus, further relations could be expected between the fields and might lead to the generation of an interfield theory.

Secondly, a stronger reason for proposing an interfield theory exists when two *fields share an interest in explaining different aspects of the same phenomenon*.[10] For example, genetics and cytology shared an interest in explaining the phenomenon of heredity, but genetics did so by breeding organisms and explaining the patterns of inheritance of characters with postulated genes. Cytology, on the other hand, investigated the location of the heredity material within the cell using microscopic techniques. Since they were both working on the problem of explaining the phenomenon of hereditary, a relation between them was expected to exist.

Furthermore, *questions arise in each field which are not answerable using the concepts and techniques of that field*. These questions direct the search for an interfield theory. For example, in genetics the question arose: where are the

---

[10] To say that two different fields share an interest in "the same phenomenon" is only to say that scientists believed and had good reasons to believe that they were dealing with the same phenomenon.

genes located? But no means of solving that question within genetics were present since the field did not have the techniques or concepts for determining physical location. Cytology did have such means.

In brief, an interfield theory is likely to be generated when background knowledge indicates that relations already exist between the fields, when the fields share an interest in explaining different aspects of the same phenomenon, and when questions arise about that phenomenon within a field which cannot be answered with the techniques and concepts of that field.

Questions about the relations between fields pose an *interfield theoretical problem*: how are the relations between the fields to be explained? The solution to an interfield theoretical problem is an interfield theory. Dudley Shapere, in discussing theoretical problems, says: "Theoretical problems call for answers in terms of ideas different from those used in characterizing the domain items.... These new ideas, moreover, are expected to 'account for' the domain..." (Shapere 1973, p. 22). Shapere's analysis is for an *intra*field theory, in our terminology, but we may extend it to our case by saying that new ideas are introduced to account for the relationships between the two different domains of the different fields. The new idea which the theory introduces gives the nature of the relations between two fields, such as the types of relations discussed in (1), (2), (3), and (4) above.[11]

Suppose a relation between fields is suspected to exist because the fields share an interest in explaining aspects of the same phenomenon. Familiar types of relations between fields can then be considered, for example, causal, part-whole, or structure-function. The most likely relation (as indicated by considerations which will not be examined in this chapter) can be chosen and particularized for the case in point. Thus, a new idea is introduced specifying the nature of the relations between fields, that is, an interfield theory is formulated.

---

[11] Shapere's analysis differs from ours in another respect. For Shapere, items that are related to one another and demand explanation make up *one* domain. In fact, "related items about which there is a problem demanding a theory as an answer" is one of the alternative definitions of a domain supplied by Shapere. Hence, once relations are seen between two domains of two different fields, Shapere would probably regard the situation as the formulation of a new single domain encompassing the other two. However, we have introduced discussion of a field and domains that are characteristic of particular fields, distinctions not explicitly discussed by Shapere. And our primary concern is with such relationships between fields. Hence, although it is possible to say that a new domain is formed as a result of the discovery of relations between different domains, we do not wish to so characterize the situation. We prefer to regard the domains of different fields as separate but related and the interfield theory as providing an explanation of the relations.

We will now turn to the examination of detailed examples of interfield theories in order to illustrate the general features of their generation just discussed and to analyze their functions in more detail.

## 5.3 THE CHROMOSOME THEORY OF MENDELIAN HEREDITY

Cytology emerged as a field in the 1820s and 1830s with improvements in the microscope and the proposal of the cell theory. By the late 1800s, as a result of their investigations of the structures within cells, cytologists asked the following question: where within the germ cells is the hereditary material located? A widely accepted answer by 1900 proposed the chromosomes (darkly staining bodies within the nuclei of cells) as the likely location. (For further discussion, see Wilson 1900; Coleman 1965; Hughes 1959.)

On the other hand, theories of heredity had been proposed in the late nineteenth century, but none had the necessary ties to experimental data to give rise to a field of heredity until the discovery of (what have come to be called) Mendel's laws in 1900. Although Mendel had worked with garden peas, noted their hereditary characteristics, crossed them artificially, and proposed a law – he formulated only one – characterizing the patterns of inheritance, he did not found a field. Genetics emerged between 1900 and 1905 with the independent discovery of Mendel's law by Hugo de Vries and Carl Correns and with the promulgation of Mendel's experimental approach by William Bateson. Although Bateson did not (for reasons too complex to examine here), other geneticists postulated (what have come to be called) genes as the causes of hereditary characteristics. (For further discussion, see Coleman 1970; Darden 1974; Dunn 1965.)

Thus, by 1903, cytology and genetics had both investigated hereditary phenomena but asked different questions about it. At least some geneticists postulated Mendelian units to account for the patterns of inheritance of observed characteristics. Cytologists, on the other hand, proposed the chromosomes as the location of the hereditary material in the germ cells. Genes were, thus, hypothetical entities with known functions; chromosomes were entities visible with the light microscope with a postulated function.

But questions arose in each field which were not answerable with the techniques and concepts of the field itself. Genetics was unable to answer the question: where are the genes located? Its techniques were those of artificial breeding which provided no way of determining physical location. Cytology was known to provide a way of investigating the cells and their contents and thus was the natural field to turn to in search of an answer to the question

**Table 5.1  Relations Between Chromosomes and Genes**

| CHROMOSOMES | GENES |
|---|---|
| ❶ Pure individuals (remain distinct, do not join) | ❶ Pure individuals (remain distinct, no hybrids) |
| ❷ Found in pairs (in diploid organisms prior to gametogenesis and after fertilization) | ❷ Found in pairs (in diploid organisms prior to segregation and after fertilization) |
| ❸ The reducing division results in one-half to gametes | ❸ Segregation results in one-half to gametes |
| ❹ *Prediction:* Random distribution of maternal and paternal chromosomes in formation of gametes | ❹ Characters from maternal and paternal lines found mixed in one individual offspring; independent assortment (often) of genes |
| ❺ Chromosome number smaller than gene number | ❺ *Prediction:* Some genes do not assort independently in inheritance; instead are linked on the same chromosome |
| ❻ Some chromosomes form chiasmata, areas of intertwining *Prediction:* An exchange of parts of chromosomes at chiasmata | ❻ More combinations of linked genes than number of chromosomes; "crossing-over" occurs |

about the location of genes. On the other hand, cytologists had no way of investigating the functioning of chromosomes in producing *individual* hereditary characteristics. Theodor Boveri had, however, investigated the loss of one (or more) *entire* chromosome(s) and the changes in many characteristics in the developing embryo that such loss produced (Boveri 1964).

In addition to there being questions in each field which could not be answered within that field, more important in the historical generation of the chromosome theory was the fact that properties of the chromosomes and genes showed striking similarities. At least three properties of chromosomes and genes had been found independently in the two fields (see Items 1, 2, and 3 of Table 5.1). Both Walter Sutton (Sutton 1903) and Theodor Boveri (1904, pp. 117–118) were struck by the remarkable similarities and were independently led in 1903 and 1904 to postulate the chromosome theory of Mendelian heredity as a result. The theory (using the modern term "gene") is the following: the genes are in or on the chromosomes. The theory solves the theoretical problem as to the nature of the relations between genes and chromosomes by introducing the new idea that the chromosomes are the physical location of the Mendelian genes. Although August Weismann had, in 1892, postulated that the chromosomes were composed of hierarchies of hereditary units, which he called "biophores, determinants, and ids," his was not a theory of Mendelian units (Weismann 1892). That the units which obeyed Mendel's law were located in or on the chromosomes was a new idea proposed by Boveri and Sutton.

The ambiguity as to whether the genes were "in" or "on" the chromosomes was resolved in favor of the "in" with further development of the theory in the hands of T. H. Morgan and his associates in the 1910s and 1920s (Morgan 1911a, 1926; Morgan et al. 1915). Thus, the relationship between genes and chromosomes postulated by the interfield theory became that of part to whole, and the theory explained the correlated properties because parts would be expected to share at least some properties of their wholes.

But the theory did more than explain properties of genes and chromosomes already known. It also functioned to predict new items for the domains of each field on the basis of knowledge of the other. For example, Item 4 of Table 5.1 is a prediction Sutton made about the behavior of chromosomes on the basis of the behavior of genes. This prediction corrected a misconception of cytologists. Mistakenly, Sutton said ([1903] 1959, p. 29) cytologists prior to the formation of the chromosome theory of Mendelian heredity had thought that the sets of chromosomes from the mother and father remained intact in their offspring and separated as units in the formation of gametes (sexual cells; in animals, eggs and sperm) in offspring. However, the independent assortment of hereditary characteristics and, therefore, the genes which cause them, led to a reexamination of the behavior of the chromosomes, with the subsequent finding that the maternal and paternal chromosomes are distributed randomly in the formation of gametes (Sutton 1903; Carothers 1913). The prediction for cytology of random segregation of chromosomes as a result of independent assortment of genes was thus substantiated.

Predictions went both ways. The knowledge from cytology of the small number of chromosomes compared to larger numbers of genes led both Boveri and Sutton to the prediction that some genes would be linked in inheritance, in other words, that exceptions to independent assortment would occur. The finding of linked genes substantiated this prediction. (See Item 5 of Table 5.1.) The finding of predictions made on the basis of the theory served to provide support for the theory. As a result, both genetic and cytological evidence provided confirmation.

Not only did the theory predict new domain items for each field, it also served to focus attention on previously known but neglected items. For example, Item 6 shows the correlation between the new finding in genetics that some genes "cross-over" (i.e., become unlinked) and the previously known chiasmata, that is, areas of intertwining between chromosomes. Chiasmata had been seen by cytologists prior to their correlation with crossing-over, but no function for them was known so they had not been considered important. With the correlation to a property of genes by Morgan (1911a), chiasmata took on a new significance and subsequent investigation showed that they were indeed areas of exchange between parts of chromosomes as predicted by the genetic evidence. This is an example of the change in relative importance of a type of domain item; an item previously considered peripheral became a center of investigation.[12]

After the formulation of the theory and its confirmation, new findings about genes raised parallel questions about chromosomes and vice versa. New types of experiments were designed using the techniques from both fields. Calvin Bridges, a coworker of T. H. Morgan, was one of the most successful practitioners of the new method of research (Bridges 1916). Thus, the theory generated a new line of research coordinating the techniques and findings of both fields.

In summary, the chromosome theory of Mendelian heredity is an interfield theory bridging the fields of genetics and cytology. It was generated to unify the knowledge of heredity found in both fields and thereby to explain the similar properties of chromosomes and genes. It functioned to focus attention on previously neglected items of the domains and to predict new items for the domains of each field. It further served to generate a new line of research coordinating the fields of cytology and genetics. Success in finding the predictions of the theory and in developing the common line of research resulted in the confirmation of the theory and the fruitful bridging of two fields of science.

## 5.4   THE OPERON THEORY AND THEORY OF ALLOSTERIC REGULATION

The chromosome theory was an important first step, eventually leading to the development of an explanation of how the genetic material acts as a carrier of information in biological systems. Once the DNA component of the chromosome was shown to carry the genetic information, then the problem of the *control* of such information emerged. The regulation of gene expression was seen to be of particular significance for an understanding of the development of organisms from embryo to adult. All cells of a multicellular organism have an identical complement of genes, but in different cells, different genes are expressed at different times. In short, differentiation occurs and must be explained. Even in unicellular organisms, gene expression varies with stages in the life cycle and, as we shall see in discussing the operon theory, with changes in the surrounding medium that affect the intracellular

---

[12] Shifts in importance of domain items were discussed by Shapere (1974a, pp. 532–533).

environment. The operon theory and the Monod–Wyman–Changeux theory of allosteric regulation are both theories of the control of gene expression: the operon theory of the control of protein levels (i.e., the quantity of a protein in a cell) and the theory of allosteric regulation of the control of protein *activity*.[13]

Biochemists became interested in one aspect of the control of protein levels, *enzyme adaptation* (later called "enzyme induction"), some fifty years after its discovery in 1900 by Dienert (1900). Dienert had described a process by which cells adjust the availability of an enzyme in response to the presence of specific metabolites (i.e., substances required for growth). This finding was later thought to suggest that gene expression is reversibly controlled by biochemical changes in the environment. For example, the bacterium *Escherichia coli* produces higher levels of the lactose-metabolizing enzymes when lactose precursors, the galactosides, are available. However, these enzyme levels are radically reduced in the absence of galactosides.

Further, transmission studies (artificial breeding and recording of characters transmitted to offspring) showed that the capacity to regulate enzyme levels in response to metabolites could be altered by mutation (heritable changes in the genes). Certain mutants in the lactose-metabolizing system within the bacterium were discovered; the mutants produce the enzymes required for the metabolism of lactose whether or not the lactose precursors are available. This mutation (*i-*, or inducer-negative) was found to be located at a site on the bacterial chromosome distinct from the sites of the genes for the lactose-metabolizing enzymes. This suggested that changes at a site somewhat distant from the genes for the lactose-metabolizing enzymes could affect the expression of those genes. Further investigation of mutants of this *i* gene and the critical experiment of Pardee, Jacob, and Monod in 1959 (Pardee et al. 1959) implicated an *i* gene product as the controlling substance responsible for the repression of the lactose-metabolizing enzymes.[14]

The control of these enzymes by an *i* gene product was incorporated into the proposal, in 1961, by Jacob and Monod (1961) of a theory of the operon, a causal theory of biochemical changes that effect specific, heritable patterns of gene expression. Two kinds of genes were postulated: *structural genes*, like the genes for the lactose-metabolizing enzymes, carry the information

---

[13] The theory of allosteric regulation discussed here is the Monod–Wyman–Changeux theory (1965). This theory of allosteric regulation is not the only conformational theory set forth to explain the regulation of protein activity; the other major account is Koshland's "induced fit" theory (1973).

[14] A more detailed examination of the genesis of the operon theory was undertaken by Schaffner (1974a, 1974c).

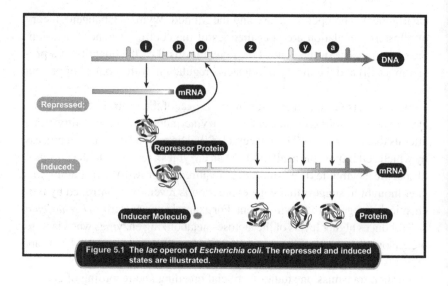

Figure 5.1 The *lac* operon of *Escherichia coli*. The repressed and induced states are illustrated.

that determines the molecular structure of enzymes, or carry the information for some proteins other than enzymes (e.g., hormones); and *regulatory genes*, of which the *i* gene represents only one type, are involved in the control of structural genes. Further, the operon theory postulates that the *lac* (lactose-metabolizing) system of *E. coli* is an *inducible* system; enzyme synthesis is induced by the presence of metabolites. Induction, that is, transcription of the lactose-metabolizing genes into a *cytoplasmic messenger* (mRNA) for protein synthesis, depends on the state of another regulatory gene called the *operator*. The structural genes whose activity is coordinately controlled, as are the lactose-metabolizing genes, form a unit of control, the *operon*. Transcription of the operon begins at the operator if the operator is not blocked by the *i* gene product, called the *repressor*. However, the repressor is not always in complex with the operator; the repressor itself is controlled by its interaction with *inducer*, in the case of the *lac* system, the galactosides. When inducer is available, the repressor binds the inducer and cannot bind the operator. And transcription of the operon proceeds. On the other hand, when inducer is absent. the repressor binds the operator and transcription is blocked (Figure 5.l).

Questions about the physical nature of the repressor, whether a polynu-cleotide like mRNA or a protein, were raised in the 1961 proposal. However, the *genetic* studies on which the proposal was based could not provide an answer to questions about the physical nature of the repressor. Subsequent to the 1961 proposal, *biochemical* findings implicated a protein repressor. Yet, the *lac* repressor was not isolated until 1966; the biochemical test used to isolate the protein was its affinity for galactosides, a property predicted by the operon theory on the basis of genetic findings. As the theory also predicted, the protein was shown to be absent or functionally impaired in the *i* mutant strains of bacteria. Finally, the repressor protein was shown to bind an operator, as the theory predicted.[15]

Significantly, the protracted failure to isolate the repressor in the five years after the 1961 proposal led to questions about the mode of interaction among repressor, inducer, and operator.[16] It was thought that a better understanding of the inducer-repressor interaction would facilitate isolation of the repressor. The characteristic biochemical pattern of activity of the repressor (a sigmoid activity curve) was seen by Monod and his colleagues to be similar to that of a class of "regulatory enzymes" and to hemoglobin, the "honorary enzyme." Thus, the operon theory served to direct new attention to an area of investigation in biochemistry, the functional similarities of a group of proteins. In addition to questions about the shared pattern of activity, questions about the possibility of shared structural features among such proteins were raised, thereby involving an area of investigation within physical chemistry, the structure or conformation of molecules.

Protein function (as revealed by a characteristic pattern of activity) was thought to be associated with protein structure as early as 1894; in that year, Fischer (1894) proposed his "lock and key" model of enzyme catalysis. In 1965, Monod, Wyman, and Changeux proposed a causal theory to relate changes in protein structure to changes in protein activity. According to this theory of allosteric regulation, the alteration of protein activity (in the case of the repressor, its affinity for the operator) is due to a reversible change in the conformation of the protein when it binds its regulatory metabolite (inducer).

The theory predicts that the regulatory protein (repressor) will have two nonoverlapping sites: one, the *active site*, has a structure complementary to the substrate (operator) and therefore binds it; another, the *allosteric site*, has

---

[15] A protein repressor was strongly suggested by the work of Bourgeois, Cohen, and Orgel (1965). The repressor was isolated by Gilbert and Müller-Hill (1966), who also showed that the repressor binds a DNA sequence, the operator (1967).

[16] On the one hand, failure to isolate the repressor led to an attempt to characterize its mode of action in the theory of allosteric regulation proposed by Monod, Wyman, and Changeux (1965). On the other hand, failure to isolate the repressor was seen by Stent (1964) as justification for the proposal of a new, competitor theory that prescribed a very different function for the *i* gene product.

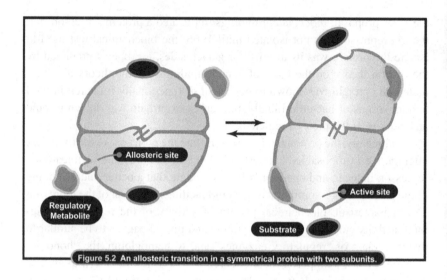

Figure 5.2 An allosteric transition in a symmetrical protein with two subunits.

a structure complementary to the regulatory metabolite (inducer) and binds it (Figure 5.2). The conformational change brought about by the formation of a protein-metabolite complex is called an *allosteric transition*, "which modifies the properties of the active site, changing one or several of the kinetic parameters which characterize the biological activity of the protein" (Monod, Wyman, and Changeux 1963, p. 307).

In both theories, the operon theory and the theory of allosteric regulation, two fields (genetics and biochemistry on the one hand and biochemistry and physical chemistry on the other) share a concern with the investigation of aspects of a phenomenon (with the operon theory, regulated gene expression and with the theory of allosteric regulation, regulated protein activity). Both theories are posed against a considerable body of background knowledge; extensive relationships among genetics, biochemistry, and physical chemistry had already been established. As a result, good reasons could be advanced for entertaining the hypotheses that patterns of gene expression are related to certain biochemical entities and processes and that the activity of proteins is related to their physicochemical conformation.

In both cases, no one field had all the concepts and techniques to answer all the questions that arose concerning the phenomenon. Genetics provided the transmission studies of the *i* mutant and characterized the active *i* gene product as a cytoplasmic substance. The physical nature of the *i* gene product was determined by a biochemical test. And the structural features of the repressor required to explain its characteristic pattern of biochemical activity could only be supplied by physicochemical techniques.

Both the operon theory and the theory of allosteric regulation are causal theories. The operon theory is an account of biochemical causes and genetic effects; specific biochemical interactions explain the patterns of inheritance characteristic of the regulated system of genes. The theory of allosteric regulation explains the characteristic pattern of activity (the function) of certain proteins by a specific and reversible sequence of structural change. Both theories supply answers to theoretical problems by introducing new ideas about the relationship among items of different domains.

## 5.5    THE FUNCTION OF INTERFIELD THEORIES

In summary, an interfield theory functions in some or all of the following ways:

a.    to solve (perhaps "correctly") the theoretical problem which led to its generation, that is, to introduce a new idea as to the nature of the relations among fields;

b.    to answer questions which, although they arise within a field, cannot be answered using the concepts and techniques of that field alone;

c.    to focus attention on previously neglected items of the domains of one or both fields;

d.    to predict new items for the domains of one or both fields; and

e.    to generate new lines of research which may, in turn, lead to another interfield theory.[17]

## 5.6    CONCLUSION

Because our examples represent significant developments in science, because they have important similarities in generation and function, and because other examples of theories sharing these characteristics may be found in other cases,

---

[17] In this discussion, we have not viewed the interfield theory as functioning to establish a new, third field with relations to the previously separate fields. We believe the situation is made unnecessarily complicated by doing so. However, just as Shapere might regard the two domains as joined (see note 11), the interfield theory could be seen as forming a new field. For example, 'cytogenetics' and 'molecular genetics' seem to be used to refer to the areas between genetics and cytology related by the chromosome theory and between modern transmission genetics and biochemistry related, in part, by the operon theory. Some scientists who used techniques from both original fields, such as Bridges did, might be considered members of the "interfield theory field." This cumbersome analysis is a possible interpretation but not the best, we think.

we have attempted to set forth the characteristics common to this type of theory, the interfield theory.

Let us now suggest why interfield theories have been ignored by philosophers of science. Philosophers have not usually discussed areas or fields of science, much less relations between them; instead, they have concentrated on theories and on the relations between those and, thus, have tended to view fields as theories. Furthermore, theories were viewed as being of the same type – interpreted axiomatic systems – and the relations between theories also were thought to take a single form – namely, derivational reduction.[18] After a derivational reduction had occurred, one theory had been "eliminated," at least in the sense that it had been explained as a deductive consequence of a more general theory. Reduction analyses were taken to provide an interpretation of the unity of science. For example, Oppenheim and Putnam (1958) interpreted the unity of science as the cumulative microreduction of theories. And progress was identified with successful reductions.

Although this overview of a tradition in philosophy of science may be simplistic in certain respects, it is sufficient to show how a concentration on reduction relations between theories would have obscured the nature of fields of science and relations between them established by interfield theories which are not reductive. An interfield theory, in explaining relations between fields, does not eliminate a theory or field or domain. The fields retain their separate identities, even though new lines of research closely coordinate the fields after the establishment of the interfield theory.

But our analysis of interfield theories does not merely serve to bring to the attention of philosophers of science a type of theory which has been ignored. In addition, our analysis suggests and provides a foundation for the further development of a conceptual apparatus for understanding the generation and function of theories. The advantage of paying attention to the developmental and functional characteristics of theories is shown by the fact that we have discovered similarities among theories that might not have been noticed otherwise. Although the relationships between genetics and biochemistry and between biochemistry and physical chemistry may be seen, at first glance, as in some way reductive, it is unlikely that anyone would claim that cytology is reduced to genetics (or vice versa) by the chromosome theory. In other words, our analysis shows important similarities between the generation and function of relationships which, on the older analysis, are in the different categories of "reductive" and "nonreductive" and would not have been seen

---

[18] See F. Suppe's Introduction (Suppe 1974) for a detailed account of this tradition in the philosophy of science (the "Received View") as well as criticisms of and alternatives to it.

as similar. It also indicates that there are important relations in science that are not reductive at all.[19]

One is led to ask: what accounted, on the older analysis, for the different ways of categorizing these two types of cases? Again, our analysis proves illuminating. Although all three cases are examples of interfield theories, the relationships between fields which they establish are not identical. The operon theory and the theory of allosteric regulation primarily establish *causal* links; the chromosome theory, on the other hand, establishes a *part–whole* relation between genes and chromosomes. We suspect that causal relations are more likely to appear reductive than part–whole relations. Of course, this speculation is only a starting point for further investigation of these and other cases.

Not only does our analysis call attention to similarities and differences overlooked in reductive analyses, it also casts doubt on the view that the unity of science is to be analyzed merely as a series of reductions, realized or potential. Provided with a new analysis of the relations between fields, it becomes natural to view the unity of science not as a hierarchical succession of reductions between theories but rather as the bridging of fields by interfield theories. The unity of science analyzed as a hierarchical classification scheme of scientific theories, graded according to generality, is precisely the picture provided by Oppenheim and Putnam (1958) as a "working hypothesis." Our preliminary analysis suggests another, new working hypothesis: unity in science is a complex network of relationships among fields effected by interfield theories.

Progress, too, receives a different analysis. Much of the progress of modern biology results from the development of interfield theories and the progressive unification (i.e., bridging) of the biological and physical sciences. With the chromosome theory, genes were associated with intracellular organelles; subsequently, with the development of the Watson–Crick model, the genes were identified with segments of DNA molecules with a specific type of structure. The operon theory provided further links between genetics and biochemistry, and the theory of allosteric regulation served as a bridge between

---

[19] Kenneth Schaffner (1974b), through examination of cases from the history of modern biology, including the operon theory, seems to be moving toward a similar view with his claim that derivational reduction is "peripheral" to developments in modern biology. We believe that our analysis of interfield theories explains the phenomena that Schaffner notes: "These principles and causally relevant entities [of the operon theory] represent both 'biological' and 'chemical' principles and entities; Jacob and Monod did not work at the strictly chemical level but at levels, 'intertwining' the biological and chemical" (1974b, p. 135). For further development of the analysis of interfield theories as an alternative to reduction, see Maull (1977).

biochemistry and physical chemistry. In sum, the chromosome theory was the first of a series of interfield theories, the operon theory and theory of allosteric regulation among them, that advanced our understanding of the relationship between the biological and physical sciences and resulted in progress in modern biology.

We have used a methodology here which ties philosophy of science not to formal logic but to the history of science, the proper subject matter of philosophy of science. But we must be clear about what our analysis has produced. We are not proposing a general analysis applicable to all varieties of theory, progress, or unity in science. Indeed, we have no reason to prejudge the still-open question as to whether there is *one* analysis applicable to all theories or instances of unification and progress. We have found a type of theory prevalent in modern biology which helps us understand one way in which unity and progress occur. Further examination of other cases from the history of biology and the history of other sciences will reveal the extent to which this analysis may be generalized.

### REFERENCES

Bourgeois, S., M. Cohen, and L. Orgel (1965), "Suppression of and Complementation among Mutants of the Regulatory Gene of the Lactose Operon of *Escherichia coli*," *Journal of Molecular Biology* 14: 300–302.

Boveri, Theodor ([1902] 1964), "On Multipolar Mitosis as a Means of Analysis of the Cell Nucleus," in B. H. Willier and J. Oppenheimer (eds.), *Foundations of Experimental Embryology*. Englewood Cliffs, NJ: Prentice-Hall, pp. 75–97.

Boveri, Theodor (1904), *Ergebnisse über die Konstitution der chromatischen Substanz des Zellkerns*. Jena: G. Fischer.

Bridges, Calvin B. (1916), "Non-disjunction as Proof of the Chromosome Theory of Heredity," *Genetics* 1: 1–52, 107–163.

Carothers, E. E. (1913), "The Mendelian Ratio in Relation to Certain Orthopteran Chromosomes," *The Journal of Morphology* 24: 487–509.

Coleman, William (1965), "Cell, Nucleus, and Inheritance: A Historical Study," *Proceedings of the American Philosophical Society* 109: 124–158.

Coleman, William (1970), "Bateson and Chromosomes: Conservative Thought in Science," *Centaurus* 15: 228–314.

Darden, Lindley (1974), *Reasoning in Scientific Change: The Field of Genetics at Its Beginnings*. Ph.D. Dissertation, University of Chicago, Chicago, IL.

Dienert, F. (1900), "Sur la Fermentation du Galactose et sur l'Accoutamance des levures à ce Sucre," *Annales de l'Institute Pasteur* 14: 138–189.

Dunn, L. C. (1965), *A Short History of Genetics*. New York: McGraw-Hill.

Fischer, E. (1894), "Einfluss der Konfiguration auf die Wirkung der Enzyme," *Berichte der deutschen chemische Gesellschaft* 27: 2985–2993.

Fruton, J. S. (1972), *Molecules and Life: Historical Essays on the Interplay of Chemistry and Biology*. New York: Wiley-Interscience.

Gilbert, W. and B. Müller-Hill (1966), "Isolation of the Lac Repressor," *Proceedings of the National Academy of Sciences* 56: 1891–1898.

Gilbert, W. and B. Müller-Hill (1967), "The Lac Operator is DNA," *Proceedings of the National Academy of Sciences* 58: 2415–2421.

Hughes, Arthur (1959), *A History of Cytology*. New York: Abelard-Schuman.

Jacob, Francois and Jacques Monod (1961), "Genetic Regulatory Mechanisms in the Synthesis of Proteins," *Journal of Molecular Biology* 3: 318–356.

Koshland, D. E. (1973), "Protein Shape and Biological Control," *Scientific American* 229: 52–64.

Kuhn, Thomas (1970), *The Structure of Scientific Revolutions*. 2nd ed. Chicago, IL: University of Chicago Press.

Lakatos, Imre (1970), "Falsification and the Methodology of Scientific Research Programmes," in Imre Lakatos and Alan Musgrave (eds.), *Criticism and the Growth of Knowledge*. Cambridge: Cambridge University Press, pp. 91–195.

Maull Roth, Nancy (1974), *Progress in Modern Biology: An Alternative to Reduction*. Ph.D. Dissertation, University of Chicago, Chicago, IL.

Maull, Nancy (1977), "Unifying Science Without Reduction," *Studies in the History and Philosophy of Science* 8: 143–162.

Monod, Jacques, J-P. Changeux, and Francois Jacob (1963), "Allosteric Proteins and Cellular Control Systems," *Journal of Molecular Biology* 6: 306–329.

Monod, Jacques, J. Wyman, and J-P. Changeux (1965), "On the Nature of Allosteric Transitions: A Plausible Model," *Journal of Molecular Biology* 12: 88–118.

Morgan, Thomas Hunt (1911a), "An Attempt to Analyze the Constitution of the Chromosomes on the Basis of Sex-Limited Inheritance in *Drosophila*," *Journal of Experimental Zoology* 11: 365–413.

Morgan, Thomas Hunt (1926), *The Theory of the Gene*. New Haven, CT: Yale University Press.

Morgan, Thomas Hunt, A. H. Sturtevant, H. J. Muller, and C. B. Bridges (1915), *The Mechanism of Mendelian Heredity*. New York: Henry Holt and Co.

Nagel, Ernest (1961), *The Structure of Science*. New York: Harcourt, Brace and World, Inc.

Oppenheim, Paul and Hilary Putnam (1958), "Unity of Science as a Working Hypothesis," in H. Feigl, M. Scriven and G. Maxwell (eds.), *Concepts, Theories and the Mind-Body Problem*. Minnesota Studies in the Philosophy of Science, v. 2. Minneapolis, MN: University of Minnesota Press, pp. 3–36.

Pardee, Arthur, Francois Jacob, and Jacques Monod (1959), "The Genetic Control and Cytoplasmic Expression of 'Inducibility' in the Synthesis of $\beta$–galatosidase by *E. coli*," *Journal of Molecular Biology* 1: 165–178.

Schaffner, Kenneth (1974a), "Logic of Discovery and Justification in Regulatory Genetics," *Studies in the History and Philosophy of Science* 4: 349–385.

Schaffner, Kenneth (1974b), "The Peripherality of Reduction in the Development of Molecular Biology," *Journal of the History of Biology* 7: 111–139.

Schaffner, Kenneth (1974c), "The Unity of Science and Theory Construction in Molecular Biology," in R. J. Seeger and R. S. Cohen (eds.), *Philosophical Foundations of Science: Proceedings of Section L, AAAS 1969*. Boston Studies in the Philosophy of Science, v. 11. Dordrecht: D. Reidel Publishing, pp. 497–533.

Shapere, Dudley (1973), Unpublished MS. Presented at IUHPS-LMPS Conference on Relations Between History and Philosophy of Science. Jyväskylä, Finland.

Shapere, Dudley (1974a), "Scientific Theories and Their Domains," in Frederick Suppe (ed.), *The Structure of Scientific Theories*. Urbana, IL: University of Illinois Press, pp. 518–565.

Stent, G. (1964), "The Operon: On Its Third Anniversary," *Science* 144: 816–820.

Suppe, Frederick (1974), *The Structure of Scientific Theories*. Urbana, IL: University of Illinois Press.

Sutton, Walter ([1903] 1959), "The Chromosomes in Heredity," *Biological Bulletin* 4: 231–251. Reprinted in J. A. Peters (ed.), *Classic Papers in Genetics*. Englewood Cliffs, NJ: Prentice-Hall, pp. 27–41.

Toulmin, Stephen (1972), *Human Understanding*. v. 1. Princeton, NJ: Princeton University Press.

Weismann, August (1892), *The Germ-Plasm, A Theory of Heredity*. Translated by W. N. Parker and H. Röonfeldt. New York: Charles Scribner's Sons.

Wilson, Edmund B. (1900), *The Cell in Development and Inheritance*. 2nd ed. New York: Macmillan.

# 6

# Theory Construction in Genetics[1]

Philosophers of science have had relatively little to say about theory construction. Theories were treated by Popper (1965) and the logical empiricists (e.g., Hempel 1966) as if they arose all at once by a creative leap of the imagination of a scientist, a process whose study was viewed as the province of the psychologist. Only after the creative leap, they agreed, were the philosopher's logical tools useful to evaluate the theory so produced. Even more historical accounts concerned with scientific change, such as Kuhn's (1970), did not discuss the way paradigms or the theories within them were constructed, except that they somehow arose in response to anomalies of their predecessors. Lakatos (1970), who proposed criteria for evaluating progressive research programs, did not discuss how the scientist constructs the program originally, although later additions, he claimed, resulted in some way from the "positive heuristic." Even Laudan (1977), who made much of the commonplace observation that science is a problem-solving activity, focused on the use of solved or unsolved problems to evaluate theories and research traditions; he did not provide an analysis of how a scientist goes about solving a problem. Thus, most of twentieth-century philosophy of science, from the logical empiricists to the most recent work, has been within the context of justification, not the context of discovery.

Dichotomizing science into these mutually exclusive contexts and concentrating on justification to the exclusion of discovery distorts the ongoing

[1] This chapter was originally published as Darden, Lindley (1980), "Theory Construction in Genetics," in T. Nickles (ed.), *Scientific Discovery: Case Studies*. Dordrecht: Reidel, pp. 151–170. It is reprinted with kind permission of Springer Science and Business Media. This research was supported by U.S. National Science Foundation Grant SOC 77-23476. For helpful comments on earlier drafts, I thank students and colleagues at the University of Maryland; participants at the Leonard Conference on Scientific Discovery at the University of Reno; and faculty at the Claremont Colleges, the University of Chicago, and the Johns Hopkins University.

process that characterizes science. A theory rarely, if ever, arises all at once in a complete form. Vague ideas about postulated explanatory factors may take on more form as new data are found and new theoretical components added. A negative result may produce a change in only one part of a theory, with the subsequent modification incorporating new ideas which fit the data better. Connections to empirically confirmed items in another field may be important in constructing part of a theory and at the same time bring a measure of justification. The processes of discovery and justification thus have complex interrelations in the development of a theory over time. Consequently, this chapter will use the phrase "theory construction," which better captures this ongoing process than do the mutually exclusive "context of discovery" and "context of justification."

A central problem in understanding this ongoing process of theory construction is to understand the various factors that play roles in it. First of all, the *domain* to be explained, that is, the phenomena to be explained by the theory (Shapere 1974a), will be one factor which, minimally, acts as a constraint. The theory must be able to account for all or most of the items of the domain. More positively than its constraining role, the domain may provide suggestions for the form the theory may take. Shapere (1974b) provided an analysis of domain indications for theory construction. But something other than the ideas in the domain must be built into the theory if the theory is to explain that domain. Shapere (1973, p. 22) characterized this difference by saying that theories must introduce *new ideas*, different from those in the domain, in order to account for that domain. Mary Hesse expressed this in linguistic terms: "the essence of a theoretical explanation is the introduction into the explanans of a new vocabulary or even of a new language" (Hesse 1966, p. 171). Carl Hempel, in discussing the problems of inductive inference, said in a similar vein: "scientific hypotheses and theories are usually couched in terms that do not occur at all in the description of the empirical findings on which they rest, and which they serve to explain" (Hempel 1966, p. 14). He continued by claiming that there could be no mechanical rules for producing the novel concepts found in the theories.[2]

Where do these new ideas, these novel concepts expressed in a new vocabulary, come from? Even if there are no mechanical rules for their production,

---

[2] "New" is ambiguous here. It may refer to a concept entirely new to the history of thought, e.g., the quantitative convertibility of a generalized energy in the late nineteenth century. Such developments are much rarer than the other sense of "new," namely, different from the concepts associated with the domain and new to this field or this type of theory. In the second case, the concept itself may be common in other sciences, e.g., material units, but it simply is different from items of the domain.

are guidelines for search obtainable? What role, exactly, do they play in the construction of a theory? In an attempt to answer these questions, this chapter will argue that either analogies or interfield connections may play the role of providing new ideas that are built into a theory according to a schema of theory construction, which will be discussed. Furthermore, a working hypothesis will be proposed: Connections to well-developed related fields are likely to be a better source of new ideas than analogies, for reasons to be given; thus, when interfield connections are available, they should be considered prior to a search for analogies in the steps of constructing a theory. In the specific historical case to be examined, I show that progress occurred in theory construction when scientists shifted from vague analogies to interfield connections as the source of new ideas.

Some scientists and philosophers have discussed the role of analogies in theory construction; fewer have discussed the connections between closely related scientific fields as providing an important source of new ideas. Before turning to a detailed examination of the historical case, we will briefly examine some of these opinions.

The philosopher, Norwood Russell Hanson (1961), in his classic work on reasoning in discovery, discussed the role of analogies. In retroductive reasoning, Hanson claimed, the scientist is puzzled by some phenomena. Then, according to Hanson, the scientist uses analogy to reason that some type of hypothesis plausibly explains the phenomena. Unfortunately, Hanson did not elaborate on this. Exactly how does the analogy function to suggest a type of hypothesis? What reasons does the scientist have for choosing one analogy rather than another? In what way do analogies make a type of hypothesis plausible?

Scientists have also indicated that analogies play a role in theory construction. G. K. Gilbert, a noted American geologist, said:

> To explain the origin of hypotheses I have a hypothesis to present. It is that hypotheses are always suggested through analogy. Consequential relations of nature are infinite in variety and he who is acquainted with the largest number has the broadest basis for the analogic suggestion of hypotheses.
>
> (Gilbert 1896, p. 2)

R. J. Blackwell, a philosopher writing in *Discovery in the Physical Sciences*, said in a similar vein:

> ... the scientist might recognize that the unsolved problematic before him bears resemblances to another problematic which has already been solved. As a result he forms the hypothesis that the same type of explanation also applies to his problem. ... This type of transformation leading to theoretical explanation we will call substitution through analogy. (Blackwell 1969, p. 179)

R. Harré (1960; 1970) provided a more detailed discussion of types of models and the ways they function in theory construction. Although it is not necessary to give all of Harré's baroque terminology in his classification of different types of models, two basic types are of interest to us here: the first to rule it out of our further discussion and the second to focus on. *Homoeomorphs,* according to Harré, have the same thing as both the *source* from which they come and the *subject* which they model. The simplest type is scale models in which the parent situation serves as both the source of information used to construct the model and as the subject modeled. Additionally, in the homoeomorph category is a type of model often used by scientists, especially modern biologists. The model is an abstraction from or an idealization of a theory; the term "theory" can often be substituted for this usage of "model." This class of homoeomorphic models does not play the role of providing new ideas in theory construction.

On the other hand, the second class, which Harré called "paramorphs," employs analogous relations and is important in constructing theories. The sources of these *analogue models,* as I will call them here, are different from the subjects which they model. Of these models, Harré said:

> The analogy that is found to hold between certain characteristics of different processes is the basis of the paramorph.... We make for ourselves conceptual models, in which something with which we are familiar or which we understand very well is used as an imaginary model of some otherwise obscure process.... The more general notion of a conceptual paramorph is a basic element in the analysis of the construction of scientific explanations.
>
> (Harré 1960, pp. 87–88)

In addition to the role of the analogical model in first constructing a theory, Harré indicated that models play a role in later steps of theory construction. Harré discussed the deployment of a model much as Hesse did elsewhere (Hesse 1966): One may explore the "neutral analogy" to see if other unexplored aspects of the model are also similar to the subject and can be used to construct additional postulates. Alternatively, Harré suggested, one may bring in a different analogue model in the further development of the theory. Finally, Harré also discussed a process that will be of interest to us in our further discussion, the use of another science in place of a model:

> The reference to another science takes the place of a description of a model.... The sciences can be arranged in a hierarchy such that an explanation of facts in one is given in terms of the description of facts in another....
>
> (Harré 1960, p. 101)

Although my points differ somewhat from those expressed by these philosophers or will develop the analyses further, we agree on the basic point that analogues or other fields of science supply important new ideas in constructing a theory.

In previous work (Darden 1976), I analyzed the role that analogies played in the construction of Darwin's theory of heredity, and from that case I extracted a general schema for theory construction. In further work done conjointly with Nancy Maull (Darden and Maull 1977; see Chapter 5, this book), we examined the generation and function of interfield theories, that is, theories that connect two scientific fields. In this chapter, I relate and extend this previous work by arguing that in the case of construction of theories of heredity, specifically, early attempts depended on vague analogies but progress occurred with the replacement of vague analogies by specific interfield connections.

In order to substantiate these claims, we now turn to an analysis of an important case from the history of biology: namely, attempts to construct a theory of heredity, beginning with Darwin's pangenesis, to the rediscovery of Mendel's laws, to William Bateson's experimental and theoretical work, to the construction of the successful theory of the gene by T. H. Morgan and his associates. The period spans some sixty years, from the 1860s to 1926.

Charles Darwin's writings have proved a rich source for discussing the role of analogies in science. Darwin used many. In previous work (Darden 1976), I examined Darwin's theory of heredity and will briefly summarize those results here. In 1868, Darwin proposed his provisional hypothesis of pangenesis, the view that hereditary units, called "gemmules," were produced all over the body, collected in the reproductive areas, and were passed on to grow into the embryo. To explain Darwin's use of analogies there, I constructed a general schema for a pattern of reasoning in theory construction (Figure 6.1).

Each item in the domain to be explained is considered as a problem to be solved. The next step is the extraction of the general form of problem embodied in that item. For example, Darwin wished to explain reversion, that is, cases in which an organism has characteristics like one of its ancestors rather than like its immediate parents. This domain item poses the general problem: How can something be present, then disappear, and then reappear in the same form at a later time? The next step, according to the schema, is the search for analogues which embody a problem with the same general form but additionally have solutions to that problem. Darwin said that many things in nature are present, then go into dormancy, then appear again: seeds in plants and dormant buds on plants were given as two analogues. Thus, "activity, dormancy, and then reactivation" is the general form of the solution found through use of analogies. The general solution is then particularized for

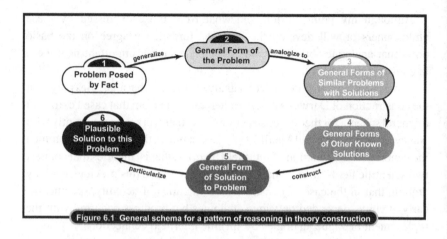

Figure 6.1 General schema for a pattern of reasoning in theory construction

the problem posed by the domain item – for example, gemmules are active, become dormant through one or more generations, and then are reactivated.

The pattern of reasoning may be used in a similar way for each of the domain items. As a result, different analogies play roles in constructing different postulates. At the early stages of theory construction, such as Darwin's, the analogies ranged widely over numerous processes known to occur in the empirical world. The use of such wide-ranging analogies provides weak plausibility to the postulates constructed using them: such processes are known to occur elsewhere. Testing is immediately in order to see if that process does occur in this case. Another nineteenth-century theory of heredity showed improvement that was provided by use of closely related knowledge to construct better postulates. Hugo de Vries, in his *Intracellular Pangenesis* of 1889, explicitly claimed to be developing and improving Darwin's provisional hypothesis of pangenesis. Among the numerous changes that de Vries made was the incorporation of knowledge from cytology, the study of cells. Darwin had proposed that gemmules circulated throughout the body, but tests had not confirmed this circulation. De Vries, in contrast, used the knowledge of the importance of the nucleus within the cell, to claim that material units, which he called "pangens," were located in the nucleus of the cell.

De Vries, like Darwin, was particularly interested in understanding reversion. His theory provided him with a framework for asking about latent pangens as the causes of reversion. The experimental program that he developed to test his theory and investigate latent pangens led him, I argued (Darden 1976), to the rediscovery of what have come to be called Mendel's laws. With that rediscovery came the transition from nineteenth-century theories of heredity with their wide-ranging domains to Mendelian genetics with its data expressed in numerical ratios of differing hereditary characteristics. The construction of theories to explain those ratios which culminated in the theory of the gene is the historical case to which we now turn.

Genetics emerged as a field in 1900 with the independent rediscovery of Mendel's results by de Vries and Carl Correns. But the theory of the gene was not formulated until some years later by T. H. Morgan and his coworkers. It is important to understand exactly what the new conceptions were that showed such promise that a new field of research emerged to investigate them (Darden 1977), yet were so limited that it seems inappropriate to call them a theory as of 1900.

Mendel's empirical results are quite familiar. When organisms differing in traits of one character were crossed, one of the traits dominated in the first generation of hybrids. The famous example was the cross between yellow and green peas, with all the $F_1$ generation being yellow peas. When the hybrids were allowed to self-fertilize, the second generation showed the traits in a proportion of 3: 1 (e.g., three yellow to one green). When plants with two different characters were crossed (e.g., pea color and height of plants), the characters behaved independently, giving a 9:3:3:1 ratio. But Mendel did not stop at merely recording these numerical analyses of the empirical data. Importantly, he also gave a minimal explanation of them by introducing the new idea that the germ cells of the hybrids differed in that each cell had the potential to produce only one of a pair of traits. Furthermore, his other independent assumptions were, first, that the types of germ cells were formed in equal numbers, and, second, that the cells combined randomly in fertilization. Bateson (1902) called the new idea "purity of the gametes." De Vries (1900) expressed his discovery by saying that the germ cells of hybrids are not hybrid but belong to one or the other of the parental types. Such assumptions explained the numerical results, as Mendel's algebraic notation showed: $A$ and $a$ represented the differing traits (Figure 6.2).

Since the presence of $A$ produced the dominant appearance, the $A + 2Aa$ combine to give the appearance $3A$ to $1a$ (Mendel 1865). Differing characters assort independently to give the $9AB:3Ab:3aB:1ab$ ratio. However, Mendel did not discuss the nature of the "*Elemente*" or "*Merkmal*" in the germ cells; he certainly did not claim that the germ cells carried differing material particles.

At the beginning of the field of genetics, there were good empirical regularities produced by the technique of artificial breeding, and the proposal that germ cells of hybrids differed with respect to the characters they later gave rise to. But there was no well-developed theory about these causes of characters. As indicated by the genetic data, the germ cells contained differing

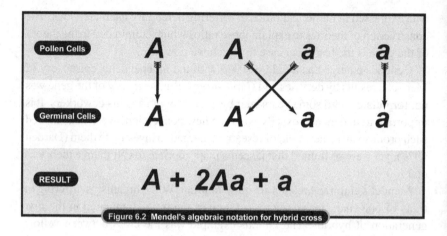

Figure 6.2 Mendel's algebraic notation for hybrid cross

black boxes which somehow caused characters. Filling in details about these black boxes and their interrelations became the chief task of genetics.

Although neither were rediscoverers of Mendel's results, William Bateson and T. H. Morgan were important figures in the development of theories in genetics. Bateson's attempts were unsuccessful; Morgan and his associates formulated the theory of the gene. The contrast between Bateson's approach and that of the Morgan school provides us with an excellent case for examining progress in methods of theory construction.

Bateson's sources of ideas about the causes of patterns of inheritance ranged widely over analogies from physics, with a predilection for forces and vibrations. His methods of theory construction resembled Darwin's in the use of numerous analogies from various sources. Morgan, on the other hand, insisted on tighter ties to empirical evidence. The source of new ideas for much of the construction of the theory of the gene was the neighboring field of cytology. By using interfield connections between postulated genes and observed chromosomes, the Morgan school constructed a theory which quickly surpassed Bateson's in explanatory power and extendibility to new data. Thus, progress resulted from abandoning the use of vague analogies and the directed search for interfield connections as guides to theory construction.

A distinction needs to be made explicitly at this point. The theory whose construction is being discussed here is an intrafield theory, completely within the field of genetics. Its successful form, as of 1926, was called by Morgan the "theory of the gene." It postulated unobservable genes in the germ cells, whose behavior served to explain the patterns of inheritance of characters determined by artificial breeding. This *intrafield* theory is different from – though closely connected to – the *interfield* chromosome theory of Mendelian heredity, which postulated that genes were in chromosomes. The interfield connections established by the chromosome theory were important in the construction of the theory of the gene; that construction is being discussed in this chapter. This discussion represents a shift of emphasis from the work on interfield theories (Darden and Maull 1977; see Chapter 5, this book) in which we emphasized the importance for both fields of the interactions between them. Here, the emphasis is on the developments within a single field. Historians and philosophers have often confused the theory of the gene and the chromosome theory because they were so closely related. For the purposes of this chapter, the theory whose construction is discussed is the theory of the gene, and the links to cytology important in that construction will be referred to as "interfield connections" or "links."

Bateson never worked out a well-developed theory of genetics. His own attempts were along quite different lines than the theory of the gene and the chromosome theory, neither of which he ever accepted. Most of his published writings show him as the empirical scientist, doing genetic experiments, coining new terminology (e.g., "genetics," "heterozygote," "homozygote"), criticizing the Morgan interpretations. He appears to be the empiricist, skeptical of theoretical leaps; however, unpublished writings and some published work (Bateson 1913) belie that interpretation. Bateson struggled to formulate his own theory of genetics in terms of vibrations, forces, and vortices. William Coleman (1970), in an excellent article, analyzed these attempts to show that Bateson was using traditions in physics in the late nineteenth century as a source for analogies to construct his theory. Coleman said:

> [Bateson's] search for an alternative to strictly atomistic, particulate, or, if you will, material bases for heredity led him to consider a diverting array of models. Of these, the most significant was one derived from popular generalizations of the physical sciences. In this rhythmic or vibratory theory of inheritance is a model whose premises are reasonably clear, whose further development is desperately obscure and whose attractions Bateson could never escape.
>
> (Coleman 1970, p. 267)

Bateson's own discussions of his hypotheses are not numerous. In a letter of 1891, he called an early view the "Undulatory Hypothesis" and briefly sketched the main idea:

> Divisions between segments, petals, etc. are *internodal* lines like those in sand figures made by sound, i.e., lines of maximum vibratory strain, while the midsegmental lines and the petals, etc. are the *nodal* lines, or places of minimum movement. Hence all the *patterns* and *recurrence of patterns* in animals and

157

> plants – hence the perfections of symmetry – hence bilaterally symmetrical variation, the completeness of repetition whether of a part repeated in a radial or linear series etc. etc. (B. Bateson 1928, p. 43)

The cell came to occupy a privileged place in Bateson's attempts at theory construction. But it was not a static structure; instead, it was a dynamic vortex:

> The cell is a vortex of chemical and molecular change. Matter is continually passing through this system. We press for an answer to the question, How does our vortex spontaneously divide? The study of these vortices is biology, and the place at which we must look for our answer is cell division.
> (Bateson 1907; quoted in Coleman 1970, pp. 274–275)

> A simple vortex, like a smoke-ring, if projected in a suitable way will twist and form two rings. If each loop as it is formed could grow and then twist again to form more loops, we should have a model representing several of the essential features of living things. (Bateson 1913, p. 40)

An understanding of living vortices, namely cells, held the key, according to Bateson, for understanding heredity and variation. Heredity was essentially like producing similar cells in cell division or symmetrical parts in a single organism. Variation, on the other hand, was like differentiation in which cells give rise to cells that differ from them. This primacy of the cell and his predilection for dynamic forces were aspects of Bateson's rejection of subcellular material structures, such as chromosomes within the nucleus of the cell, as the important locus of activity. Not only did he reject the chromosome theory, he also was quite skeptical of interpretations of Mendelism as requiring discrete material units. On numerous occasions, Bateson stressed that there was no evidence to show that the Mendelian factors were material substances (Bateson 1913, p. 268).

> We are in the state in which the students of physical science were, in the period when it was open to anyone to believe that heat was a material substance or not, as he chose. (Bateson 1902, pp. 2–3)

These arguments left open the possibility that Bateson's dynamic approach was correct.

Bateson's use of analogies to physics was reflected in the terms he used to report data. An important example on which we will focus was the discovery by Bateson and coworkers of exceptions to the 9:3:3:1 Mendelian ratios (Bateson et al. 1905). These anomalies were called by Bateson either "coupling" if the dominant traits tended to be inherited together more often than expected, or "repulsion" if the dominants were found together less often. The "intensity" of coupling or repulsion varied for different pairs of characters. The example reported in 1905 was for partial coupling in pollen shape and colors of flowers in sweet peas.

By examining this terminology and additional statements that Bateson made, one may infer the form of a theory of heredity that Bateson might have proposed. "Powers" within the swirling vortex of the germ cells produce characters during development (Bateson 1913, p. 268). These powers or forces may attract or repel each other with varying intensity in the formation of the germ cells (a denial of independent assortment) with a resultant distortion of the 9:3:3:1 ratio of characters produced. Alternatively, the powers may not influence each other, giving the usual Mendelian ratio.

But this new idea of coupling and repulsing powers did not prove to be fruitful and was abandoned by Bateson in favor of an alternative hypothesis: selective reduplication of certain cells. For instance, instead of equal numbers of germ cells with *AB, aB, Ab, ab* as Mendel proposed, Bateson suggested unequal gametic ratios (e.g., 7*AB*:1*Ab*:1*aB*:7*ab*) produced by selective reduplication of the cells containing *AB* and *ab*. Thus, he denied that equal numbers of the different types of germ cells were produced. Instead, a disproportionate number of certain types formed and combined in fertilization to give rise to the anomalous ratios. Bateson and Punnett (1911) called this the "reduplication hypothesis."

By abandoning coupling and repulsion in favor of cellular reduplication, Bateson made progress in his method of theory construction. He abandoned vague analogies from the physical sciences which produced a theory difficult, if not impossible, to test. In place of those analogies, he substituted a process from the related field of cytology – cell division – as a means of constructing an explanation of the genetic data. Unfortunately, Bateson was not appealing to any specific cytological data about selective reduplication of cells. Ideally, in theory construction, one can find empirically confirmed information in a related field with relevant similarities such that it can be used to solve one's problem. But the ideal may not obtain since scientists in the related field may simply not have looked in the right place or at the right things. In that event, one can predict phenomena in the related field on the basis of its presumed relation to the theory being constructed. Thus, Bateson's hypothesis predicted selective reduplications. But the cytologists had found none.

Another way in which the reduplication hypothesis failed was in its extendibility. It became entirely unwieldy when extended from the coupling of two characters to the coupling of three or more. The numbers of different types of cells that had to be produced by selective reduplications became quite large. Since extendibility to new cases determines the acceptability of a new theory, it should act as a demand in theory construction: choose the analogies or interfield connections which will provide ideas for theory development in the light of additional cases. Bateson failed to take this demand into account and his method of theory construction was the poorer for it.

The successful use of the interfield links between genetics and cytology, begun by Boveri (1904) and Sutton (1903), came to fruit in the hands of Morgan and his associates between 1910 and 1926. Unlike Bateson, they did not focus on the level of the whole cell and its divisions for sources of new ideas to explain the genetic data. Instead, they made use of knowledge of the subcellular level – specifically, knowledge about the chromosomes inside the nucleus of the cell. Their postulations were often made on the basis of known cytological processes or properties; thus, their constructions had a measure of empirical support that Bateson's lacked. Also, additional information about chromosomes provided ideas to Morgan and his associates for constructing additional postulates of the theory of the gene; in other words, the interfield links provided ideas for extending the theory.

In 1903 and 1904, Walter Sutton and Theodor Boveri independently postulated the chromosome theory of Mendelian heredity. Both scientists were struck by the similarities in the known properties of chromosomes and (what came to be called) Mendelian genes: (1) they both consisted of pure individuals which did not join or contaminate each other; (2) both genes and chromosomes were found in pairs; and (3) in the formation of the sex cells, each set is reduced by one half. On the basis of these similarities, Boveri and Sutton proposed that the genes were in or on the chromosomes and made predictions for both cytology and genetics on the basis of knowledge from the other field. Using the knowledge that the chromosome number was much smaller than the gene number, they both predicted that some genes would be carried on the same chromosome and would therefore be linked in inheritance. (For more discussion of this case, see Darden and Maull 1977; see Chapter 5, this book.)

Morgan began his career with work in embryology and became an outspoken proponent of the use of experiments in biology. Prior to 1910, Morgan was a critic of both Mendelism and the chromosome theory. But his own and others' experimental results caused him to change his views. (For a discussion, though I think probably not the complete explanation of this change, see Manier 1969; Allen 1978.) As shown in Carlson (1971) and Roll-Hansen (1978), Morgan's students, Sturtevant, Bridges, and Muller, were important contributors to the development of the theory of the gene, often bringing Morgan around to their points of view. Detailed examination of the step by step construction of the theory of the gene is outside the scope of this chapter (see Darden 1991). This discussion concentrates only on the explanation of the deviations from the 9:3:3:1 ratios, namely the postulation of linkage.

One of Morgan's criticisms of the chromosome theory was that the prediction by Sutton and Boveri of characters inherited together had been found in only a few instances (Morgan 1910a, p. 489). The most striking work for indicating the importance of the chromosomes in heredity, Morgan conceded, was the finding of connections between differing chromosomes and sex determination as discovered by Nettie Stevens and investigated by E. B. Wilson (Brush 1978a). The experimental work that marked the turning point in Morgan's views about Mendelism and the chromosome theory was Morgan's discovery of what came to be called sex-linked characteristics. As Morgan searched for mutations among *Drosophila* cultures, a white-eyed male appeared among the normal red-eyed flies. Subsequent breeding experiments showed white and red eyes to be linked to sex (Morgan 1910b). Morgan quickly discovered other characters linked both to sex and to each other. But the linkage was not complete or, in Bateson's terminology, there were variations in the strengths of coupling. If all the genes were carried on and remained on a single intact chromosome, then one expected complete linkage of the characters. A problem thus became to explain the partial linkage.

Morgan made excellent use of recent cytological work to construct an explanation for partial linkage. After discussing Bateson's hypothesis, Morgan said:

> In place of attractions, repulsions and orders of precedence, and the elaborate systems of coupling, I venture to suggest a comparatively simple explanation based on results of inheritance of eye color, body color, wing mutations and the sex factor for femaleness in *Drosophila*.          (Morgan 1911b, p. 384)

Morgan continued by postulating that the factors (genes) for those characters were located linearly along the chromosome. Citing the work of Janssens (1909), which showed the intertwining of chromosomes during the formation of germ cells, Morgan proposed that some of the homologous chromosomes exchanged parts, thereby producing crossing-over of linked factors. Morgan concluded:

> Instead of random segregation [independent assortment] in Mendel's sense we find "associations of factors" that are located near together in the chromosomes. Cytology furnishes the mechanism that the experimental evidence demands.
> (Morgan 1911b, p. 384)

Thus, Morgan, in contrast to Bateson, did not search for analogies from the physical sciences; instead, he used empirical knowledge from the related field of cytology and the postulated physical relation of genes to chromosomes to construct an explanation for the genetic data. The data presented the problem: How are factors partially linked to each other? The cytological evidence of linear chromosomes, which sometimes intertwine, provided the solution: The genes are linked linearly to each other and sometimes parts of the homologous linkage groups cross-over.

By 1915, when Morgan, Sturtevant, Muller, and Bridges wrote their now-classic *The Mechanism of Mendelian Heredity,* Morgan's explanation of linkage had been markedly confirmed and extended to new cases. Linkage groups had been associated with the other, non-sex chromosomes; the number of linkage groups corresponded to the number of chromosomes. Bridges (1914) had found an abnormal fly which showed corresponding anomalies in both the chromosomes and the patterns of inheritance. Sturtevant (1914) constructed a linear map of the relative positions of genes using the numerical crossing-over data. When the data showed less crossing-over than predicted, once again the chromosomes provided a way of constructing the explanation: double crossing-over occurred or one cross-over interfered with the formation of another (Morgan et al. 1915, pp. 63–64).

In 1926, Morgan gave an elegant presentation of the theory of the gene:

> The theory states that the characters of the individual are referable to paired elements (genes) in the germinal material that are held together in a definite number of linkage groups; it states that the members of each pair of genes separate when the germ-cells mature in accordance with Mendel's first law, and in consequence, each germ-cell comes to contain one set only; it states that the members belonging to different linkage groups assort independently in accordance with Mendel's second law; it states that an orderly interchange – crossing over – also takes place, at times, between the elements in corresponding linkage groups; and it states that the frequency of crossing over furnishes evidence of the linear order of the elements in each linkage group and of the relative position of the elements with respect to each other.
>
> (Morgan 1926, p. 25)

Note particularly that nothing is said about chromosomes in this statement of the theory of the gene. The interfield connections to cytology were used in its construction. Furthermore, since the information from cytology was already empirically confirmed, it provided support for the new theory. The theory of the gene, apart from the chromosome theory, contained new ideas, genes associated in linkage groups. The connections to cytology played the same role as analogies in theory construction, but they were a much more powerful source of new ideas since they supplied information about physically related processes.

However, connections to cytology were not sufficient to fill in all the details of the black boxes, the genes. Although the postulation that the genes were in the chromosomes made plausible the assumption that the genes were material particles, it would be the task of new interfield connections established by molecular biology to illuminate the rest of the black box (see Chapter 4, this book).

A hypothesis which emerges from this case study and which needs to be tested in others is that progress occurs in theory construction with the abandoning of vague analogies in favor of interfield connections to provide new ideas. At this point, a more detailed discussion of "vague analogies" as opposed to "interfield connections" is needed. A critic of this hypothesis might ask: What is the difference? In both cases, one is postulating similarities to factors in other sciences. In your example, for instance, similarities to electricity are just as good a source of new ideas as similarities to the behavior of chromosomes. How is one to be judged better than the other, the critic continues, except with hindsight as to which produced the successful theory?

My reply concedes something to this objection: Analogies and interfield connections do have properties in common. First, a relation of similarity exists between the postulate constructed and either the analogue or the information from another field. Second, both may serve as sources of new ideas. Third, in both instances, assuming the analogy has been drawn to a known thing or process, the use of either analogies or interfield connections brings some plausibility to the postulate constructed using it: A similar thing or process is known to occur elsewhere. Finally, the analogy and interfield connections may both play a role in developing further postulates to extend the theory, especially if the analogy is a more detailed one than the Bateson example. In Hesse's terms, the neutral analogy, or the areas of unexplored relationships, may be used in further development of the theory. Similarly, other information from the related field may be brought in to construct additional postulates.

Despite these commonalities in the roles played by analogies and interfield connections in theory construction, I would reply to the critic that there are also important differences that make the interfield connections better. It is much weaker to say that, since a *similar type of factor* has been found in other sciences, such a type may be operative here (appeal to analogy) than to say that the *actual factor* found in a related science behaves in a similar way to the postulated factors as a result of a *physical* relationship between them (interfield connections). Instances of this weaker plausibility provided by analogies are common in the history of science. In postulating that powers were coupling or repulsing, Bateson was appealing to analogies in the way Newton did when he proposed that there might be chemical and life forces similar to mechanical ones. Similarly, natural philosophers of the eighteenth century postulated different fluids to explain electricity, heat, and nerve impulses. It is interesting to ask what justification exists for this appeal to similar types of explanatory factors. One speculation is that a weak unity of science thesis is operating: Different phenomena should be explicable by similar types of factors; if one type, such as forces or fluids, has been successful in one field, try it in others.

On the other hand, one might postulate a stronger unity of science directive: Instead of postulating analogous factors, see if one and the same factor may be the operative one. For example, instead of postulating a vital fluid analogous to an electrical fluid, check to see if actual electromagnetic forces are operating in living things. If Bateson had been appealing to detailed empirical knowledge about coupling and repulsing forces known to occur in cells, then he would have been using interfield connections rather than analogies. Since his analogies were not even specifically drawn from similar types of successful explanatory factors found elsewhere but were merely suggestive of some sort of "power" or "force" or "vibration" which could "couple" and "repulse," I have called them "vague analogies."

Interfield connections postulate a relationship to empirically confirmed information in a related field in the ideal case or make a prediction of the existence of such a relation in the weaker case. The interfield connection, in contrast to an analogy, is itself a scientific hypothesis that postulates a *physical relation*[3] between the entities or processes in the neighboring field and in the theory under construction. Thus, in contrast to analogies, one expects *empirical investigation* of the interfield connection, with the result that such a connection may be confirmed or disconfirmed. Also, since there is a physical relation, the payoff for theory construction may be greater: One has good reason to believe that unexplored properties will have relations to each other; there is no similar basis for expectations about neutral analogy being useful. For example, the physical relation postulated in the case of the theory of the gene was between hypothetical genes and observed chromosomes: The genes were claimed to be parts of chromosomes. Since parts usually do

---

[3] "Physical relation" may not be the most general way of expressing the types of relations between fields which are scientific hypotheses. It fits this case very well since genes were claimed to be physically located on the chromosomes. But, if this analysis is extended, e.g., to relations between psychology and neurophysiology, then some term more general than "physical" would be desirable.

what their wholes do, then most of the properties of chromosomes provided potential information about genes. One is not merely importing knowledge from cytology as a heuristic device to construct the theory of the gene in such a way that that knowledge may be discarded once the theory is constructed. Instead, one is postulating a physical relation which may be confirmed or disconfirmed. Other kinds of physical relations besides part-whole obtain in other cases (e.g., causal, structure–function, or even identity) if one and the same factor is claimed to be operative (Darden and Maull 1977, p. 49; see Chapter 5, this book).

An important question about the use of analogies or interfield connections in theory construction is how the theorist chooses the ones to use. With respect to analogies, I have little to say in answer to that difficult question. Perhaps a quantitative measure of the amount of positive (and/or neutral) analogy would be useful. One would like one's analogue to have numerous properties similar to the subject. Furthermore, more areas of neutral analogy, properties that are not yet known to be similar or dissimilar, would provide more potential for further development.

With respect to interfield connections, I can provide more in the way of answer to the question about choice of a related field. Harré suggested that a reductive hierarchy provides interscience explanations. Certainly, the search for underlying structure has proved fruitful in the development of science; however, detailed examination of actual interfield relations shows that they come in more varieties than the identities necessary for reductive accounts. Entities and properties are not always being identified with entities and properties at a lower level, a necessary condition for reduction. What the lower level is may not be easily decidable: Is genetics with its observable, large-scale characteristics and unobserved genes the lower level, or is cytology with its microscopically observable chromosomes the lower? Neither is a candidate to be reduced to the other. Science is more reticulate and less hierarchical, so one needs directives other than use the field at the next lower level. (For further discussion of the interfield analysis as an alternative to reduction, see Maull 1977.)

Thus, in addition to choosing fields which may provide information about underlying structure, other reasons exist for choosing another field for constructing interfield relations. In searching for fruitful interfield connections to use in constructing an explanation of data within a subject field, one may have background knowledge that another field is related if it has supplied ideas for other explanations within the subject field. More strongly, one may know that scientists in another field have been investigating the same phenomenon using differing techniques; for example, cytologists used microscopic techniques to investigate questions about heredity while geneticists used breeding techniques. One would expect some relation to be established between the results in the two fields. Temporal relations may also indicate which fields are related. As Stephen Brush (1978b) discussed, in order to explain a domain of geological phenomena, certain interfield connections with astronomical theory about the formation of planets were postulated, since the astronomical theory dealt with phenomena temporally prior to the geological.[4]

I have proposed the hypothesis that interfield connections are better than analogies as a source of new ideas. Let me now draw on the differences between the two to provide evidence for that claim. First, the use of interfield connections produces a more plausible theory if the interfield connections have been empirically confirmed. One is not merely appealing to processes known to occur somewhere; one is appealing to known processes that one claims have physical relations to the postulated factors or processes. Second, with regard to extendibility, if the related field does not already provide needed information, then further empirical work in that field can be done. Presumably with analogous relations, one already knows the properties of the analogue. But with interfield connections and the physical relationship they postulate, predictions for the related field can be made and tested in order to find additional information. Thus comes both the means of further theory construction and independent evidence from another field for the constructed theory. Finally, using interfield connections in the construction of a theory provides a unification of that theory with other fields of science at the outset. In contrast to the account of unification of science by reduction of one completed theory by another completed theory (and the attendant difficulty of finding adequate identity relations without modifying the theories), this account shows that unification can be accomplished in the very way the theory is constructed, not "after the fact." The new theory constructed with interfield connections is already connected to another theory. Since interfield connections play these important roles of providing empirical evidence for the theory and of unifying it with other fields, they are not merely heuristic devices, useful as a source of ideas but then eliminable after theory construction. Instead, they are themselves important scientific hypotheses: Things are connected in these ways.

My thesis should not be interpreted to be stronger than it is. I am not claiming that interfield links *must* be used in the construction of any adequate

---

[4] It has been suggested to me that in some parts of psychology and in the social sciences where related fields are not well developed, analogies may be the preferable source of new ideas. This seems plausible and needs to be investigated.

explanatory theory. Nor am I claiming that whenever they are used, a successful (i.e., subsequently confirmed) theory will necessarily be produced. In fact, I doubt that any necessary and sufficient schemas for constructing correct theories will be found. But I am claiming that in the case examined, progress occurred with the use of interfield connections in theory construction, and this analysis may apply to other past or future instances.

## REFERENCES

Allen, Garland (1978), *Thomas Hunt Morgan*. Princeton, NJ: Princeton University Press.
Bateson, Beatrice (1928), *William Bateson, Naturalist*. London: Cambridge University Press.
Bateson, William (1902), *Mendel's Principles of Heredity – A Defense*. Cambridge: Cambridge University Press.
Bateson, William (1913), *Problems of Genetics*. New Haven, CT: Yale University Press.
Bateson, William, R. C. Punnett, and E. R. Saunders (1905), "Further Experiments on Inheritance in Sweet Peas and Stocks: Preliminary Account," *Proceedings of the Royal Society* 77. Reprinted in R. C. Punnett (1928), *Scientific Papers of William Bateson*, v. 2. Cambridge: Cambridge University Press, pp. 139–141.
Bateson, William and R. C. Punnett (1911), "On Gametic Series Involving Reduplication of Certain Terms," *Journal of Genetics* 1. Reprinted in R. C. Punnett (1928), *Scientific Papers of William Bateson*, v. 2. Cambridge: Cambridge University Press, pp. 206–215.
Blackwell, R. J. (1969), *Discovery in the Physical Science*. Notre Dame, IN: University of Notre Dame Press.
Boveri, T. (1904), *Ergebnisse über die Konstitution der chromatischen Substanz des Zellkern*. Jena: G. Fischer.
Bridges, Calvin B. (1914), "Direct Proof Through Non-disjunction that the Sex-linked Genes of *Drosophila* are Borne by the X-chromosome," *Science*, N. S., 40: 107–109.
Brush, Stephen (1978a), "Nettie M. Stevens and the Discovery of Sex Determination by Chromosomes," *Isis* 69: 163–172.
Brush, Stephen (1978b), "A Geologist Among Astronomers: The Rise and Fall of the Chamberlin-Moulton Cosmogony," *Journal for the History of Astronomy* 9: 1–41.
Carlson, Elof A. (1971), "An Unacknowledged Founding of Molecular Biology: H. J. Muller's Contributions to Gene Theory, 1910–1936," *Journal of the History of Biology* 4: 149–170.
Coleman, William (1970), "Bateson and Chromosomes: Conservative Thought in Science," *Centaurus* 15: 228–314.
Darden, Lindley (1976), "Reasoning in Scientific Change: Charles Darwin, Hugo de Vries, and the Discovery of Segregation," *Studies in the History and Philosophy of Science* 7: 127–169.
Darden, Lindley (1977), "William Bateson and the Promise of Mendelism," *Journal of the History of Biology* 10: 87–106.
Darden, Lindley (1991), *Theory Change in Science: Strategies from Mendelian Genetics*. New York: Oxford University Press.

Darden, Lindley and Nancy Maull (1977), "Interfield Theories," *Philosophy of Science* 44: 43–64.

Darwin, Charles (1868), *The Variation of Animals and Plants under Domestication.* New York: Orange Judd & Co.

Gilbert, G. K. (1896), "The Origin of Hypotheses, Illustrated by The Discussion of a Topographic Problem," *Science*, N. S., 3: 1–13.

Hanson, Norwood Russell ([1961] 1970), "Is There a Logic of Scientific Discovery?" in H. Feigl and G. Maxwell (eds.), *Current Issues in the Philosophy of Science.* New York: Holt, Rinehart and Winston. Reprinted in B. Brody (ed.) *Readings in the Philosophy of Science.* Englewood Cliffs, NJ: Prentice Hall, pp. 620–633.

Harré, R. (1960), *An Introduction to the Logic of the Sciences.* London: Macmillan.

Harré, R. (1970), *The Principles of Scientific Thinking.* Chicago, IL: University of Chicago Press.

Hempel, C. G. (1966), *Philosophy of Natural Science.* Englewood Cliffs, NJ: Prentice-Hall.

Hesse, Mary (1966), *Models and Analogies in Science.* Notre Dame, IN: University of Notre Dame Press.

Janssens, F. A. (1909), "La theorie de la chiasmatypie," *La Cellule* 25: 389–411.

Kuhn, Thomas (1970), *The Structure of Scientific Revolutions.* 2nd ed. Chicago, IL: University of Chicago Press.

Lakatos, Imre (1970), "Falsification and the Methodology of Scientific Research Programmes," in I. Lakatos and A. Musgrave (eds.), *Criticism and the Growth of Knowledge.* Cambridge: Cambridge University Press, pp. 91–195.

Laudan, Larry (1977), *Progress and Its Problems.* Berkeley, CA: University of California Press.

Leatherdale, W. H. (1974), *The Role of Analogy, Model and Metaphor in Science.* New York: American Elsevier.

Manier, E. (1969), "The Experimental Method in Biology, T. H. Morgan and the Theory of the Gene," *Synthese* 20: 185–205.

Maull, Nancy (1977), "Unifying Science without Reduction," *Studies in the History and Philosophy of Science* 8: 143–162.

Mendel, Gregor ([1865] 1966), "Experiments on Plant Hybrids," translated from German and reprinted in Curt Stern, and Eva Sherwood (eds.), *The Origin of Genetics, A Mendel Source Book.* San Francisco, CA: W. H. Freeman, pp. 1–48.

Morgan, Thomas Hunt (1910a), "Chromosomes and Heredity," *The American Naturalist* 44: 449–496.

Morgan, Thomas Hunt (1910b), "Sex-Limited Inheritance in *Drosophila*," *Science* 32: 120–122.

Morgan, Thomas Hunt (1911b), "Random Segregation versus Coupling in Mendelian Inheritance," *Science* 34: 384.

Morgan, Thomas Hunt (1926), *The Theory of the Gene.* New Haven, CT: Yale University Press.

Morgan, Thomas Hunt, A. H. Sturtevant, H. J. Muller, and C. B. Bridges (1915), *The Mechanism of Mendelian Heredity.* New York: Henry Holt and Company.

Popper, Karl (1965), *The Logic of Scientific Discovery.* New York: Harper Torchbooks.

Punnett, R. C. (ed.) (1928), *Scientific Papers of William Bateson.* Cambridge: Cambridge University Press, 2 vols.

Roll-Hansen, N. (1978), *"Drosophila* Genetics: A Reductionist Research Program," *Journal of the History of Biology* 11: 159–210.

Shapere, Dudley (1973), Unpublished MS. Presented at IUHPS-LMPS Conference on Relations Between History and Philosophy of Science, Jyväskylä, Finland.

Shapere, Dudley (1974a), "Scientific Theories and Their Domains," in F. Suppe (ed.), *The Structure of Scientific Theories.* Urbana, IL: University of Illinois Press, pp. 518–565.

Shapere, Dudley (1974b), "On the Relations Between Compositional and Evolutionary Theories," in F. J. Ayala and T. Dobzhansky (eds.), *Studies in the Philosophy of Biology.* Berkeley, CA: University of California Press, pp. 187–204.

Stern, C., and E. Sherwood (eds.) (1966), *The Origin of Genetics, A Mendel Source Book.* San Francisco, CA: W. H. Freeman.

Sturtevant, A. H. (1913), "The Linear Arrangement of Six Sex-linked Factors in *Drosophila,* as Shown by Their Mode of Association," *Journal of Experimental Zoology* 14: 43–59.

Sutton, Walter (1903), "The Chromosomes in Heredity," *Biological Bulletin* 4: 231–251.

Vries, Hugo de ([1889] 1910), *Intracellular Pangenesis.* Translated by C. S. Gager. Chicago, IL: Open Court.

# 7

## Relations Among Fields in the
## Evolutionary Synthesis[1]

### 7.1 INTRODUCTION

The synthetic theory of evolution is a multilevel theory that serves to synthesize knowledge from fields at different levels of organization. It provides a solution to the problem of the origin of species. Biologists attempted to solve this problem for years, during which time the key fields emerged and developed to the point that the synthesis was possible. Prior to the synthesis, debate occurred about what components were necessary to solve the problem and alternative theories were proposed. Had any of the prior theories been correct, then the fields that exist within evolutionary studies would have been different. Which fields exist is a contingent fact about the nature of the world and the way we study it. Which fields are synthesized to solve certain problems is contingent on the nature of the solution and the stage of development of various fields at the time the solution is proposed.

Relations among levels of organization, fields of study, and the problem of the origin of species are the subjects of this chapter. The focus is on the synthetic theory of evolution as proposed by Theodosius Dobzhansky in 1937, the multiple levels within it, the knowledge from different fields that it synthesized, and its various predecessors. The final section discusses the concept of a synthetic theory and contrasts that with previously studied interfield theories.

[1] This chapter was originally published in Bechtel, William (ed.) (1986), *Integrating Scientific Disciplines*. Dordrecht: Martinus Nijhoff Publishers. It is reprinted with kind permission of Springer Science and Business Media. This research was supported by the General Research Board of the University of Maryland. Thanks to John Beatty and William Bechtel for comments on an earlier draft.

## 7.2 THE SYNTHETIC THEORY

Theodosius Dobzhansky published the seminal book outlining the evolutionary synthesis in 1937: *Genetics and the Origin of Species*. Although others contributed to the synthesis, this discussion focuses primarily on Dobzhansky's 1937 book. In it, he clearly distinguished between evolution (change of species) and the causal mechanisms of evolutionary change. It is worth quoting him at length on this topic:

> The theory of evolution asserts that the beings now living have descended from different beings which have lived in the past; that the discontinuous variation observed at our time-level, the gaps now existing between clusters of forms, have arisen gradually, so that if we could assemble all the individuals which have ever inhabited the earth, a fairly continuous array of forms would emerge; that all these changes have taken place due to causes which now continue to be in operation and which therefore can be studied experimentally.
>
> (Dobzhansky 1937, p. 7)

Dobzhansky thus stated the three components that are usually referred to as common descent, gradualism, and uniformitarianism. The explanation of common descent and the gaps now existing between forms was the key problem that he addressed in his book. Note that adaptations were not mentioned. Since Darwin combined the explanation of adaptations and the origin of new forms in his theory of natural selection, these two problems are often lumped together. Although Dobzhansky advocated selection as a means of producing adaptive change, it was not the only mechanism he postulated for the production of new forms. Furthermore, the other theories we will consider were directed more to solving the problem of the origin of new forms than to the explanation of adaptations.

Dobzhansky clearly stated the three levels that constitute the "mechanisms of evolution." These constitute the "synthetic theory of evolution," although Dobzhansky did not use that term. First:

> Mutations and chromosomal changes are...the first stage, or level, of the evolutionary process, governed entirely by the laws of the physiology of individuals.

Second, the populational level:

> A mutation may be lost or increased in frequency...without regard to the beneficial or deleterious effects of the mutation. The influences of selection, migration, and geographical isolation then mold the genetic structure of populations

171

into new shapes in conformity with the secular environment and the ecology, especially the breeding habits, of the species. This is the second level of the evolutionary process, on which the impact of the environment produces historical changes in the living population.

Third, the species level:

> Finally, the third level is a realm of fixation of the diversity already attained on the preceding two levels.... A number of mechanisms encountered in nature (ecological isolation, sexual isolation, hybrid sterility, and others) guard against a ... fusion [due to interbreeding].          (Dobzhansky 1937, pp. 12–13)

Knowledge from three levels of organization was synthesized: the level of genes and chromosomes, the level of the population, and the level of the species. These levels constitute an inclusive hierarchy: Genes are parts of chromosomes, which make up individual organisms, which make up populations, which make up species. Dobzhansky as the geneticist focused on the organism's genes and chromosomes; the individual organism was not accorded a separate level in his hierarchy. Additionally, a direction of causal influence characterizes the relations among the levels: mutational changes provide raw material for changes in populations; populational changes, in turn, are necessary for the production of new species.

Different fields provided the key knowledge about mechanisms operating at each level: Mendelian genetics for mutations; cytology for chromosomal abnormalities; mathematical population genetics and experimental and field studies of populations for the populational level. Finally, the study of isolating mechanisms in speciation was a new area of study that emerged as a result of the synthesis. Dobzhansky was aware that the final level was the least studied as of 1937: "The origin and functioning of the isolating mechanisms constitute one of the most important problems of the genetics of populations" (Dobzhansky 1937, p. 14). It is clearly too simplistic to say that the synthetic theory synthesized only Mendelism and Darwinism; numerous other components were involved. Additionally, a new level was established as separate from the previously existing ones. Dobzhansky drew on extensive knowledge from these numerous fields as well as delineating problems for the new level of organization. These he combined into one comprehensive explanation of the origin of new species.

It is sometimes asked why the synthesis did not occur shortly after the rediscovery of Mendel's laws in 1900. The stage of development of the fields at the time of the synthesis was crucial. Mendelism by 1937 was a very different field than Mendelism in 1900. By 1937, Dobzhansky could draw on extensive knowledge of mutations, modifying genes, genes that produce sterility – none of which was available in 1900. Thus, fields must be seen as developing entities, providing shifting information at different times to problem-solving occurring in other fields. The stage of development determines whether appropriate knowledge is available. We will now examine alternative explanations of the origin of new species prior to the synthetic theory and see how the various fields were related within those theories.

### 7.3 OTHER THEORIES, OTHER LEVELS

We will begin with Darwin's theory of natural selection, examine Hugo de Vries's mutation theory, and the objections of some of the early Mendelians to Darwinian selection. This analysis will show how prior theories focused on different levels of organization as the key to the origin of species. Had any of the alternatives been correct, then the fields supplying knowledge would have been different. Levels themselves are not just "given" but must be discovered to exist as causally efficacious.

In the *Origin of Species* of 1859, Darwin laid out his theory, which explained the origin of new species and the origin of adaptations that organisms exhibited. Darwin argued that his theory of natural selection was sufficient to explain the origin of all new species. The theory may be stated as a deductive argument:

I. In all organic forms, heritable variations occur.
II. Organic forms multiply at a greater rate than their sources of sustenance; therefore, there is a struggle for existence.
Conclusion 1: Those organisms that possess variations advantageous in the struggle for existence will tend to survive.
Conclusion 2: A permanent and adaptive change in organic forms will be effected, as over time, the new forms replace the old.

(modified from Vorzimmer 1970)

The levels of organization found in Darwin's theory are the level of variation in individual organisms and the population in which selection occurs. No separate level of isolating mechanisms to produce new species was postulated; in fact, Darwin (1859) emphasized the continuity between varieties and species. Thus, if we were still operating with Darwin's theory, we would have fields for the study of variation and the field of population studies but no separate study of speciation mechanisms. That a separate level existed in order to move from the population to the species and that separate study of isolating mechanisms was needed was one of the additions of the synthesis.

Darwin amassed much empirical evidence for the existence of variations, both in domesticated forms and in nature. However, in the *Origin* he did not have a theory to explain the origin or inheritance of variations. Darwin attempted to supply such a theory in 1868 – his unsuccessful provisional hypothesis of pangenesis. He postulated that cells produced hereditary units called "gemmules." The gemmules varied in quantity or quality, and modified cells produced modified gemmules. Since Darwin believed that most hybrids were intermediate in form between the two parental forms, hybrids were postulated to produce hybridized gemmules. The gemmules circulated throughout the body and collected in the reproductive areas, were passed on to offspring, and grew into cells in the developing embryo (Darwin 1868, Ch. 27). Darwin's theory of pangenesis was thus both particulate and accounted for blending inheritance. (Unfortunately, particulate and blending are often contrasted as two different types of hereditary theories. To use later terminology, blending at the phenotypic level may or may not be correlated with discrete units at the genotypic.)

Darwin's theory of pangenesis was not confirmed when tested (Galton 1871). The problem of variation remained unsolved in the late nineteenth century. Hugo de Vries took up Darwin's theory of pangenesis and modified it. As a result of his experiments on variations, de Vries discovered that Darwin was wrong: Hybrid forms do not produce hybrid units; they carry one or other of the pure parental units in an unmodified form (de Vries 1900; Darden 1976; but see Kottler 1979). This is Mendel's law of segregation, namely that hereditary units separate or segregate in a pure form in the formation of pollen and egg cells so that each sex cell receives one or the other of a pair of units for a trait. After de Vries (and Correns) published this finding in 1900, the field of Mendelism began.

Even though Mendelism emerged as a field in 1900, as late as 1908, Vernon Kellogg in *Darwinism Today* said that one of the chief problems in biology was the "origin, the causes and the primary control" of variations (Kellogg 1908, p. 30). The relation between Darwinian selection (with its stress on small, frequently occurring "individual differences") and the Mendelian law pertaining to particulate (in de Vries's formulation), nonblending, discontinuous variations was not readily apparent. Kellogg iterated objections to Darwinian natural selection, most of which focused on problems of variation: the problem of the size of variations, that is, whether "continuous variations" could provide the raw material for selection or whether "discontinuous variations" played a role; the problem of the swamping of unusual variations by interbreeding; the problem of producing a sufficient number of similar variants at a given time to alleviate the swamping problem; the problem that many variations that distinguished species showed no utility (Kellogg 1908, Ch. 3). Darwin had focused much attention on adaptations; subsequent work had turned up many instances of traits whose adaptive significance was not apparent. In discussing alternatives to Darwinism, Kellogg suggested that a number of people had come to the view: "Why may not variation be the actual determinant factor in species-forming, in descent?" (Kellogg 1908, p. 34).

Although de Vries rediscovered Mendel's law of segregation, he did not see it as applicable to the problem of the origin of species. Kellogg prominently featured de Vries's mutation theory of 1901–1903 as one of the alternatives to Darwinian selection that did in fact make the origin of certain types of variations and the origin of new species coincide. De Vries proposed that progressive mutations arose in organisms and were of such a magnitude that they gave rise to new, true-breeding species in a single generation. According to de Vries, progressive mutations did not segregate, nor were they necessarily adaptive. Selection served merely to eliminate the most harmful; thus, the numerous seemingly nonadaptive variants could be explained. He based his theory primarily on his findings from experiments with the *Oenothera,* the evening primrose. The mutations appearing in the evening primrose were due to its being in a mutating period, de Vries believed. Similarly, de Vries thought, other organisms would go through such periods and thus sufficient mutations would be produced (de Vries 1909–1910).

If de Vries had been right that a new species could arise with the formation of a single mutation in a hereditary unit, then the fields of genetics and evolutionary studies would not be separate. If the study of how variations arise had been, as Kellogg had suggested, the actual determinant factor in forming new species, then the study of mutations and their inheritance would have been the key to the origin of new species. No synthesis of Darwinism and Mendelism would have been required; no separation of genetic studies and evolutionary studies would have occurred. But de Vries's theory turned out to be incorrect; the evening primrose was not an appropriate model organism since it has unusual chromosomal mechanisms. An understanding of the origin of species did require more than the study of genetics alone.

## 7.4 THE FIELD OF GENETICS

William Bateson is rightly called the founder of genetics, for he pursued Mendelian inheritance and its relation to the discontinuous variations that he had long advocated as the key to evolutionary change. Mendelism emerged as a separate field in 1900 with the rediscovery of the law of segregation. A field may be characterized as having the following elements: a central problem that characterizes the subject matter of the field; a set of facts related to the problem; concepts, laws, and theories that aid in solving the central problem; and techniques and methods for studying the phenomena or analyzing the concept. (Sometimes fields may also be characterized as having goals providing expectations as to how the problems are to be solved; however, the goals are often difficult to identify in the historical record.)

The emergence of the field of Mendelian genetics in 1900 was marked by both continuity with the past and change from that past. Some of the elements of the new field had been present in biology prior to 1900; others came into being with the emergence. The problem of heredity was an old problem: What are the patterns of resemblance of offspring and parents and how are they to be explained? It had puzzled people for centuries. Moreover, the qualitative phenomena that made up the domain of the new field had been known. Offspring sometimes resemble grandparents rather than parents, thus exhibiting "reversions." The techniques of hybridization or artificial breeding – the main experimental technique of the new field – had a long history. Similarly, the appeal to material units as explanatory factors in heredity was not new.

However, even though these elements had existed separately, they had not been related to one another as they were once genetics emerged as a separate field. The problem of heredity existed but how was it to be solved? Many types of reversions had been noted, but little quantitative analysis and careful steps of crossings had been carried out. Hybridization had been primarily used for practical purposes of artificial breeding. Material units of heredity had been postulated, but no way of investigating them experimentally had been found.

These separate elements came together in a new way with the discovery of quantitative ratios in types of characters in hybrid crosses. When a yellow pea was crossed with a green pea, all the hybrids were yellow; but when the hybrids were crossed, the result was three yellow to one green. Such data was explained by postulating a unit, variously called a "pangen," a "factor," an "allelomorph," and later "gene," that occurred in pairs which segregated in the formation of sex cells that combined randomly to produce the ratios. Investigations of the generality of these hybridization ratios occupied Bateson and others in the period immediately following 1900.

The new field of genetics that thus came into being may be characterized in the following way. Its central problem was the explanation of patterns of inheritance of characteristics. The technique of artificial breeding was used to investigate the characteristics. The laws were Mendel's laws of segregation and independent assortment, and the gene was the primary theoretical concept of the new field.

Mendelian traits were seen by Bateson and others as representing "discontinuous" variations. These were contrasted with the smaller-scale differences that graded into each other called "continuous variations." The biometricians were seen as the proponents of Darwinism. They advocated evolutionary change by means of small "individual differences," as Darwin had called them, or "continuous variations," as they came to be called (Provine 1971). It took a number of years of work within Mendelism before quantitative variations were explained as due to multiple, interacting Mendelian factors; the two types of variation were thereby shown not to be fundamentally different. Thus, the early disputes between Mendelians and Darwinians were resolved as the result of extension of the scope of Mendelism.

The studies of mutation carried out by T. H. Morgan and his students working on the fruit fly *Drosophila* provided the most extensive modifications of Mendelism after 1900 and provided much of the knowledge on which Dobzhansky was able to draw by 1937. Mutations found in *Drosophila* were of a much smaller scale than de Vries's mutations; they served well as raw material for a more gradual process of evolutionary change. Modifying genes (one gene modified the effect of another) allowed even smaller-scale changes. The fact that one gene can affect more than one character provided Dobzhansky with the ability to explain seemingly non-adaptive traits – they might be caused by genes that produced other traits that were adaptive (Dobzhansky 1937, p. 29). The discovery of genes that produced sterility helped to alleviate the concern that the Mendelian differences were not sufficient to account for species differences (Dobzhansky 1937, p. 264). The extensive development of the chromosome theory of Mendelian heredity, linking Mendelian genes to chromosomes, provided key information for Dobzhansky to use in explaining larger-scale differences between varieties and species. The Morgan school's development of this theory and the subsequent work by cytologists in the 1920s and 1930s added substantial new information. In short, subsequent developments of Mendelism in the period from 1900 to 1937 were crucial to the utilization of Mendelism and cytological studies in the evolutionary synthesis.

Similar stories can be told for the stages of development of mathematical population genetics as well as the experimental and field studies of populations (Provine 1971; Mayr and Provine 1980; Mayr 1982). We will not dwell on the details here except to point out that the stages of development by 1937 were also crucial to the possibility of a synthesis occurring.

177

## 7.5   THE SYNTHETIC THEORY VERSUS INTERFIELD THEORIES

Nancy Maull and I discussed a type of theory that we called "interfield theories" (Darden and Maull 1977; see Chapter 5, this book). These theories relate two fields by providing relations among components of the two fields. For example, the chromosome theory of Mendelian heredity was an interfield theory that related the fields of genetics and cytology by postulating that genes were parts of chromosomes. Interfield theories may postulate relations other than part-whole ones: they may propose a structure–function relation or specify the physical nature of an entity or process. Interfield theories function to solve problems that arise within a field but cannot be solved with the techniques and concepts available in that field. The interfield theory serves as a bridge between two previously separate fields that may have been working on the same problem from different perspectives. Predictions may be made for both fields on the basis of the other; thus, the relation is reciprocal. By postulating a physical relation among entities or processes in two fields, interfield theories thus provide a kind of unity of science.

We failed in that chapter to stress the importance of the stage of development of a field in determining to what it can be linked. However, in our examples, we discussed genetics in an early stage being linked to cytology, then genetics at a much later stage being linked to biochemistry to produce the developments in the operon theory of gene regulation. Thus, implicitly, we were aware of the changing nature of fields and the effect of such changes on what kinds of interfield connections are possible.

The synthetic theory differs somewhat from the previously studied interfield theories since it is a multifield theory. It postulates a process by which new species are formed. This process depends on changes at multiple levels. An understanding of those changes is provided by multiple fields. The stage of development of those fields was important in providing adequate information so that the synthetic theory could be formulated. The problem of the origin of species did not arise within one of the fields and thus initiate a search for information in another field to solve it. Instead, it was an old problem that had numerous proposed solutions, some involving different fields and levels of organization than those integrated by the synthetic theory.

The synthetic theory not only related knowledge already present in existing fields, it also postulated the need for a new area of study at a higher level of organization, namely at the species level. Thus, the theory did not merely build bridges among existing fields. In addition to integrating knowledge from other fields, it added a new component with new research problems – the study of isolating mechanisms in the formation of new species. The solution of an

178

old problem may thus not only require relating existing knowledge but also necessitate the emergence of a new area of study.

A causal relation exists among the mechanisms studied by the separate fields of the synthetic theory. Gene mutation and recombination provide the raw material for population changes, which, in turn, are necessary for the formation of new species. Because of these hierarchical, causal relations among the fields, the relations among the fields are not reciprocal as in the interfield theory cases. In Dobzhansky's formulation of the theory, knowledge about gene mutations provided information about population changes, but knowledge about selection processes did not provide new knowledge for the geneticist about mutations. Mendelian genetics proceeded relatively independently of population genetics, but population genetics was dependent on new findings of mutational processes in genetics. Knowledge was merely taken from genetics to be used in the postulation of a new process; the genes themselves were not connected to any new entities as they were when they were connected to chromosomes or to DNA. Similarly, population processes, such as selection and migration, can continue whether or not isolation has occurred. The asymmetry of the causal relations among the fields thus makes their relations less reciprocal than in previously studied interfield theories.

Since genetics can proceed without taking into account the population level, it retains more relative independence than do the other fields in the synthesis. Dobzhansky was wrestling with the question of the relations of fields within the synthesis when he said:

> It should be reiterated that genetics as a discipline is not synonymous with the evolution theory nor is the evolution theory synonymous with any subdivision of genetics. Nevertheless, it remains true that genetics has so profound a bearing on the problem of the mechanisms of evolution that any evolution theory which disregards the established genetic principles is faulty at its source.
>
> (Dobzhansky 1937, p. 8)

Genetics was clearly a key component of the theory, but it remained relatively independent. It was not affected by the higher levels in the way that it affected them, in Dobzhansky's formulation of the theory. (From a more contemporary viewpoint, it might be argued that higher levels do influence the lower. For example, selection may influence the evolution of gene mutation mechanisms. However, instances of influence of selection on the evolution of genetic processes were not part of the original synthetic theory proposed by Dobzhansky; our slim knowledge of the evolution of genetic mechanisms is due to post-1937 developments.)

## 7.6  CONCLUSION

The synthetic theory is a multifield, multilevel theory that postulates a well-integrated causal process combining mechanisms from the different fields to solve the old problem of the origin of species. It established the study of isolating mechanisms as an important new area of study, while relating the already developed fields of Mendelian genetics, mathematical population genetics, and experimental and field studies of populations. The stages of development of these fields were crucial to their roles in the evolutionary synthesis. The new synthetic theory that was formed has provided the basis for evolutionary studies from the time it was proposed to the present. It has been one of the truly remarkable syntheses of twentieth-century science.

### REFERENCES

Darden, Lindley (1976), "Reasoning in Scientific Change: Charles Darwin, Hugo de Vries, and the Discovery of Segregation," *Studies in the History and Philosophy of Science* 7: 127–169.

Darden, Lindley (1977), "William Bateson and The Promise of Mendelism," *Journal of the History of Biology* 10: 87–106.

Darden, Lindley (1980), "Theory Construction in Genetics," in Thomas Nickles (ed.), *Scientific Discovery, Case Studies*. Dordrecht: Reidel, pp. 151–170.

Darden, Lindley and Nancy Maull (1977), "Interfield Theories," *Philosophy of Science* 44: 43–64.

Darwin, Charles ([1859] 1966), *On the Origin of Species, A Facsimile of the First Edition*. Cambridge, MA: Harvard University Press.

Darwin, Charles (1868), *The Variation of Plants and Animals under Domestication*. 2 vols. New York: Orange Judd and Co.

Dobzhansky, Theodosius (1937), *Genetics and the Origin of Species*. New York: Columbia University Press.

Galton, Francis (1871), "Experiments in Pangenesis," *Proceedings of the Royal Society (Biology)* 19: 393–404.

Kellogg, Vernon (1908), *Darwinism Today*. New York: Henry Holt.

Kottler, Malcolm (1979), "Hugo de Vries and the Rediscovery of Mendel's Laws," *Annals of Science* 36: 517–538.

Mayr, Ernst (1982), *The Growth of Biological Thought*. Cambridge, MA: Harvard University Press.

Mayr, Ernst and William Provine (eds.) (1980), *The Evolutionary Synthesis*. Cambridge, MA: Harvard University Press.

Provine, William (1971), *The Origin of Theoretical Population Genetics*. Chicago, IL: University of Chicago Press.

Vorzimmer, Peter (1970), *Charles Darwin, The Years of Controversy: The Origin of Species and its Critics 1859–1882*. Philadelphia, PA: Temple University Press.

Vries, Hugo de ([1900] 1966), "The Law of Segregation of Hybrids." Translated from German and reprinted in C. Stern and E. Sherwood (eds.), *The Origin of Genetics: A Mendel Sourcebook*. San Francisco, CA: W. H. Freeman, pp. 107–117.

Vries, Hugo de ([1903–4] [1909–10] 1969), *The Mutation Theory*, 2 vols., translated from German by J. B. Farmer and A. D. Darbishire. New York: Kraus Reprint Co.

# 8

# Selection Type Theories[1]

8.1 INTRODUCTION

This chapter discusses abstract characterizations of selection theories. Finding such abstractions is a task in a larger research program, which is based on the assumption that some scientific theories are representative of types of theories that solve types of problems. Natural selection, clonal selection for antibody production, and selective theories of higher brain function are examples of selection type theories. Selection theories solve adaptation problems by specifying a process through which one thing comes to be adapted to another thing.

When Darwin elaborated his theory of natural selection in 1859, he provided a new type of theory for explaining adaptation problems. Others, such as Burnet (1957) in immunology, argued by analogy from Darwinian natural selection for selection processes in other fields. One analysis of analogy is that two analogues share a common abstraction (Genesereth 1980). Thus, an analysis of natural selection and its analogues aids in the development of an abstraction for selection theories. The selection literature in philosophy of biology includes several discussions of natural selection that provide help in this task. The abstraction for natural selection extracted from these discussions will then be used in analyzing clonal selection for antibody production in immunology and selective theories for higher brain function in

[1] This chapter was originally published as Darden, Lindley and Joseph A. Cain (1989), "Selection Type Theories," *Philosophy of Science* 56: 106–129. This work was supported by a Special Research Assignment from the College of Arts and Humanities at the University of Maryland. Thanks to Michael Bradie, David Casey, Marjorie Grene, Joel Hagen, Pamela Henson, Joshua Lederberg, Elisabeth Lloyd, Elizabeth Napier, James Platt, Frederick Suppe, and two anonymous referees for their comments.

neurobiology.[2] Before beginning the analysis of selection theories, however, we will briefly discuss the concept of an abstraction.

### 8.2 THE CONCEPT OF AN ABSTRACTION

An abstraction provides a schematic outline that can be filled in or "instantiated" to give an actual theory. It delineates typical components of theories of a given type. Although little work has been done in history and philosophy of science on abstract types of theories, some is suggestive.

Hanson (1961), in discussing reasoning in discovery, suggested that one might reason that problematic phenomena can be explained by a "type" of hypothesis. Hanson said little about what constitutes a type of hypothesis, but he gave an example: Newton's theory of gravitation is an inverse-square type of theory. Despite its lack of development, Hanson's idea has appeal: Find types of hypotheses proposed in the past and analyze the nature of the puzzling phenomena to which the type applied. Use this knowledge of a type of hypothesis in future instances of theory construction. In Darden (1987), this method of theory construction is referred to as using the history of science for finding "compiled hindsight."

Suppe (1979) suggested that investigations of types of theories is an important research task in philosophy of science and mentioned two types that Shapere (1974b, 1977) discussed: compositional and evolutionary. Compositional theories are theories that explain by appealing to component parts, such as atomic theory. Evolutionary theories claim that change over time has occurred, such as biological evolution, stellar evolution, and theories concerning the evolution of the chemical elements. It is important to distinguish selection theories, which we will discuss, from evolutionary theories, on which Shapere focuses. Darwinian natural selection is one explanation for biological evolution, but other theories have been proposed to explain it, such as inheritance of acquired characters and orthogenesis (i.e., development toward an end driven by some sort of internal dynamic without regard to

---

[2] A note about terminology is useful at the outset. In using the term "natural selection" in contrast to clonal selection and neuronal selection, we do not mean to imply that the latter two are examples of "artificial" selection. We are following accepted usage in restricting natural selection to the theory that explains adaptations arising through evolutionary time. Although the immune system and brain evolved, the theories about them under discussion explain adaptations within a single individual during its life to combat disease or to learn from experience. Also, we are using "adaptation" in a more general way than the restricted sense of the product formed as a result of natural selection; see, for example, Gould and Vrba (1982).

environmental conditions). Furthermore, selection theories, such as the clonal theory of antibody production, can also be used to explain adaptive processes that are not evolutionary.

Working at a less abstract level, Kitcher (1981) suggested that theories can be viewed as providing schematic argument patterns. He developed instructions for instantiating such schemas. His examples included Newtonian laws and a pattern provided by Darwin for explaining the evolution of a particular species. Kitcher's "schemas" are less abstract than the level at which we will discuss selection type theories. His cases showed the use of a single schema within one domain to explain numerous, similar instances. Thus, for Kitcher, using the same schema within a single domain provided a kind of explanatory unification for that area of science. Our discussion of selection theories will be for analogous theories found in separate fields. Use of the same type of theory merely shows that similarly structured theories can be used to explain different domains; it does not unify the areas in the strong sense discussed by Kitcher. However, if one has a realistic interpretation of theories, then finding the same abstract type of theory in numerous domains may be taken as an indication that nature uses similar processes in diverse phenomena. Such a conclusion might be regarded as a weaker form of unity in science than that discussed by Kitcher.

Brandon (1980) also used the language of "schema" and "instantiation." In discussing the principle of natural selection, he regarded the principle of natural selection as a "schematic law." The schema may be "cashed out" in terms of differences in particular traits that apply "to particular populations under particular environmental conditions" (Brandon 1980, pp. 432–433).

Despite this previous work, historians and philosophers of science have not taken as a primary task finding types of problems and types of theories to solve them.[3] To find other previous work that seeks theory types and attempts to construct abstractions of them, we must turn to another field.

---

[3] The question may be raised of how our analysis relates to axiomatic (e.g., Williams 1970) and semantic (e.g., Beatty 1980; Thompson 1983; Lloyd 1984) analyses of evolutionary theory. Our analysis is compatible with these approaches but has different motivations and aims. Darden was motivated to undertake this analysis as a result of her work on analogy and theory construction (Darden 1982a, 1983). The aim here is not to use or provide a general analysis of the structure of all scientific theories. It is an open question whether all theories can be used to form abstractions; furthermore, it is likely that some theories are unique and not examples of recurring types. Instead, the aim is to provide one theory formation strategy, namely, invoking a theory type. Numerous other strategies for theory construction exist (see, e.g., Darden 1980; see Chapter 6, this book; 1982a, 1986). The idea of instantiating variables in different ways to produce different theories also has some similarity to Schaffner's (1980) analysis of theories of the middle range as a series of overlapping, temporal models. Clonal selection is one of Schaffner's examples, although he does not focus on its nature as a selection type theory. None of these previous analyses, however,

Work on reasoning by analogy and attempts to implement analogical reasoning in computer systems using techniques in artificial intelligence (AI) have produced additional examples of abstract theory types. Gentner (1983) discussed central force systems, such as the solar system and the Bohr model of the atom. Greiner (1985) constructed an abstraction for flow problems, for example, water and electric flow. Holyoak and Thagard (1995) investigated wave type theories, including water waves and sound waves.

The analysis of two analogous things as sharing a common abstract structure (Genesereth 1980) has proved to be fruitful in AI. Methods for forming abstractions and for instantiating them are an active area of research, not only in the analogy work but also in research on problem-solving more generally.[4] Philosophers, insofar as they have considered abstractions, have tended to think of one level of abstraction of a particular theory, such as a formal set of equations embodied in the theory. But methods of knowledge representation and of implementing reasoning strategies in AI have led to the idea of multiple levels of abstraction (Korf 1985).

Imagine a scientist confronted with the problem of explaining how one thing comes to be adapted to another thing – how an immune system comes to be adapted to fighting an invading antigen or how a brain becomes adapted to responding to a type of signal. History of science provides the compiled hindsight that selection theories have been useful in solving adaptation problems (Darden 1982a, 1987). So, our imaginary scientist asks: What should be the case in the problem situation if a selection theory is to provide a useful solution? What needs to be specified in order to construct a selection theory? In order to answer this scientist, our task is to construct an abstraction for selection theories. By instantiating the abstraction, our imaginary scientist may be able to generate testable hypotheses.

Several methods for forming abstractions exist. An abstraction can be constructed from one example by replacing constants with variables. Alternatively, more than one example with the same structure can be used, as in analogical reasoning. Abstracting involves pruning away content and leaving essentials (or "typicals," if one isn't looking for necessary conditions). Comparison of two examples is a good way to locate what is most important. Another way of forming abstractions is to consider several discussions of

shares with ours the desire to construct abstractions that capture structure shared by *analogous* theories in different fields. Nonetheless, analyzing theories as abstract schemas or models that can be instantiated in different ways may be a fruitful idea to pursue because it has been arrived at from quite different perspectives by a number of philosophers.

[4] For an algorithmic approach to adaptation, see Holland (1975). For a comparison of Holland's adaptive theories and selection theories, see Darden (1983).

abstract components of a theory and put complementary parts together in a piecemeal way. (For additional methods, see Darden 1987.)

Our task now is to proceed to develop an abstraction for selection type theories. We will draw on recent discussions in the philosophy of biology literature to develop an abstraction for natural selection. The construction of that abstraction will be constrained by our desire to make it applicable, possibly at a slightly higher level of abstraction, to clonal selection and selective theories of brain function. After stating the abstraction for natural selection, we will instantiate it in the other two analogous cases.

### 8.3   THE PROCESS OF NATURAL SELECTION

A selection process may be broken down into a series of steps from which a more abstract characterization can be developed, as follows:

(A) First are the preconditions before a selective interaction. These include a set of individuals that vary among themselves. Also, the individuals must be in an environment with critical factors that provide a context for the ensuing interaction.

(B) The actual step of selection involves an interaction between individuals and their environment. Because they vary, different individuals will interact differently.

(C) Several types of effects result from the differential interactions. In the short range, individuals benefit or suffer. If the individuals can be located in a hierarchy (e.g., gene, organism, group), then there may also be short-range effects of sorting at other levels.

(D) Longer-range effects may follow the short-range effects of the interaction, such as increased reproduction of individuals with certain variations or reproduction of something associated with those individuals.

(E) Even longer-range effects may also occur, such as accumulation of benefits through numerous generations to produce a lineage of individuals.

Our task now is to discuss these steps in more detail and develop an abstraction with appropriate variables.

### A.   Preconditions for Natural Selection

Natural selection operates on a population of varying individuals in a given location, such as the classic example of the peppered moths in a region with trees darkened by industrial pollution. This case illustrates the following preconditions for selection: the set of individuals making up a population, the presence of the variation, and the environment.

186

Although variation is often discussed as a precondition for selection (e.g., Lewontin 1970; Sober 1984), the other preconditions that we list have received less attention. In addition to varying, the individuals on which selection acts must *share* some features. If the population is an interbreeding population, then the organisms, of course, share many genes and their associated characteristics. We will assume that a set of individuals is delineated by specifying features that they share, not by enumerating its members. Thus, our abstraction has as a first precondition that a set of individuals exists. "Group" in Hull's (1980, 1981) sense might be a better word than "set"; groups tend to be spatiotemporally localized entities and their members are considered part of the group because of their location (Hull 1980, p. 314). But using "group" would introduce confusion with the different issue of whether "group selection" exists. The term "class" might avoid introducing perhaps extraneous connotations from set theory, but Kitcher (1987, p. 186) argued that it introduces confusions of its own.

Hull stresses that selection "can act only on spatiotemporally localized entities" (Hull 1980, p. 314). Sober makes a related claim when he argues that a "common causal influence" operates in a selective process:

> Think of two populations living on opposite sides of the universe, each driven by its own internal dynamic. Suppose that one group ends up outreproducing the other. If the two populations are not subject to a common causal influence, then it makes no sense to view them as participating in the same selection process. (Sober 1984, p. 274, footnote 40)

In order to make more explicit the idea, which is perhaps implicit in the views of Hull and Sober, we will state that the set of individuals must be "in an environment *E*." This phrase has the advantage of allowing for further elaboration of the nature of the environment, as we will shortly see.

Much can be said about the nature of variation within the set. Sober (1984) reduced his discussion of selection to the simplest case: Individuals either have or do not have a single variant property *P*. Darden (1982a) uses the phrase "array of variants," which implies a range of variations within a set. She also discussed the scale of the variants relevant to natural selection. The amount of difference between variants has been a matter of dispute historically; Darwin's small-scale individual differences and de Vries's (1903–1904) large-scale mutations are examples. Another issue that may arise within a discussion of the preconditions for selection is how the variants are produced. But, as long as preexisting diversity exists, the details of the mechanism of its production can be omitted from a selection type theory.

A key precondition for selection that has received comparatively little discussion in the philosophy of biology is the nature of the environment. Brandon (1990) recognized this omission too. Cain stressed the role that critical environmental factors play in creating conditions for a selective interaction (Cain and Darden 1988). The components of the environment that affect the nature of the struggle for existence will be referred to as "critical factors." What factors in the environment are critical determines what variant properties are relevant to surviving (or otherwise benefiting from) the selective interaction. Thus, the fit between variant and environment (or, in a finer-grained analysis, the fit between the variant property and the critical factor) is the key aspect of the next step, the selective interaction.

## B. Selective Interaction

Much of the early discussion in philosophy of biology about the nature of selection focused on the *effects* of selection, namely, differential survival and reproduction. If no account can be given of selection except in terms of the effects, then the specter of tautology is raised: "survival of the fittest" becomes "survival of those that survive."

Lewontin, in his now classic 1970 article discussing the steps in selection and the properties of evolving entities, stated: "different phenotypes have different rates of survival and reproduction in different environments" (Lewontin 1970, p. 1). But, we may ask, *why* is reproduction differential? Lewontin lists three properties of entities that evolve: variation, reproduction, and heritability. A fourth component, one involving interaction during the selection step, must be added to this list to explain how variation gives rise to differential reproduction. In an earlier attempt at producing an abstraction for natural selection, Darden characterized the selection step common to artificial selection and Darwinian natural selection: selection of a subset of variants occurs by an agent selecting according to a criterion (Darden 1982a). But "agent" brings inappropriate negative analogy from the artificial to the natural case.

A better characterization of the selection step can be constructed using ideas from Hull (1980) and Sober (1984). Especially important are the ideas of *interaction* and the *causal role* of the variant property. Their ideas have to be augmented by more detail about the role of the *critical factor* in the environment. Hull recognized the nexus of the selective event:

> Reproduction by itself is sufficient for evolution of sorts, but not evolution through natural selection. In addition, certain entities must interact causally with their environments in such a way as to bias their distribution in later generations.
> (Hull 1980, p. 317)

Sober (1984) stressed the *causal* nature of the selective interaction even more than Hull did. Sober advanced us on our way to producing an abstraction for natural selection by providing somewhat helpful characterizations of group and genic selection:

> [There] is group selection for groups that have some property P if (and only if)
> 1) Groups vary with respect to whether they have P, and
> 2) There is some common causal influence on those groups that make it the case that
> 3) Being in a group that has P is a positive causal factor in the survival and reproduction of organisms.
>
> The analysis of genic selection goes in parallel. There is selection for possessing the gene P if (and only if)
> 1) Organisms vary with respect to whether they have P, and
> 2) There is a common causal influence on those organisms that makes it the case that
> 3) Possessing the gene P is a positive causal factor in the survival and reproduction of organisms.
>
> (Sober 1984, p. 280)

Sober stressed the relation between the possession of a property and the causal role it plays in producing a benefit. Sober provided neither a similar analysis for organismic selection nor an abstraction for natural selection that is independent of hierarchical levels. We may attempt to put variables into Sober's accounts:

There is $X$ selection for $Y$s that have some $X$-level property $P$ if (and only if)

(1) $Y$s vary with respect to whether they have $P$
(2) There is some common causal influence on $Y$s that makes it the case that
(3) $Y$s possession of $P$ is a positive causal factor in the survival and reproduction of $Z$s.

$X$ is the hierarchical level considered as the level of selection. $Y$s are the objects that vary. $Z$ is the level that Sober called the "benchmark." A benchmark is an object that benefits or suffers in virtue of having properties that are selected for (Sober 1984, p. 270). However, the relation of $Y$ to $X$ and $Z$ is unclear. For the genic selection example, $Y = Z =$ organisms. But in the group selection case, $Y$ is not equivalent to $Z$: $Y =$ groups, but $Z =$ organisms. $Y$s are the variants, but Sober was unclear about whether the variation is to be located at the level of causal influence ($Y$) or at the benchmark ($Z$). For simplicity, initially we are going to assume that the variant, the interactor, and that which directly benefits or suffers can be represented by the same variable. Alternative, lower-order abstractions can be constructed that differentiate these three variables.

Sober's analysis introduced complexities because he was grappling with the issue of effects at various hierarchical levels. Since Lewontin's (1970) analysis of selection at different levels, others have also tried to analyze selection and its effects at different levels in structural or functional hierarchies (e.g., Hull 1980; Arnold and Fristrup 1982). In the abstraction we are constructing, we wish to avoid most of the complexity the hierarchical perspective on selection introduces. However, we will use one distinction used by Vrba and Gould (which parallels a similar distinction in Sober): sorting and selecting.

> In Darwinian theory, evolutionary change is the product of sorting (differential birth and death among varying organisms within a population). Sorting is a simple description of differential representation; it contains, in itself, no statement about causes. As its core, Darwinism provides a theory for the causes of sorting – natural selection acting upon organisms in the "struggle for existence." However, other processes (genetic drift, for example) produce sorting as well; thus, the two notions – sorting and selection (a favored theory for the cause of sorting) – are quite distinct and should be carefully separated.
>
> (Vrba and Gould 1986, p. 217)

Vrba and Gould's definition of selection makes use of their distinction between sorting and selecting and also shows their localization of the selective interaction at the hierarchical level of the variant property:

> We suggest that individuals are subject to selection at any level for the characters they possess only if these characters are heritable and *emergent*, and if they interact with the environment to cause sorting ... : *Selection encompasses those interactions between heritable emergent character variation and the environment that cause differences in rates of birth or death among varying individuals.* (Vrba and Gould 1986, p. 219; emphasis theirs)

The single most important omission from the previous discussions is a neglect of the role that environments play in selective interactions. Although Hull and Vrba and Gould allude to the interactions with the environment, no analysis is given of conditions in an environment that are relevant to the selective interaction. Sober did not explicitly mention the role of the environment in his schematic discussions of group and genic selection (Sober 1984, p. 280). Consideration of his steps shows that he packed the role of the environment into "common causal influence" and into his analysis of how a variant property can be a "positive causal factor" in selection. We analyze the environment in terms of the critical factor that makes a given property causally relevant to

190

the selective interaction. The critical environmental factor and the interaction of the variant property with it are the keys to the subsequent effect of the selective interaction.

In summary, the main points about the selective interaction are the following: First, an explicit recognition that a step in the temporal process occurs between the formation of variants and the effect of differential reproduction; second, that this step involves *interaction*; third, the interaction can be analyzed either in terms of an individual and its environment or, in a finer-grained way, in terms of the *variant property* and the *critical factor* in the environment; fourth, the variant property plays a *causal role* in the interaction; and, finally, effects can be manifested at levels other than the level of the interaction.

## C.  Effect

Sober used the language of "benefit and suffer" to characterize the differential effects of the selective interaction. Although such language sounds somewhat anthropomorphic and will require a value judgment as to what is a benefit, it is appropriately abstract for our purposes for a number of reasons. In natural selection, the interaction may produce an effect that only later results in differential survival and reproduction. Thus, the process should have a step for specifying such an intermediate effect. Also, the benefits in the many different instances of selection events will be of many different types, for example, escaping predators, obtaining food, defending territory. "Benefit" is a neutral term for characterizing numerous types of positive effects.

Most important, the neutral language of "benefit" is desirable for constructing an abstraction that can apply at different levels in an organizational hierarchy within an evolutionary theory context. It is even more desirable for our current purpose of constructing an abstraction that can be instantiated in other selection type theories. We would be happy with a less anthropomorphic and value-laden sounding term than "benefit," but we haven't yet found one at an appropriately abstract level.

## D. and E.   Longer-Range Effects

We have carefully delayed adding differential reproduction to our steps in a selection process until these later steps. As others have argued, defining selection as differential reproduction masks many of the important features of selection. For artificial selection, it may be very important to separate the benefit from differential reproduction. The breeder might select on a criterion of low seed output in grapes; hence, decreased reproductive capacity actually results in more grape plants being grown. Nonetheless, for natural

191

selection, differential reproduction as a result of the differential interaction is the important consequence that makes a difference in evolutionary change.

Differential replication is an integral part of Hull's analysis of selection, as is the even longer-range consequence of the production of a lineage. Hull defined a lineage as "an entity that changes indefinitely through time as a result of replication and interaction" (Hull 1980. p. 327).

We have discussed steps A–E and will turn to a construction of a more abstract characterization of them, attempting to be more precise and use variables in place of many of the words of the looser preliminary list.

### 8.4   ABSTRACTION FOR NATURAL SELECTION

In considering what aspects of the various discussions should be preserved in an abstraction for selection theories, the following components are important. Darden's idea of an "array of variants" focused attention on the fact that one needs to specify a set of things that vary. Implicit in that idea is that the individuals in the set share many properties but vary in one or more. For simplicity, Sober suggested a single variant property $P$. Hull stressed causal interaction. Sober's primary emphasis was on the "common causal influence" operating in the selective event and the role of the variant property as a "positive causal factor." He provided appropriate neutral language of the effect being some sort of "benefit." Vrba and Gould provided the distinction between selection at a given level in a hierarchy and sorting at other levels. To these discussions we have added the concept of a critical factor in the environment.

Drawing on this previous work, we now attempt to develop our abstraction. A typical selection process can be analyzed into a group of preconditions such that if they obtain, then a selective interaction occurs, resulting in a series of effects. A selection process typically occurs if:

### A.   *Preconditions*

i.  A set of $Y$s exists and
ii. $Y$s vary as to whether they have property $P$ and
iii. $Y$s are in an environment $E$ with critical factor $F$

### B.   *Interaction*

iv. $Y$s, in virtue of possessing or not possessing $P$, interact differently with environment $E$ and
v. critical factor $F$ affects the interaction such that

192

## C.   Effect

vi. the possession of *P* causes *Y*s with *P* to benefit and those without *P* to suffer.

(vi′. This causal interaction may have the concomitant effect of sorting *Z*s.)

## D.   Longer-Range Effect

vii. *C* may be followed by increased reproduction of *Y*s with *P* or reproduction of something associated with *Y*s.

## E.   Even Longer-Range Effect

viii. *D* may be followed by longer-range benefits.

To give some substance to these abstract statements, let's consider the classic case of the peppered moths. (i) A population or set of peppered moths exists. (ii) The moths vary in the property of color, dark or light. (iii) Moths are preyed upon by birds and live in woods near coal-burning plants where trees are darkened by soot. (iv) Moths, by virtue of possessing or not possessing dark color, interact differently with birds. (v) The darkness of the trees affects the interaction such that (vi) the possession of dark color causes moths with dark color to benefit and those without dark color to suffer. (This causal interaction may have the concomitant effect of sorting genes in the moth population.) (vii) Those moths with dark color tend to produce more offspring, which inherit dark color.

The abstraction is a simplification in a number of explicit ways. In Condition (i), the existence of a set is specified, but common properties that define the set are not explicitly given, though they could be. Condition (ii) is the simplest case in which *Y*s differ by only one property. If the way in which *Y*s differ is unknown, or *Y*s differ in more than one property, this condition could be altered accordingly. Condition (ii) considers the range of the variant property, which here is stated in the simplest possible way. Having or not having *P* could be altered to reflect other possibilities, such as a continuous gradation of some sort or quantitative differences. (For a discussion of instantiating details about variation, see Darden 1982a.)

Specifying an environment and a single factor within it that is relevant to the selective interaction (Conditions iv and v) may sometimes be difficult. How to delineate the relevant "environment" from everything else may be a formidable task. Limiting discussion to a single environmental factor *F* is a simplification. In the moth example, the environment includes the nature of the predator as well as environmental conditions, for example, color of trees.

Thus, sometimes the effects of the critical factor are mediated through an agent, such as a predator.

Condition (iv) of interaction is separated from the causal Condition (vi) because it is possible to have different interactions without having differential benefits. For example, if toads have tongues of different lengths, then those with longer tongues can catch insects at a greater distance. But if insects are sufficiently plentiful, then all toads get adequate food despite this differential interaction; no relative benefit or suffering results. Some critical factor in the environment, Condition (v), must produce a situation that makes $P$ causally relevant to benefiting.

With some trepidation, we have chosen to follow others and include causal language in our abstraction. We are not here providing an analysis of the "cause" in our abstraction. We are merely stating that in a typical selection process, the interaction between variant property $P$ and the environmental factor $F$ *causes* those variants with $P$ to benefit. Explaining the way differences in interaction cause differences in benefit will depend on case-specific information. We realize the statistical nature of the process and the implicit *ceteris paribus* clause: Sometimes those with $P$ will be in appropriate conditions to benefit, but other times something else will befall them and they won't actually benefit.

"Benefit" and "suffer" are vague, but this vagueness has the advantage of allowing instantiation in numerous ways in different specific cases. Benefit could be interpreted as contributing to survival and reproduction (with inheritance of the beneficial property), but it need not. Benefit can be measured instead in terms of immediate effects on $Y$s. Theories lacking a reproduction step might be called "election" rather than "selection" theories.[5]

The parenthetical remark in (vi') about the concomitant effect of a selective interaction sorting $Z$s makes use of Vrba and Gould's distinction between sorting and selecting. Our abstraction clearly separates $Y$s that are selected and $Z$s that are sorted but not directly selected. Whether sorting in this sense actually occurs depends on whether there are hierarchical features in the particular case.

The variable $Y$ is used in three different ways in our abstraction: in (A) as a variator, in (B) as an interactor, and in (C) as the benefitor. In a finer-grained analysis, one might propose three related but separate variables, which need not be restricted to a single hierarchical level.

---

[5] This terminology was used in Lederberg (1959). Shapere and Edelman (1974, pp. 202–204) explicitly separated reproduction and differential reproduction in their "criteria of natural selection."

## 8.5 CLONAL SELECTION THEORY FOR ANTIBODY FORMATION

The goal of this work is to develop an abstraction for selection type theories. Thus far, we have developed an abstraction for natural selection. Now we will consider another successful selection theory, which comes from immunology. In 1955, Jerne proposed the natural selection theory for antibody formation. Jerne's theory was subsequently modified by Burnet (1957) to the theory of clonal selection.

The problem is to explain how the immune system forms antibodies that are able to deactivate many different types of foreign substances (called "antigens") that invade the body while not attacking the body's own substances. Jerne, in reflecting on the components of his theory, said:

> ... three mechanisms must be assumed: (1) a random mechanism for ensuring the limited synthesis of antibody molecules possessing all possible combining sites, in the absence of antigen, (2) a purging mechanism for repressing the synthesis of such antibody molecules that happen to fit auto-antigens, and (3) a selective mechanism for promoting the synthesis of those antibody molecules that make the best fit to any antigen entering the animal.
>
> (Jerne 1966, p. 301)

Thus, Jerne proposed several mechanisms: one for producing variants, another for negative selection against any antibodies that would attack the body's own substances, and a third positive selective mechanism for increasing production of antibodies against an invading antigen. The positive selective mechanism works, according to Jerne, by a specific, circulating antibody attaching to an invading antigen, the complex of antigen-antibody being engulfed by a cell, and that cell (or perhaps another one that is signaled) beginning to produce more of the specific antibody molecule. Jerne, writing before the details of protein synthesis had been worked out in molecular biology, admitted that the idea that a protein molecule entering a cell could signal the cell to produce more molecules of that kind is an "unfamiliar" notion (Jerne 1955, pp. 849–850).

Although Burnet (1957) endorsed much of Jerne's theory, he objected to the mechanism of production of antibodies and proposed an alternative at the level of the cell rather than the molecule. A similar change in level from molecule to cell was proposed (apparently independently) by Talmadge (1957). (For further discussion of Talmadge's role, see Burnet 1957, p. 67; Talmadge 1986, p. 7; Ada and Gustav 1987, p. 65). Jerne (1966, p. 308), in a retrospective on the theory in 1966, gave credit to Burnet and does not mention Talmadge.

Of Jerne's theory, Burnet said:

> Its major objection is the absence of any precedent for, and the intrinsic unlike-
> lihood of the suggestion, that a molecule of partially denatured antibody could
> stimulate a cell, into which it had been taken, to produce a series of replicas
> of the molecule. . . . it would be more satisfactory if the replicating elements
> essential to any such theory were cellular in character *ab initio* rather than
> extracellular protein which can replicate only when taken into an appropriate
> cell.                                                              (Burnet 1957, p. 67)

Thus, Burnet changed the level of organization of the unit that replicates
from molecule to cell. Furthermore, he claimed that the cell is the unit of
variation; one cell produces one type of antibody. He proposed somatic mutation as the mechanism of the production of variants. In addition to making
the cell the unit of variation and of reproduction, he also made it the unit
of interaction; the lymphocyte cell has a reactive site that is "equivalent" to
the antibody it produces. The reaction between the reactive site and antigen
serves as a signal to the cell to produce a clone of cells like itself. The cloned
cells then release free antibodies of the specific type to attack the invading antigen. (For additional discussion, see Lederberg 1959; Schaffner 1980,
pp. 69–71.)

Jerne and Burnet's theories illustrate additional instances of selection type
theories that may be analyzed to yield an abstraction. The differences in their
theories show that the variables can be instantiated in different ways either to
improve a theory or to yield a competing alternative. In this case, the level of
organization – molecule or cell – provides an important point of difference in
the ways the variables are instantiated.

Jerne's and Burnet's theories attempted to solve an adaptation problem:
How is the immune system's defense against an invading antigen produced,
or, more specifically, how are antibodies against an antigen formed? Fighting
invaders is an adaptive response of the immune system.

By comparing Burnet's theory to our previous abstraction, we can develop
an abstraction for Burnet's clonal selection theory. A selection process occurs
if:

## A.  *Preconditions*

   i.  A set of lymphocyte cells exist and
  ii.  these cells vary as to the reactive sites on their cell surfaces.
 iii.  These cells are in the bloodstream with invading antigens.

## B.  Interaction

iv.  The cells, in virtue of possessing or not possessing certain reactive sites, interact differently with the antigens and

v.  the properties of the antigen affect the interaction such that

## C.  Effect

vi.  the possession of a particular reactive site causes cells with such a site to be activated and those without such a site not to be activated.

## D.  Longer-Range Effect

vii.  Activated cells proliferate in clones thereby producing more cells of their type; the cloned cells release free antibodies of a given type that attack the antigen.

## E.  Even Longer-Range Effect

viii.  More cells of that type will be present even after the invading antigen has been eliminated and the system returns to "normal"; thus, the immune system has acquired the ability to respond more quickly in the future if an antigen of that type invades.

The selection event involves a differential activation. The effect is a differential amplification of cells that has the concomitant effect of producing more antibodies of the given type.

### 8.6  SELECTION THEORIES FOR BRAIN FUNCTION

A number of rather speculative theories for explaining various aspects of brain functioning have been proposed that are selection type theories, including theories by Conrad (1976), Changeux (1985), and Edelman and Mountcastle (1978). Here, we will focus on Edelman's theory.

The adaptation problem is to explain increasing efficiency of the brain in processing a given type of signal. First, how does the brain accommodate new types of signals that have never before been encountered? Second, how does the brain refine a response pattern when it encounters a type of signal repeatedly? (Edelman and Mountcastle 1978, p. 55). Edelman proposed a theory to solve this adaptation problem; he called it the "group-selective theory of higher brain function":

> The basic idea is that the brain is a *selective* system that processes sensory motor information through the temporally coordinated interactions of collections or repertoires of functionally equivalent units, each consisting of a small group of neurons [brain cells].   (Edelman and Mountcastle 1978, p. 52; emphasis his)

The units of variation are groups of neurons. Groups having nearly identical functions are part of a *repertoire*, in Edelman's terminology (p. 62), or a *set* of variants, in our terminology. Functionality of groups of neurons is defined by their response to signals. Although all the groups in a repertoire share the property of being able to respond to a given signal, these groups vary in their internal structural organization. In Edelman's terminology (and used similarly in molecular biology), the groups are degenerate, not redundant: Groups with different intrinsic connections respond to identical stimuli. "Degenerate groups are isofunctional, but nonisomorphic . . . redundant groups are isofunctional and isomorphic" (Edelman and Mountcastle 1978, p. 59, Fig. 3).

The variant groups are produced as a result of both genetic factors and "epigenetic" factors, which occur during the embryological and early development of the brain. Such factors include "selective stabilization" of synapses (connections between neurons) discussed by Changeux (Edelman and Mountcastle 1978, pp. 88–89). But even after these selection events, which give rise to structurally different neuron groups, much degeneracy remains, according to Edelman. This degeneracy provides the variability for the selective interactions that occur between the neuron groups and the signal input throughout the life of the organism.

By some means that Edelman did not specify, the variation in internal structure of each particular group determines how well it will respond to a given stimulus. As a result of the variation in structure, some neuron groups are more easily or more often stimulated than others. That stimulation has the effect of either reinforcing a particular neuron group or somehow selecting against the other groups in that repertoire, thus altering their potential for subsequent stimulation by similar signals. The accumulation of such alterations has the longer-range effect of building up a secondary repertoire of groups. In other words, the reinforced groups have the potential to respond more efficiently in the future.

Edelman's work is obviously only in the early stages of theory construction. Many details remain to be specified. Constructing an abstraction following the pattern from the natural selection and clonal selection theories is a method that Edelman himself recognized as valuable (see his Table 8.1: "Some Characteristics of Selective Systems," Edelman and Mountcastle 1978, p. 91). The following is our attempt at instantiating our abstraction with Edelman's theory.

A selection process occurs if:

## A.  *Preconditions*

  i. A set of degenerate neuron groups exist that respond to a given signal (i.e., an isofunctional repertoire exists), and
 ii. neuron groups vary in intrinsic connections and thus in their abilities to respond to a given signal, and
iii. neuron groups are in regions of the brain that receive a given sensory signal with a particular spatiotemporal configuration such that

## B.  *Interaction*

 iv. neuron groups, in virtue of their different intrinsic connections, interact differently with the sensory signal, and
  v. the particular spatiotemporal configuration of the signal affects the interaction such that

## C.  *Effect*

 vi. the possession of certain intrinsic connections causes neuron groups with such connections to be stimulated and/or those without such connections to be inhibited.
(vi'. This causal interaction has the concomitant effect of sorting individual neurons.)

## D.  *Longer-Range Effect*

vii. Stimulation of neuron groups causes reinforcement of those groups.

## E.  *Even Longer-Range Effect*

viii. A secondary repertoire of committed neuron groups develops for responding more efficiently to signals similar to those previously encountered.

| Table 8.1 Summary of Instantiations for Selection Type Theories | | | | |
|---|---|---|---|---|
| **Variable** | **Abstraction** | **Natural Selection** | **Clonal Selection** | **Neuron Group Selection** |
| *Y* | variant<br>interactor<br>benefitor | peppered moths<br>peppered moths<br>peppered moths | lymphocytes<br>lymphocytes<br>lymphocytes | neuron groups<br>neuron groups<br>neuron groups |
| set of *Y*s | set of variants | population of moths | population of lympho-<br>cytes | repertoire of neuron<br>groups |
| *P* | variant property | surface color | structure of reactive<br>sites | pattern of intrinsic<br>connections |
| nature of *P* | range of variation | discrete (light or dark) | continuous variation<br>in structure of active<br>site | continuous variation<br>in possible<br>connections |
| *E* | environment | woods darkened by<br>soot where birds<br>prey on moths | bloodsteam with<br>antigens | region of brain<br>receiving given input<br>signal |
| *F* | critical factor | color of trees | antigen structure | spatiotemporal config-<br>uration of signals |
| *Z* | level of sorting | genes influencing<br>surface color | antibody molecules | individual neurons |
| benefit | metric for measuring<br>differential effect | survival of moths | lymphocyte activation | stimulation |
| longer-range<br>effect | | reproduction of moths | cloning of activated<br>lymphocytes | amplification<br>(reinforcement) |
| even longer-<br>range effect | | lineage adapts | antigen removed,<br>secondary response<br>develops | amplified groups<br>become committed,<br>more efficient pro-<br>cessing of similar<br>signals |

Instantiating the variables in our abstraction for Edelman's theory shows the vagueness in his theory as to the nature of the relation between the variations and the critical factor in the signal that interacts with the variant groups. In other words, what is it about the intrinsic connections that causes some groups to have a better "match" (Edelman and Mountcastle 1978, p. 56) to a signal? In terms of our variables, the *Y*s have in common the ability to process a given signal, but Edelman's analysis has not yet provided values to instantiate *P* and *F*; more specifically, what property of the intrinsic connections (*P*) causes them to process a given signal configuration (*F*) in a better or worse way? Edelman told us too little about the relation between neuron connections and signal configurations to allow us to see why some would "fit" better than others. Our abstraction suggests further theory construction needs to be done on this point.

Table 8.1 shows the instantiations for the variables in the abstraction extracted from the three cases. It shows the analogies between the three

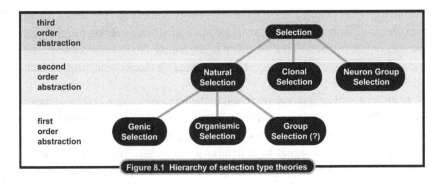

Figure 8.1 Hierarchy of selection type theories

theories and the components of Edelman's theory that we have used to instantiate his theory in our abstraction.

## 8.7 CONCLUSION

The abstraction extracted from the characterizations for natural selection proved of sufficiently high order for the immunology and neuronal cases, except that the longer-range effects were not necessarily reproduction. A disjunction, such as "reproduction or amplification or reinforcement," might serve unless some more abstract term for these payoffs can be found. Table 8.1 shows ways of instantiating the variables in the different cases. Figure 8.1 shows the hierarchy of relations among the various theories.

Consider once again an imaginary scientist who has a problem to solve. Our results can now provide guidance in theory construction. First, a problem must be identified as an adaptation problem. Making such an identification may not be an easy task. It involves determining that a process occurs over time such that something becomes better adapted to another. One explanation of something's becoming better adapted is that it is a consequence of a selective process.

The term "adaptation" has received a lot of attention in the philosophy of biology literature, and we wish to skirt most of those issues here. However, it is useful to relate what has been called "engineering fitness" (see, e.g., Burian 1983) and our usage of "adaptation" in an adaptation problem. A deer may be fit in the engineering sense because it can run swiftly: it is adapted to escaping predators. Running swiftly is property $P$ in our abstraction; predator speed is the critical environmental factor $F$. The interaction between $P$ and $F$ may cause the deer to benefit. Many such interactions lead to the population of deer becoming better adapted: they run more swiftly and escape predators. Thus, two senses of adaptation are present: the one called "engineering fitness" that focuses on the relation of the organism *being* adapted to its environment, and the sense of population *becoming* better adapted by the accumulation of such adapted properties.

Both senses of adaptation are found in the immune system and brain cases as well. The fit between a given antibody and an antigen is the "engineering fitness" that contributes to the body's *becoming* adapted to fight that antigen. The recognition of a signal by a neuron group is the "match" (Edelman's term) that corresponds to "engineering fitness"; reinforcement of that group with numerous such matches produces adaptation in the brain, which *becomes* better able to respond more efficiently to such types of signals.

Adaptation problems can be identified in two different types of situations – by observing a *process* or observing an *adapted product*. First, a process may appear to be an adaptation process because something is changing so as to better cope with an environmental condition. Alternatively, all that may be noted is the prevalence of some property with good engineering fitness. Either finding is a reason to consider instantiating a selection type theory as an explanation. If a selection theory is applicable, then the adaptive process or the presence of the adaptation will be explained by a process that chooses individuals that have preexisting properties that display relative engineering fitness in a given environmental situation.

Historically, another type of theory has been proposed to solve adaptation problems: instructive theories. Instructive theories lack the set of variants prior to interaction with the environment. Instead, a mechanism exists for receiving instructions from the environment in order to construct the adapted form, in an engineering sense; the waste of nonadapted variants is thus avoided. But biologists have not been very successful in proposing instructive theories with such mechanisms. The theory for the inheritance of acquired characteristics is an example of an instructive theory in which a mechanism has never been successfully developed for acquiring adapted characters. The template theory of antibody formation (Pauling 1940) did provide a mechanism by which the antigen served as a template (provided instruction) for antibody formation; however, it was disproved once the details of protein synthesis were discovered in molecular biology.

In the cases of evolutionary adaptation and antibody formation, selection theories rather than instructive theories have been successful. Although numerous additions have been made to the clonal selection theory to explain antibody formation, a selection process remains central (Golub 1981, p. 9). As Edelman notes, "It is clear from both evolutionary and immunological theory ... that in facing an unknown future, the fundamental requirement for successful adaptation is preexisting diversity" (Edelman and Mountcastle 1978, p. 56). Such diversity is a precondition for a selection process. (For further discussion of instructive theories, see Edelman and Mountcastle 1978; Darden 1982a, 1987; Piattelli-Palmarini 1986.)

Sometimes a problem may be incorrectly classed as an adaptation problem. What is now called "enzyme induction" was originally called "enzyme adaptation" (Schaffner 1974a). When, for example, lactose is introduced in a medium with certain bacteria, the bacteria begin synthesizing enzymes to utilize it. It looked as if the bacteria were acquiring a new character to adapt to the new environment. However, this case was explained in terms of induction (or derepression) of genes already present to begin producing the needed enzymes. Thus, induction processes share features with adaptation processes and must be distinguished from them in order to determine what kinds of theories to invoke. In induction, a very specific preprogrammed response is initiated; it is a less creative process than are selective and instructive processes, which can give rise to new types of adapted forms.

Once an adaptation problem has been identified, then our analysis suggests considering a selection theory as an explanation for it. Instantiating the variables in our abstraction is one way of constructing a selection theory. Table 8.1 provides the blanks to be filled, and the abstraction in Section 8.4 provides the outline of the theory. Some variables may be instantiated in different ways (e.g., at different levels of organization) to produce alternative hypotheses, which can then be tested to confirm or disconfirm them.

Using a selection theory in a new case provides a measure of "plausibility by analogy" (Darden 1976, p. 145). By invoking a type of process that has been confirmed to apply in other biological cases, a measure of plausibility is provided for the new hypothesis – it is like processes known to occur elsewhere. But "plausibility by analogy" does not substitute for direct empirical evidence. Of course, a new theory will require testing against the data in its own domain. Instantiating a type of theory, even a previously successful type, is no guarantee that a theory will be confirmed in a new case.

This work demonstrates that at least one abstraction for an important type of theory can be constructed. The extent to which other types of problems and types of theories can be found and abstracted is an important area for future work for historians and philosophers of science. Such work holds the promise of providing "compiled hindsight" (Darden 1987) from the history of science, which may prove useful to current researchers in constructing new theories and, perhaps, providing insights into types of processes common in the natural world in which we live.

203

REFERENCES

Ada, G. L. and N. Gustav (1987), "The Clonal Selection Theory," *Scientific American* 257: 62–69.

Arnold, A. J. and K. Fristrup (1982), "The Theory of Evolution by Natural Selection: A Hierarchical Expansion," *Paleobiology* 8: 113–129.

Ayala, F. J. and T. Dobzhansky (eds.) (1974), *Studies in the Philosophy of Biology*. Berkeley, CA: University of California Press.

Beatty, John (1980), "What's Wrong with the Received View of Evolutionary Theory?" in P. D. Asquith and R. N. Giere (eds.), *PSA 1980*, v. 2. East Lansing, MI: Philosophy of Science Association, pp. 397–426.

Brandon, Robert N. (1980), "A Structural Description of Evolutionary Theory," in P. D. Asquith and R. N. Giere (eds.), *PSA 1980, v. 2*. East Lansing, MI: Philosophy of Science Association, pp. 427–439.

Brandon, Robert N. (1990), *Adaptation and Environment*. Princeton, NJ: Princeton University Press.

Burian, Richard (1983), "Adaptation," in Marjorie Grene (ed.), *Dimensions of Darwinism*. Cambridge: Cambridge University Press, pp. 287–314.

Burnet, F. M. (1957), "A Modification of Jerne's Theory of Antibody Production Using the Concept of Clonal Selection," *The Australian Journal of Science* 20: 67–69.

Cain, Joseph A. and Lindley Darden (1988). "Hull and Selection," *Biology and Philosophy* 3: 165–171.

Changeux, J. P. (1985), *Neuronal Man: The Biology of Mind*. New York: Pantheon, Random House.

Conrad, M. (1976), "Complementary Molecular Models of Learning and Memory," *BioSystems* 8: 119–138.

Darden, Lindley (1976), "Reasoning in Scientific Change: Charles Darwin, Hugo de Vries, and the Discovery of Segregation," *Studies in the History and Philosophy of Science* 7: 127–169.

Darden, Lindley (1980), "Theory Construction in Genetics," in T. Nickles (ed.), *Scientific Discovery: Case Studies*. Dordrecht: Reidel, pp. 151–170.

Darden, Lindley (1982a), "Artificial Intelligence and Philosophy of Science: Reasoning by Analogy in Theory Construction," in T. Nickles and P. Asquith (eds.), *PSA 1982*, v. 2. East Lansing, MI: Philosophy of Science Association, pp. 147–165.

Darden, Lindley (1983), "Reasoning by Analogy in Scientific Theory Construction," in R. S. Michalski (ed.), *Proceedings of the 1983 International Machine Learning Workshop*. Urbana, IL: Department of Computer Science, University of Illinois, pp. 32–40.

Darden, Lindley (1986), "Reasoning in Theory Construction: Analogies, Interfield Connections, and Levels of Organization," in P. Weingartner and G. Dorn (eds.), *Foundations of Biology*. Vienna, Austria: Holder-Picher-Tempsky, pp. 99–107.

Darden, Lindley (1987), "Viewing the History of Science as Compiled Hindsight," *AI Magazine* 8 (2): 33–41.

Darwin, Charles ([1859] 1966), *On the Origin of Species, A Facsimile of the First Edition*. Cambridge, MA: Harvard University Press.

Edelman, Gerald and V. Mountcastle (1978), *The Mindful Brain: Cortical Organization and the Group Selective Theory of Higher Brain Function*. Cambridge, MA: MIT Press.

Genesereth, Michael (1980), "Metaphors and Models," *Proceedings of the First Annual National Conference on Artificial Intelligence*. Menlo Park, CA: American Association for Artificial Intelligence, pp. 208–211.

Gentner, Dedre (1983), "Structure Mapping – A Theoretical Framework for Analogy," *Cognitive Science* 7: 155–170.

Golub, E. (1981), *The Cellular Basis of Immune Response*, 2nd ed. Sunderland, MA: Sinauer Associates.

Gould, Stephen J. and Elisabeth S. Vrba (1982), "Exaptation – A Missing Term in the Science of Form," *Paleobiology* 8: 4–15.

Greiner, Russell (1985), "Learning by Understanding Analogies," Ph.D. Dissertation, Department of Computer Science, Stanford University, Stanford, CA.

Hanson, Norwood Russell ([1961] 1970), "Is There a Logic of Scientific Discovery?" in H. Feigl and G. Maxwell (eds.), *Current Issues in the Philosophy of Science*. New York: Holt, Rinehart and Winston. Reprinted in B. Brody (ed.), *Readings in the Philosophy of Science*. Englewood Cliffs, NJ: Prentice-Hall, pp. 620–633.

Holland, John H. (1975), *Adaptation in Natural and Artificial Systems*. Ann Arbor, MI: University of Michigan Press.

Holyoak, Keith J. and Paul Thagard (1995), *Mental Leaps: Analogy in Creative Thought*. Cambridge, MA: MIT Press.

Hull, David (1980), "Individuality and Selection," *Annual Review of Ecology and Systematics* 11: 311–332.

Hull, David ([1981] 1984), "Units of Evolution: A Metaphysical Essay," in U. L. Jensen and R. Harré (eds.), *The Philosophy of Evolution*. Brighton: Harvester Press, pp. 23–44. Reprinted in R. N. Brandon and R. M. Burian (eds.), *Genes, Populations, and Organisms: Controversies over the Units of Selection*. Cambridge, MA: MIT Press, pp. 142–160.

Jerne, Niels K. (1955), "The Natural-Selection Theory of Antibody Formation," *Proceedings of the National Academy of Sciences (USA)* 41: 849–857.

Jerne, Niels K. (1966), "The Natural Selection Theory of Antibody Formation: Ten Years Later," in J. Cairns, G. S. Stent, and J. D. Watson (eds.), *Phage and the Origins of Molecular Biology*. Cold Spring Harbor, NY: Cold Spring Harbor Laboratory of Quantitative Biology, pp. 301–312.

Kitcher, Philip (1981), "Explanatory Unification," *Philosophy of Science* 48: 507–531.

Kitcher, Philip (1987), "Ghostly Whispers: Mayr, Ghiselin, and the 'Philosophers' on the Ontological Status of Species," *Biology and Philosophy* 2: 184–192.

Korf, R. E. (1985), "An Analysis of Abstraction in Problem Solving," in J. J. Pottmyer (ed.), *Proceedings of the 24th Annual Technical Symposium*. Gaithersburg, MD: Washington, DC Chapter of the ACM, June 20, 1985, pp. 7–9.

Lederberg, Joshua ([1959]1961), "Genes and Antibodies," *Science* 129: 1649–1653. Reprinted with a postscript in *Stanford Medical Bulletin* 19: 53–61.

Lewontin, Richard C. (1970), "The Units of Selection," *Annual Review of Ecology and Systematics* 1: 1–18.

205

Lloyd, Elizabeth (1984), "A Semantic Approach to the Structure of Population Genetics," *Philosophy of Science* 51: 242–264.

Pauling, Linus (1940), "A Theory of the Structure and Process of Formation of Antibodies," *Journal of the American Chemical Society* 62: 2643–2657.

Piattelli-Palmarini, M. (1986), "The Rise of Selection Theories: A Case Study and Some Lessons from Immunology," in W. Demopoulos and A. Marras (eds.), *Language Learning and Concept Acquisition: Foundational Issues*. Norwood, NJ: Ablex Publishing Co., pp. 117–130.

Schaffner, Kenneth (1974a), "Logic of Discovery and Justification in Regulatory Genetics," *Studies in the History and Philosophy of Science* 4: 349–385.

Schaffner, Kenneth (1980), "Theory Structure in the Biomedical Sciences," *The Journal of Medicine and Philosophy* 5: 57–97.

Shapere, Dudley (1974b), "On the Relations Between Compositional and Evolutionary Theories," in F. J. Ayala and T. Dobzhansky (eds.), *Studies in the Philosophy of Biology*. London: Macmillan, pp. 187–201.

Shapere, Dudley (1977), "Scientific Theories and Their Domains," in F. Suppe (ed.), *Structure of Scientific Theories,* 2nd ed. Urbana, IL: University of Illinois Press, pp. 518–562.

Shapere, Dudley and Gerald Edelman (1974), "A Note on the Concept of Selection," in F. J. Ayala and T. Dobzhansky (eds.), *Studies in the Philosophy of Biology*. London: Macmillan, pp. 202–204.

Sober, Elliott (1984), *The Nature of Selection*. Cambridge, MA: MIT Press.

Suppe, Frederick (1979), "Theory Structure," in P. Asquith and H. Kyburg (eds.), *Current Research in Philosophy of Science*. East Lansing, MI: Philosophy of Science Association, pp. 317–338.

Talmadge, D. W. (1957), "Allergy and Immunology," *Annual Review of Medicine* 8: 239–256.

Talmadge, D. W. (1986), "The Acceptance and Rejection of Immunological Concepts," *Annual Review of Immunology* 8: 239–256.

Thompson, Paul (1983), "The Structure of Evolutionary Theory: A Semantic Approach," *Studies in the History and Philosophy of Science* 14: 215–229.

Vrba, Elisabeth S. and Stephen J. Gould (1986), "The Hierarchical Expansion of Sorting and Selection: Sorting and Selection Cannot Be Equated," *Paleobiology* 12: 217–228.

Vries, Hugo de ([1903–04] [1909–1910] 1969), *The Mutation Theory*. 2 vols. Translated by J. B. Farmer and A. D. Darbishire. New York: Kraus Reprint Company.

Williams, Mary (1970), "Deducing the Consequences of Evolution: A Mathematical Model," *Journal of Theoretical Biology* 29: 343–385.

# 9

# Strategies for Anomaly Resolution

## Diagnosis and Redesign[1]

### 9.1 INTRODUCTION

Understanding the growth of scientific knowledge has been a major task in philosophy of science. No successful general model of scientific change has been found; attempts were made by, for example, Kuhn (1970), Toulmin (1972), Lakatos (1970), and Laudan (1977). A different approach is to view science as a problem-solving enterprise. The goal is to find both general and domain-specific heuristics (i.e., reasoning strategies) for problem solving. Such heuristics produce plausible but not infallible results (Nickles 1987; Thagard 1988).

Viewing science as a problem-solving enterprise and scientific reasoning as a special form of problem solving owes much to cognitive science and artificial intelligence (AI) (e.g., Langley et al. 1987). Key issues in AI are representation and reasoning. More specifically, AI studies methods for representing knowledge and methods for manipulating computationally represented knowledge. From the perspective of philosophy of science, the general issues of representation and reasoning become how to represent scientific theories and how to find strategies for reasoning in theory change. Reasoning

[1] The chapter was originally published as Darden, Lindley (1992), "Strategies for Anomaly Resolution," in R. Giere (ed.), *Cognitive Models of Science*, Minnesota Studies in the Philosophy of Science, v. 15. Minneapolis, MN: University of Minnesota Press, pp. 251–273. It is reprinted here with permission of the University of Minnesota Press. This work was supported by a General Research Board Award from the Graduate School of the University of Maryland and by the U.S. National Science Foundation Grant RII-9003142. The AI implementation was done by Dale Moberg and John Josephson with help from Dean Allemang at the Laboratory for Artificial Intelligence Research at Ohio State University. Thanks to John Josephson and Dale Moberg for their helpful comments on an earlier draft of this chapter.

in theory change is viewed as problem solving, and implementations in AI computer programs provide tools for investigating methods of problem solving. This approach is called "computational philosophy of science" (Thagard 1988).

Huge amounts of data are now available in online databases. The time is now ripe for automating scientific reasoning, first, to form empirical generalizations about patterns in the data (Langley et al. 1987); then, to construct new explanatory theories; and, finally, to improve them over time in the light of anomalies. Thus, the development of computational models for doing science holds promise for automating scientific discovery in areas where large amounts of data overwhelm human cognitive capacities (Schaffner 1986a; Morowitz and Smith 1987). The goal is to devise good methods for doing science, whether or not the methods are ones actually used by humans. The goal is not the simulation of human scientists but rather the making of discoveries about the natural world, using methods that extend human cognitive capacities. Thus, computational models allow exploration of methods for representing scientific theories, as well as methods for reasoning in theory formation, testing, and improvement. Such exploration holds the promise of making philosophy of science an experimental science. The models can be manipulated to allow "what if" scenarios to be explored; that is, experiments on scientific knowledge and reasoning can be done.

The reasoning method to be discussed in this chapter occurs in the context of resolving anomalies for theories. When a theory makes a prediction that fails, an empirical anomaly for the theory results. There are strategies that can be followed systematically to resolve the anomaly. The next section discusses prior work by philosophers of science on anomalies. Then systematic strategies for anomaly resolution will be proposed. Reasoning in anomaly resolution may be compared to a diagnostic reasoning and therapy planning task: find the ailing part of the theory and propose a way to fix it. The analogy between diagnostic reasoning and reasoning in anomaly resolution allows AI work on "model-based diagnosis" (e.g., Davis and Hamscher 1988) to be used to build computational models of theory change. The final sections describe a pilot AI program to implement an episode of anomaly resolution. A portion of Mendelian genetic theory is represented and methods for localizing a failing component of the theory are demonstrated. Finally, extensions to the current implementation are suggested.

To put the examination of strategies for anomaly resolution into a broader framework of research on scientific reasoning, consider the idea that there are stages in the development of a theory and strategies for making those

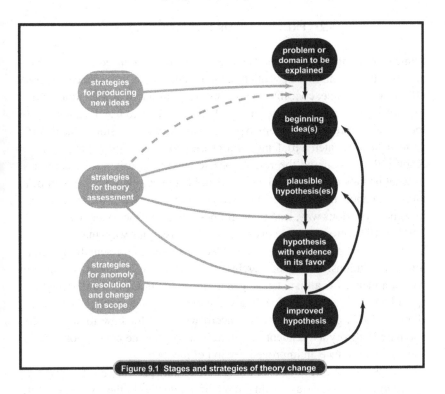

Figure 9.1 Stages and strategies of theory change

changes. Figure 9.1 is a diagram of stages and strategies of theory change. An episode may be considered to begin with a problem to be solved (problem-finding strategies are an issue not addressed here). Then, new ideas get built into plausible theories (i.e., hypotheses) which are tested and improved over time. Reasoning strategies guide these changes. The types of strategies include (a) strategies for producing new ideas, (b) strategies for theory assessment, and (c) strategies for anomaly resolution and change of scope. Change of scope can be (1) an expansion of the theory to explain items originally outside its domain (Shapere 1974a) of phenomena, (2) expansion to explain new items just discovered, or (3) specialization to exclude anomalies and preserve adequacy within a narrower domain. Anomaly resolution strategies are the focus in this chapter. Strategies for producing new ideas and strategies for theory assessment will be briefly mentioned. (All three types of strategies are the subject of Darden 1991.)

To make it easy to recognize the names of specific strategies in this discussion, they will be **boldfaced**.

## 9.2 PREVIOUS WORK ON ANOMALIES

Somewhat surprisingly, methods for anomaly resolution received compar-atively little attention in twentieth-century philosophy of science. Popper (1965), for example, concentrated on falsifying instances as indicators of the inadequacy of a theory, but he gave no hints as to how to use the anomaly to localize and correct the problem to produce an improved version of the theory. Instead, he advocated a trial and error process in "conjecturing" hypotheses. Kuhn (1970) discussed the role of puzzle solving in driving the activities in what he called "normal science." Also, he argued, the accumulation of a number of anomalies (puzzles?) sometimes provoked crises; such anomalies, in some mysterious way, lead to the proposal of a new theory or "paradigm." Laudan (1977) proposed a set of categories for classifying ways that anomalies become solved problems, but he provided no strategies for actually generat-ing such solutions. Instead of searching for a method of localizing a problem within a theory (as a first step to generating a solution), Laudan argued for spreading the blame for an anomaly evenly among the parts of the theory (Laudan 1977, p. 43). Laudan's concern was only for how to weight the anomaly in theory assessment, not with how to generate new hypotheses to solve it and produce an improved version of the theory.

My discussion of the role that an anomaly may play in *generating alter-native hypotheses* contrasts with the role of anomalies in theory assessment, which was the focus of the philosophers of science mentioned previously. One step in anomaly resolution is localization; potential sites of failure within the theoretical components need to be found. The idea of localization of problems within a theory or theoretical model in the light of an anomaly has received some attention (e.g., Glymour 1980; Nickles 1981; Darden 1982b; Wimsatt 1987). However, finding one or more plausible locations is only the first step in resolving an anomaly. How the new hypotheses are generated is the next stage of the process of anomaly resolution. Lakatos (1976) suggested heuristics for use in mathematics to improve conjectures in light of counter-instances; how-ever, when he (1970) discussed scientific reasoning, he spoke only vaguely of domain-specific heuristics associated within particular research programs. Shapere (1974a) suggested that simplifications made in the early stages of the-ory development were likely areas for hypothesis formation, once anomalies arise. Wimsatt (1987) suggested how mechanical and causal models might aid in forming hypotheses for resolving anomalies that arise for them; his analysis extended that of Hesse (1966). She suggested that unexplored areas of an analogy used in the original construction of a theory might function in forming hypotheses to resolve an anomaly at a later stage.

210

Researchers in AI are developing methods for determining which parts of a complex, explanatory system are involved in the explanation of particular data points. Their techniques for doing "credit assignment" are relevant to the problem of localizing plausible sites for modification (or even all possible sites, given an explicit representation of all knowledge in the system), given a particular anomalous data point (Charniak and McDermott 1985, p. 634). The localization of an anomaly in a theoretical component can be compared to diagnosing a fault in a device or a disease in a patient. Thus, AI methods for diagnostic reasoning can be applied to anomaly resolution (Darden 1990). After one or more components have been identified as potential sites where the theory may be failing, then changing the components is like a redesign task (Karp 1989, 1990). Reasoning in design involves designing something new to fulfill a certain function, in light of certain constraints. Redesigning theoretical components involves constructing a component that will account for the anomaly, with the constraints of preserving the unproblematic components of the theory and producing a theory that satisfies criteria of theory assessment. Especially important criteria in anomaly resolution are systematicity and lack of ad hocness. It is important that the new theoretical component be systematically connected with the other theoretical components and not be merely an ad hoc addition that serves only to account for the anomaly.

Not surprisingly, AI research has shown that implementing methods for improving faulty modules is more difficult than the first step of localizing the problem. Creatively constructing new hypotheses is, in general, a more difficult task than discovering the need for a new hypothesis. To better understand the reasoning in anomaly resolution, we need to distinguish (a) the problem of localization, and (b) the problem of generating a new hypothesis to account for the anomaly. This chapter discusses each of these issues.

## 9.3   STRATEGIES FOR ANOMALY RESOLUTION

I am combining what I call "strategies for anomaly resolution" and "strategies for change of scope." What is inside and what is outside the domain of a theory may be open to debate. Thus, a given item, such as an experimental result, might be considered an anomaly for the theory because the item is inside the domain to be explained by that theory. Alternatively, the same item, it might be argued, is not an anomaly for the theory because that item is outside the scope of the domain of the theory. Perhaps some other theory is expected to account for it. If theory construction begins with a large domain to be explained

211

and then anomalies arise, narrowing the scope of the domain may aid in anomaly resolution. If a proposed hypothesis is very general and an anomaly arises for it, then one strategy is to **specialize the over-generalization** and exclude the anomalous item from the domain. On the other hand, if theory construction begins with a hypothesis that applies to a very narrow domain, then theory change may occur with expansion of the domain, by including items originally not part of it. Because of this close relationship between the scope of the domain and the identification of an anomaly, strategies for anomaly resolution and change of scope are closely related and are considered together here.

The term "anomaly" usually refers to a problem posed by data within the theory's domain that the theory cannot explain. Often an anomaly is generated when a prediction fails to be confirmed. However, theories may face other kinds of problems besides those posed by empirical anomalies. A theory may be incomplete, even though no empirical anomaly indicates that it is incorrect. (For more discussion of incorrectness versus incompleteness, see Shapere 1974a; Leplin 1975). In addition to problems of incorrectness or incompleteness, a theory may face conceptual problems of various kinds, such as determining the nature of a newly proposed theoretical entity. Discussing general strategies for resolving all the kinds of problems that a theory may face would make this section too lengthy. Hence, the focus here will be on empirical anomalies, due to failed predictions, that seem to show that a theory is incorrect.

A general strategy for anomaly resolution entails several stages. Table 9.1 shows four primary stages: (1) **confirm the anomalous data**, (2) **localize the problem**, (3) **resolve the anomaly**, and (4) **assess the resulting theory**. The following subsections discuss these stages of anomaly resolution.

### 9.3.1   Confirm Anomalous Data or Problem

Before efforts are made to resolve an anomaly, the correctness of the anomalous data needs to be confirmed. If the data is wrong, then no anomaly exists and the subsequent steps in anomaly resolution need not be taken. If experimental error can be blamed for a failed prediction, then the anomaly is resolved without further work or change in the theory.

### 9.3.2   Localize the Anomaly

If the anomaly can be localized outside the theory, then the theory will not need to be modified. One way to localize the anomaly elsewhere is to argue that it

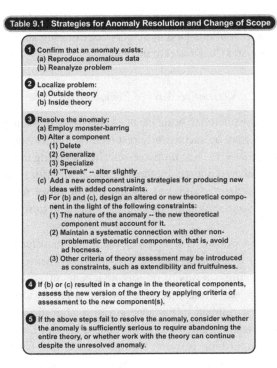

Table 9.1  Strategies for Anomaly Resolution and Change of Scope

**1** Confirm that an anomaly exists:
  (a) Reproduce anomalous data
  (b) Reanalyze problem

**2** Localize problem:
  (a) Outside theory
  (b) Inside theory

**3** Resolve the anomaly:
  (a) Employ monster-barring
  (b) Alter a component
    (1) Delete
    (2) Generalize
    (3) Specialize
    (4) "Tweak" -- alter slightly
  (c) Add a new component using strategies for producing new ideas with added constraints.
  (d) For (b) and (c), design an altered or new theoretical component in the light of the following constraints:
    (1) The nature of the anomaly -- the new theoretical component must account for it.
    (2) Maintain a systematic connection with other non-problematic theoretical components, that is, avoid ad hocness.
    (3) Other criteria of theory assessment may be introduced as constraints, such as extendibility and fruitfulness.

**4** If (b) or (c) resulted in a change in the theoretical components, assess the new version of the theory by applying criteria of assessment to the new component(s).

**5** If the above steps fail to resolve the anomaly, consider whether the anomaly is sufficiently serious to require abandoning the entire theory, or whether work with the theory can continue despite the unresolved anomaly.

is outside the scope of the domain of the theory. Thus, the phenomenon is not an anomaly for the theory after all. Another way of localizing the problem without requiring theory change is to show that the instance is an unusual one; while it is inside the domain of the theory, it does not represent a typical phenomenon. This method, which I call "monster-barring," is discussed in the next subsection.

If the anomaly is to be localized as a failing within the theory, then various methods exist for forming hypotheses about which component(s) is failing. One method for localization is to represent the theory as typical steps in typical processes. Localization is achieved by determining which step failed in the anomalous situation. The anomaly is localized in the failing step. Localization may require additional information not supplied by the experiment that generated the anomaly. Such additional information is needed to determine which steps in the process occurred and which did not. That information may have to be generated by additional experiments to detect which steps were reached and which were not. This kind of localization is analogous to a diagnostic reasoning task. It is like localizing a faulty module in a device by seeing, for example, which module has electricity coming in but puts none out.

Queries about inputs and outputs to the steps of the process aid in localization. (For an AI discussion of this kind of diagnostic reasoning about devices, see Sembugamoorthy and Chandrasekaran 1986.)

### 9.3.3   Alternative Ways of Removing an Anomaly

If an anomaly is considered to be inside the scope of the domain of the theory, then the anomaly may be successfully resolved by a theory in at least three ways (listed in Table 9.1): 3(a) **monster-barring**, showing that the anomaly is an exception that does not require theory change; 3(b) **altering an existing component** of the theory; and 3(c) **adding one or more new theoretical components**. This subsection discusses each of these in turn.

9.3.3.1. MONSTER-BARRING. I have taken the term "monster-barring" from Lakatos (1976). It is a strategy that he introduced for mathematical reasoning. **Monster-barring** is a way of preserving a generalization in the face of a purported exception: If the exception can be barred as a monster – that is, shown not to be a threat to the generalization after all – then it can be barred from necessitating a change in the generalization. Lakatos was concerned with distinguishing between legitimate exception-barring instances and illegitimate barring of instances that really did require a theory change. The strategy I label "monster-barring" is what he called a legitimate instance of "exception barring." (His "local counter-examples" would be included in what I call "model anomalies," to be discussed later. I use "model" to indicate that the anomaly is not really an exception but is an exemplary case that will, itself, serve as a model for other such cases. However, because Lakatos was not discussing scientific theories, the term "model" would have been less appropriate for the mathematical, "local" counter-examples that he discussed. Anomalies that serve to falsify an entire theory would correspond, I think, to what Lakatos called "global counter-examples.")

Two kinds of monster anomalies are possible: unique ones and those that belong to classes. Unique ones can often be difficult to account for. More interesting monsters are ones that occur often and, thus, can be seen as instances of a malfunction class. Both kinds of monsters, unique ones and malfunction classes, are barred from necessitating a change in the theory. They are explained or, better, explained away. An account is given of what went wrong in the normal process so as to produce the anomaly.

9.3.3.2. ALTER A COMPONENT OF THE THEORY. To have a term to contrast to "monster" anomalies, I introduced the term "model" anomalies, by analogy with "model" organisms (Darden 1990). Model anomalies require a change in the theory. The model anomalies then are shown to be normal, that is, not actually anomalous at all. They serve as models of normal types of processes that are commonly found. Model anomalies are resolved either by changing an existing theoretical component or by adding a new one or both (e.g., one component may be specialized and an additional new component added).

Table 9.1 divides strategies for changing a theoretical component into two categories: 3(b) **alter a component**, and 3(c) **add a new component**. Changing a component already present is usually an easier task than adding an entirely new component. A number of different strategies exist for altering an existing component. They are listed in Table 9.1, Step 3(b) and will now be discussed.

**Deleting a component** is obviously an easy kind of change to make. If the deleted component explains other items in the domain in addition to the anomalous ones leading to its deletion, then some other component(s) will have to be modified or added to account for those items. If deleting a component leaves a problematic "hole" in the representation of the theory, then one or more other components may have to be added to replace the deletion.

Another method for slightly changing a theoretical component in the light of an anomaly is to **generalize or specialize** the component. Generalization and specialization are methods of modifying a working hypothesis that have been extensively used in studies of induction and concept learning in AI (Mitchell 1982; Dietterich et al. 1982). **Generalization** expands the scope of a hypothesis; **specialization** narrows the scope. (Relations among generalization, abstraction, and simplification are discussed in Darden 1987.)

If bold, general, simplifying assumptions marked the beginning stages of theory construction, then **specialization** and **complication** will be likely strategies to use as anomalies arise. If the theory was constructed originally in a conservative way, carefully specialized to apply to a narrow domain, then **generalization** will be a way of expanding the scope of its domain, even if no specific anomaly is at issue.

A more systematic strategy for anomaly resolution is, at the outset, to consider explaining the problematic data in both the most general and the most specialized ways consistent with the data. Then, alternative hypotheses – varying along a spectrum from general to specific – become candidates for future development. This method of systematically considering a range of hypotheses, from the most general to the most specific, is called the "version space" method of hypothesis formation in AI (Mitchell 1982).

The version space is the space of all hypotheses between the most general and the most specific that account for a given set of data. Then, refinements are made in light of new data points or anomalies. Exceptions to generalizations are resolved by adding conditions to make a general hypothesis more specific. A new instance not covered by a specific hypothesis drives the formation of a more general one by dropping conditions from the specific one. I doubt that scientists typically engage in such systematic generation of alternative hypotheses; perhaps they should consider doing so.

**Tweaking** is a term for the strategy of **changing a component slightly** to account for an anomalous or a new instance. Schank (1986, p. 81) used the term similarly, when he suggested explaining anomalies by invoking past explanation patterns and changing the patterns slightly to apply to the new situation. However, he was not discussing explanations in science. I am using the term as an eclectic class of strategies for ways of changing a theoretical component. "Tweaking" strategies are strategies for making slight changes in the theory that do not fit into any of the more specific ways of making slight changes discussed previously. An example is slightly changing the parameters in a quantitative model to account for a quantitative anomaly that is just a little off from the predicted value.

9.3.3.3. ADD SOMETHING NEW. The strategies for altering a component produce slightly new hypotheses. However, those strategies may all prove to be inadequate and a new component of the theory may be needed. In such a case, strategies for producing new ideas will need to be invoked. Such strategies include **reasoning by analogy** (Hesse 1966; Darden 1982a; Darden and Rada 1988b; Holyoak and Thagard 1989); **reasoning by postulating an interfield connection** (Darden and Maull 1977; see Chapter 5, this book; Darden and Rada 1988a); **reasoning by postulating a new level of organization** (Darden 1978); **reasoning by invoking an abstraction** (Darden 1987; Darden and Cain 1989; see Chapter 8, this book); **reasoning by conceptual combination** (Thagard 1988); and **abductive assembly** of a new composite hypothesis from simpler hypothesis fragments (Josephson, Chandrasekaran, Smith, and Tanner 1987). Detailed discussion of these strategies would take us too far afield (several are discussed in Darden 1991, Ch.15). If some components of the theory are not candidates for modification, then a consistent and systematic relation with them must be maintained while adding new components. This consistency is a constraint not present in the use of strategies for producing new ideas when an entirely new theory is constructed.

**Table 9.2   Criteria for Theory Assessment**

1. internal consistency and absence of tautology
2. systematicity
3. clarity
4. explanatory adequacy
5. predictive adequacy
6. scope and generality
7. lack of ad hocness
8. extendibility and fruitfulness
9. relations with other accepted theories
10. metaphysical and methodological constraints, e.g., simplicity
11. relation to rivals

### 9.3.4   Assessing the Hypotheses to Resolve the Anomaly

Hypotheses proposed as modified theoretical components have to be evaluated, using the criteria of theory assessment. Such criteria are listed in Table 9.2; they will not be discussed in detail here (see Darden 1991, Ch. 15). The stages diagram in Figure 9.1 can be used to represent anomaly resolution. The anomaly is the problem to be solved. The strategies for localization, coupled with strategies for altering a component or the strategies for producing new ideas, provide one or more hypotheses as candidates for the modified theory components. Then the criteria for theory assessment are used in evaluating the hypotheses, with the added constraint that the new components must be compatible with the unmodified components; the criterion of a systematic relation among all the components of the theory ("systematicity") becomes important. Also especially important are the criteria of explanatory adequacy, predictive adequacy, and the lack of ad hocness (listed in Table 9.2). The new component added to resolve the anomaly should improve the theory in accordance with these criteria. What counts as a legitimate addition to the theory and what is an illegitimate ad hoc change may be a matter of debate, especially when a new component is first proposed to resolve an anomaly. Additional work will be necessary to test the explanatory and predictive adequacy of the newly proposed component and the altered theory of which it is a part.

### 9.3.5   Unresolved Anomalies

All of these strategies for resolving an anomaly may fail. In such a case, scientists working on the theory will have to decide whether the anomaly is sufficiently serious to require abandoning the entire theory or whether it can be shelved as a problem requiring resolution, while work on other parts of the theory continues. Again, no decisive criteria may be present to choose among these alternatives. Even if the entire theory is to be abandoned, the anomaly may well provide a pointer to the components of the theory most at fault and provide hints as to what a new theory should contain in order to avoid having the same anomaly.

### 9.4 REPRESENTATION AND IMPLEMENTATION OF ANOMALY RESOLUTION IN GENETICS

A subset of these strategies for anomaly resolution has been investigated in an AI system. Revising scientific knowledge is analogous to redesigning tools or other devices. Scientists make use of anomalies in diagnosing faults in a theory and then propose fixes for those faults. The perspective of "theory as device" allows AI representation and reasoning techniques to be applied to the representation of scientific theories and to the simulation of reasoning in anomaly resolution. A pilot AI system has been implemented to represent a portion of Mendelian genetics and simulate the process of resolving one monster anomaly for it.

### 9.4.1 *Representation of a Scientific Theory*

At least some scientific theories can be represented by a series of steps in a mechanistic process. Representing a theory as such a series of steps in a normal process provides a "schematic flow diagram" of the steps in a normal case. Mendelian genetics lends itself to this kind of representation. A series of steps in a normal hereditary process can be used to represent a normal Mendelian segregation process. The steps in a normal process of Mendelian segregation can best be illustrated by an example (Figure 9.2). If a pure yellow variety of pea (AA) is crossbred with a pure green variety (aa), in the next generation, all the peas will be hybrid yellow (symbolized by Aa; yellow is called "dominant"). If two hybrid yellows are mated (Aa x Aa), then the next generation produces the ratio of three yellow to one green (symbolized by AA + 2Aa + aa; AA and Aa both appear yellow). The ratio of 3:1 is produced given that (i) the Aa of the hybrid separate or segregate during the formation of germ cells (gametes) in a pure (i.e., nonblended) way; (ii) the two types of germ cells (A and a) form in equal numbers; (iii) the fertilization process is random; and (iv) all the types of zygotes are equally viable.

218

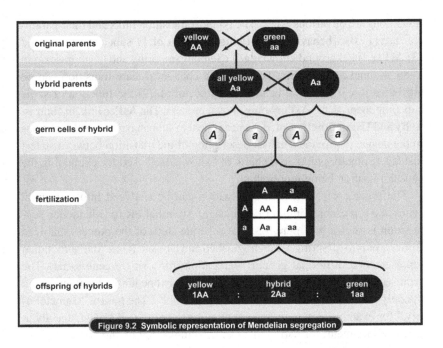

Figure 9.2 Symbolic representation of Mendelian segregation

This way of representing the process of Mendelian segregation can easily be put into a computational form using the functional representation (FR) language. FR was created to represent the functioning of devices. The goal in designing the language was to support the problem-solving activities of troubleshooting and of predicting changes in function in a device when there was a change in its components, structure, or constituent behaviors (Sembugamoorthy and Chandrasekaran 1986). The language has been used to represent both concrete devices (e.g., bodily systems in medical applications; manufactured devices and manufacturing processes for engineering applications) and abstract "devices" (e.g., plans and computer programs) (Sticklen 1987; Chandrasekaran, Josephson, and Keuneke 1986; Chandrasekaran, Josephson, Keuneke, and Herman 1989; Allemang 1990).

To construct a Functional Representation (FR), the system (the normal segregation process) must be analyzed into a sequence of states, with steps leading from state to state. The separate steps of the process to be represented are distinguished and labeled. The computational representation of the states supports operations that enable the system to assess (with the aid of a human user) whether the state has been entered. The transitions between states in an FR are typically considered causal links and the nature of the causal responsibility for the transition is specified. In the FR language, the

transitions are of various types. The types of transition links and their param-
eter are (1) **By** {behavior}, (2) **Using function of** {a subcomponent}, and
(3) **AsPer** {background knowledge} (Sembugamoorthy and Chandrasekaran
1986; Keuneke 1989). In other words, a change of state transition can be
attributed to (1) **By** some behavior that occurred, (2) the function of some
subcomponent, or (3) **AsPer** some known process. The **AsPer** link, in contrast
to **By** and **Using function of**, does not appeal to subbehaviors or subfunctions
in the device to provide a causal description of the transition between states;
instead, it merely points to a chunk of "knowledge" used for explaining the
causal transition being represented.

The process of Mendelian segregation can be analyzed into sequential
states and represented in the FR language. An initial FR for Mendelian segregation is shown in Figure 9.3. The separate steps of the process that lead
from parents to offspring are distinguished and labeled. It is a "flat" functional
representation of normal genetic segregation (for a single gene locus). The
representation is called "flat" because it is at only one hierarchical level, and
the only kind of transitions specified is **AsPer** links. The phrase "GametePu-
rity," for example, refers to one chunk of our conjectured knowledge about
germ-cell (gamete) formation that can be summarized by the rule that genes
separate (segregate) completely or "purely" when gametes form; there are
no intermediate or blended genes, only pure parental types. The other **AsPer**
links point off to additional chunks of knowledge according to which the
state transitions occur. A way to extend the representation and connect it to
representation of underlying cytological processes would be to change the
**AsPer** links to **By** or **Using function of**. This ability to package hierarchical
relations is a strength of the FR language that has not yet been exploited for
the Mendelian segregation representation.

An FR, once constructed, supports simulation. The representation can be
"run" by providing an input, the first state, and noting the output, the final
state. The final state (or perhaps some intermediate one) can be interpreted as
a prediction made by the theory. For example, the FR of normal Mendelian
segregation predicts that, given fertile hybrid parents that are mated, offspring
will be produced in the ratio of three dominant to one recessive. This qualitative simulation provides a test of the adequacy of the representation and
also allows simulation of the failure of a step in an anomalous case. Such
simulation abilities are useful during anomaly resolution for showing that the
hypothesized failing step produces the anomaly.

The analogy of a theory to a device is a useful one. The theory is a device
with component parts; the parts have functions that combine to support the
functions of the device as a whole. A theory has the functions of explanation

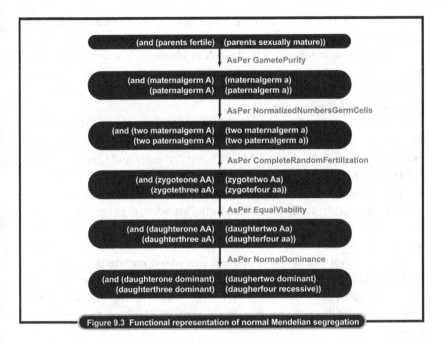

Figure 9.3 **Functional representation of normal Mendelian segregation**

and prediction. If the theory encounters an anomaly, then the theory is compared to a faulty device. Some part is failing and that theoretical component needs to be localized and shown to be capable of producing the fault. The FR language is a particularly useful computational tool for representing a theory and for supporting reasoning in anomaly localization (i.e., fault diagnosis).

### 9.4.2   Anomaly Resolution: Localization

An instance of the localization of a monster anomaly in Mendelian genetics has been implemented in a simple form. The normal Mendelian segregation process predicts that, when two hybrid (Aa) forms are crossed, a 3:1 ratio will be produced. For example, if hybrid yellow mice (Aa) are crossed (Aa + Aa), then three yellow to one non-yellow mice are, on average, predicted to be the ratios for the occurrence of those colors in the litters. The symbolic representation for the outcome is the following: AA + 2Aa + aa. Because those with A are dominant, this equation leads to the prediction of three yellow to one non-yellow.

When Cuénot did such a cross in 1905, he did not find the expected 3:1 ratios (Cuénot 1905). Instead, he found 2:1 ratios: 2Aa + 1aa, with no AA. Various hypotheses were proposed to account for the anomaly (Darden 1990,

1991, Ch. 8). The only one to be discussed here is the hypothesis that eventually proved to be the correct one: namely, that the AA combination is not a viable embryo or, in other words, the AA combination is a lethal gene combination. The localization of the failing step in normal segregation – namely, the "Equal Viability" step in Figure 9.3 – has been demonstrated, using the FR of Mendelian segregation.

The FR language supports diagnostic reasoning to localize faults in devices. The representation can automatically be converted into a diagnostic classification hierarchy (Bylander and Mittal 1986; Bylander and Chandrasekaran 1987). This hierarchy is a directed acyclic graph produced by a recursive descent parsing of the functional representation. The transition links (e.g., **AsPer**) serve as the basis for constructing "knowledge groups" whose values, along with the control messages (normally, an "establish and refine" message), govern the fault localization process. According to the establish-and-refine control strategy, only those nodes whose knowledge group satisfies an "establish" threshold, and which have received a "refine" message, will be pursued; other nodes are "rejected," and a saving in search effort is then obtained by not pursuing their subnodes (Moberg and Josephson 1990).

In the present implementation, the knowledge group generates queries that are asked of the human user of the program. Automatic use of the diagnostic system would involve a database that can be suitably queried for data, allowing the knowledge group node to establish or reject. The generation of useful and appropriate queries is the most significant contribution of the diagnostic hierarchy to the anomaly resolution process. The user's answers permit the system to localize a potential site of fault in the theory and thus to provide focus for the computationally costly process of theory redesign (Goel and Chandrasekaran 1989; Moberg and Josephson 1990).

The operation of the diagnostic hierarchy can be illustrated by tracing through the process of localizing the fault that produces the monster anomaly of 2:1 ratios. The system takes the FR and makes a classification hierarchy. To localize the fault in the hierarchy, the system queries the user about which states have been reached. Those queries are illustrated in Table 9.3. First, the need to engage in diagnostic reasoning is established by asking whether fertile sexually mature organisms have mated, and whether a three dominant to one recessive ratio was observed. The human user replies that the initial state (fertile and mature organisms mating) was entered, but that the final state of 3:1 ratios of a trait in the offspring population was not observed. This reply establishes the existence of an anomaly. Next, the system refines its diagnostic hypothesis about what part of the theory might plausibly be mistaken. At

| Table 9.3 List of Diagnostic Queries |
| --- |

Queries generated by the diagnostic hierarchy (the system's queries are in *italics*; the human user's answers are in **boldface**; explanatory comments are in [brackets]):

*parents fertile and parents sexually mature?* **yes**
[the initial state was reached]

*daughterone dominant, daughtertwo dominant, daughterthree dominant, daughterfour recessive?* **no**
[the final state was not reached; therefore there is an anomaly]

*daughterone AA, daughtertwo Aa, daughterthree aA, daughterfour aa?* **no**
[the next-to-final state was not reached, so the system proceeds to query about the previous one]

*zygote AA, zygote Aa, zygote aA, zygote aa?* **yes**
[this state was reached, so the problem is localized in the link "AsPer EqualViability"]

present, the system traverses the classification hierarchy from the final state toward the initial state, inquiring at each point whether the intermediate state was entered. In this case, localization occurs when the system is informed that four zygotes were produced but that no AA offspring were produced. Note in Figure 9.3 that "**AsPer** EqualViability" was the link before the failed step. This information establishes that the causal behavior referred to as "equal zygote viability" (**AsPer** EqualViability) is at fault (Moberg and Josephson 1990).

This pilot project shows that the causal process specified by part of the theory of the gene can be represented using the FR language. It also demonstrates how diagnostic reasoning about theory faults is supported by the availability of a compiler for a diagnostic classification hierarchy. The diagnostic queries carry out a search for knowledge needed to localize a fault. This search may reveal the need for additional experimental work in order to answer the queries. Historically, when Cuénot found the 2:1 anomaly in 1905, techniques did not exist to discriminate among the states of germ-cell formation, zygote formation, and development of zygotes into mature offspring. Later techniques provided experimental access to confirm that small embryos formed (i.e., zygotes formed), but that they did not develop into mature offspring (Castle and Little 1910; Kirkham 1919). The queries from such a diagnostic localization system could direct experimental work to discriminate between states and to determine which state was reached and which failed.

### 9.4.3 Extensions to the Current Implementation

A full implementation of all the anomaly resolution strategies discussed in Section 9.3 requires more than just strategies for localization. At the beginning of the process of anomaly resolution, queries are needed to try to resolve the anomaly without localizing it in the theory. In other words, the earlier steps in Table 9.1 need to be included in an implementation. Also, the difference between monster and model anomalies needs to be exploited by the system. In a monster anomaly case, the normal process is not questioned but its application to a particular anomalous instance is questioned. As in the Cuénot case, the normal process of segregation was not changed but its application to the hybrid yellow mice case was altered. The anomaly was resolved by showing that the AA combination was not viable. The FR representation could easily be extended to show that "breaking" the "EqualViability" link for AA forms would produce 2:1 ratios. In other words, the FR simulation capabilities could be used to show that the hypothesized failing state, if it is disabled, does in fact produce the resultant anomaly.

In model anomaly cases, not only localization is needed but also the theory (i.e., the representation of the normal process) must be changed. Such anomaly-directed redesign includes both proposing new hypotheses (i.e., new theoretical components) and testing to see that the old success of the theory is retained while the anomalous result is removed. The model anomaly resolution task can be formulated as an information-processing task:

> Given a representation of theory in the form of a series of causal steps that produce a given output and an anomaly (a failed predicted output) for that theory, construct a modified theory that no longer has the given anomaly and that retains previous explanatory successes.

This task is not uniquely specified in that many modified theories may be constructed that retain successes and eliminate the anomaly. Some of the different directions for theory modification are associated with different strategies. A goal for future work is to investigate alternative programs for theory redesign with respect to a sequence of anomalies and selection of strategies. The present implementation efforts have been directed toward finding a good representation for a theory's content and localizing potential failing sites. Such localization for a model anomaly case makes the redesign task simpler; however, producing the new hypotheses needed for redesign is a difficult task, given current AI techniques (Moberg and Josephson 1990). Such implementation will require considerable further effort and analysis.

The representation of a theory's central causal processes does not yet take advantage of several features of the FR language that would provide considerable power. The pilot project has not yet used a full analysis of components, structure, and behavior of the genetic system, nor the provision for simulation of normal genetic processes. An FR normally involves a hierarchy whose vertical organization provides a way of grouping details of state sequence, subcomponents, and knowledge. These groups then organize process sequences at different levels of detail. These in turn are the parts that can be selected as loci of failure and thus become targets for potential redesign. The strategy of using knowledge from another level of organization to form hypotheses at a given level is one strategy to be investigated. Decisions about hierarchical organization and abstraction will be made to facilitate localization and redesign. These will be investigated in the next phase of this research project.

## 9.5 CONCLUSION

This chapter has proposed that computational methods from artificial intelligence are fruitful for representing and reasoning about scientific theories. Sequential strategies for anomaly resolution suggest that quasi-algorithmic reasoning processes can be used systematically in theory refinement. A pilot project to implement anomaly resolution in an AI system has been discussed. This work shows the promise of the analogy between anomaly resolution in science and diagnosis and redesign in AI. Computational approaches provide a way for philosophers to do experiments on strategies for theory change.

### REFERENCES

Allemang, Dean (1990), *Understanding Programs as Devices*, Ph.D. Dissertation, Department of Computer and Information Sciences, The Ohio State University, Columbus, OH.

Bylander, T. and B. Chandrasekaran (1987), "Generic Tasks for Knowledge-Based Reasoning: The 'Right' Level of Abstraction for Knowledge Acquisition," *International Journal of Man-Machine Studies* 28: 231–243.

Bylander, T. and Sanjay Mittal (1986), "CSRL: A Language for Classificatory Problem Solving and Uncertainty Handling," *AI Magazine* 7(3): 66–77.

Castle, W. E. and C. C. Little (1910), "On a Modified Mendelian Ratio among Yellow Mice," *Science* 32: 868–870.

Chandrasekaran, B., John Josephson, and Anne Keuneke (1986), "Functional Representation as a Basis for Generating Explanations," *Proceedings of the IEEE Conference on Systems, Man, and Cybernetics*, Atlanta, GA, pp. 726–731.

Chandrasekaran, B., John Josephson, Anne Keuneke, and David Herman (1989), "Building Routine Planning Systems and Explaining Their Behaviour," *International Journal of Man-Machine Studies* 30: 377–398.

Charniak, Eugene and Drew McDermott (1985), *Introduction to Artificial Intelligence*. Reading, MA: Addison-Wesley.

Cuénot, Lucien (1905), "Les Races Pures et Leurs Combinaisons Chez Les Souris," *Archives de Zoologie Expérimentale et Générale* 4 Serie, T. 111: 123–132.

Darden, Lindley (1978), "Discoveries and the Emergence of New Fields in Science," in P. D. Asquith and I. Hacking (eds.), *PSA 1978*, v. 1. East Lansing, MI: Philosophy of Science Association, pp. 149–160.

Darden, Lindley (1982a), "Artificial Intelligence and Philosophy of Science: Reasoning by Analogy in Theory Construction," in T. Nickles and P. Asquith (eds.), *PSA 1982*, v. 2. East Lansing, MI: Philosophy of Science Association, pp. 147–165.

Darden, Lindley (1982b), "Aspects of Theory Construction in Biology," in *Proceedings of the Sixth International Congress for Logic, Methodology and Philosophy of Science.* Hanover: North Holland Publishing Co., pp. 463–477.

Darden, Lindley (1987), "Viewing the History of Science as Compiled Hindsight," *AI Magazine* 8(2): 33–41.

Darden, Lindley (1990), "Diagnosing and Fixing Faults in Theories," in J. Shrager and P. Langley (eds.), *Computational Models of Scientific Discovery and Theory Formation.* San Mateo, CA: Morgan Kaufmann, pp. 319–346.

Darden, Lindley (1991), *Theory Change in Science: Strategies from Mendelian Genetics.* New York: Oxford University Press.

Darden, Lindley and Joseph A. Cain (1989), "Selection Type Theories," *Philosophy of Science* 56: 106–129.

Darden, Lindley and Nancy Maull (1977), "Interfield Theories," *Philosophy of Science* 44: 43–64.

Darden, Lindley and Roy Rada (1988a), "Hypothesis Formation Using Part-Whole Interrelations," in David Helman (ed.), *Analogical Reasoning.* Dordrecht: Reidel, pp. 341–375.

Darden, Lindley and Roy Rada (1988b), "Hypothesis Formation Via Interrelations," in Armand Prieditis (ed.), *Analogica.* Los Altos, CA: Morgan Kaufmann, pp. 109–127.

Davis, Randall and Walter C. Hamscher (1988), "Model-Based Reasoning: Troubleshooting," in H. E. Shrobe (ed.), *Exploring Artificial Intelligence.* Los Altos, CA: Morgan Kaufmann, pp. 297–346.

Dietterich, Thomas G., B. London, K. Clarkson, and G. Dromey (1982), "Learning and Inductive Inference," in Paul R. Cohen and E. Feigenbaum (eds.), *The Handbook of Artificial Intelligence*, v. 3. Los Altos, CA: Morgan Kaufmann, pp. 323–511.

Glymour, Clark (1980), *Theory and Evidence.* Princeton, NJ: Princeton University Press.

Goel, Ashok and B. Chandrasekaran (1989), "Functional Representation of Designs and Redesign Problem Solving," in *Proceedings of the Eleventh International Joint Conference on Artificial Intelligence*, Detroit, MI, pp. 1388–1394.

Hesse, Mary (1966), *Models and Analogies in Science.* Notre Dame, Indiana: University of Notre Dame Press.

Holyoak, Keith J. and Paul Thagard (1989), "Analogical Mapping by Constraint Satisfaction," *Cognitive Science* 13: 295–355.

Josephson, J., B. Chandrasekaran, J. Smith, and M. Tanner (1987), "A Mechanism for Forming Composite Explanatory Hypotheses," *IEEE Transactions on Systems, Man, and Cybernetics,* SMC-17: 445–454.

Karp, Peter (1989), *Hypothesis Formation and Qualitative Reasoning in Molecular Biology*, Ph.D. Dissertation, Stanford University, Stanford, CA. (Available as a technical report from the Computer Science Department: STAN-CS-89-1263.)

Karp, Peter (1990), "Hypothesis Formation as Design," in J. Shrager and P. Langley (eds.), *Computational Models of Scientific Discovery and Theory Formation*. San Mateo, CA: Morgan Kaufmann, pp. 275–317.

Keuneke, Anne (1989), *Machine Understanding of Devices: Causal Explanation of Diagnostic Conclusions*, Ph.D. Dissertation, Department of Computer and Information Science, The Ohio State University, Columbus, OH.

Kirkham, W. B. (1919), "The Fate of Homozygous Yellow Mice," *Journal of Experimental Zoology* 28: 125–135.

Kuhn, Thomas (1970), *The Structure of Scientific Revolutions*. 2nd ed. Chicago, IL: University of Chicago Press.

Lakatos, Imre (1970), "Falsification and the Methodology of Scientific Research Programmes," in I. Lakatos and Alan Musgrave (eds.), *Criticism and the Growth of Knowledge*. Cambridge: Cambridge University Press, pp. 91–195.

Lakatos, Imre (1976), *Proofs and Refutations: The Logic of Mathematical Discovery*. J. Worrall and E. Zahar (eds.). Cambridge: Cambridge University Press.

Langley, Pat, Herbert Simon, Gary L. Bradshaw, and Jan M. Zytkow (1987), *Scientific Discovery: Computational Explorations of the Creative Process*. Cambridge, MA: MIT Press.

Laudan, Larry (1977), *Progress and Its Problems*. Berkeley, CA: University of California Press.

Leplin, Jarrett (1975), "The Concept of an *Ad Hoc* Hypothesis," *Studies in the History and Philosophy of Science* 5: 309–345.

Mitchell, Tom M. (1982), "Generalization as Search," *Artificial Intelligence* 18: 203–226.

Moberg, Dale and John Josephson (1990), "Appendix A: An Implementation Note," in J. Shrager and P. Langley (eds.), *Computational Models of Scientific Discovery and Theory Formation*. San Mateo, CA: Morgan Kaufmann, pp. 347–353.

Morowitz, Harold and Temple Smith (1987), "Report of the Matrix of Biological Knowledge Workshop, July 13–August 14, 1987," Santa Fe, NM: Santa Fe Institute.

Nickles, Thomas (1981), "What Is a Problem That We May Solve It?" *Synthese* 47: 85–118.

Nickles, Thomas (1987), "Methodology, Heuristics, and Rationality," in J. C. Pitt and M. Pera (eds.), *Rational Changes in Science*. Dordrecht: Reidel, pp. 103–132.

Popper, Karl (1965), *The Logic of Scientific Discovery*. New York: Harper Torchbooks.

Schaffner, Kenneth (1986a), "Computerized Implementation of Biomedical Theory Structures: An Artificial Intelligence Approach," in Arthur Fine and Peter Machamer (eds.), *PSA 1986*, v. 2. East Lansing, MI: Philosophy of Science Association, pp. 17–32.

Schank, Roger C. (1986), *Explanation Patterns: Understanding Mechanically and Creatively*. Hillsdale, NJ: Lawrence Erlbaum.

Sembugamoorthy, V. and B. Chandrasekaran (1986), "Functional Representation of Devices and Compilation of Diagnostic Problem-solving Systems," in J. Kolodner and C. Reisbeck (eds.), *Experience, Memory, and Reasoning*. Hillsdale, NJ: Lawrence Erlbaum Associates, pp. 47–73.

Shapere, Dudley (1974a), "Scientific Theories and Their Domains," in F. Suppe (ed.), *The Structure of Scientific Theories*. Urbana, IL: University of Illinois Press, pp. 518–565.

Sticklen, J. (1987), *MDX2: An Integrated Medical Diagnostic System*, Ph.D. Dissertation, Department of Computer and Information Science, The Ohio State University, Columbus, OH.

Thagard, Paul (1988), *Computational Philosophy of Science*. Cambridge, MA: MIT Press.

Toulmin, Stephen (1972), *Human Understanding*, v. 1. Princeton, NJ: Princeton University Press.

Wimsatt, William (1987), "False Models as Means to Truer Theories," in Matthew Nitecki and Antoni Hoffman (eds.), *Natural Models in Biology*. New York: Oxford University Press, pp. 23–55.

# 10

## Exemplars, Abstractions, and Anomalies

### Representations and Theory Change in
### Mendelian and Molecular Genetics[1]

10.1   INTRODUCTION

This chapter discusses representation of scientific theories and reasoning in theory change. A scientific theory may be represented by a set of concrete exemplary problem solutions. Or, alternatively, a theory may be depicted in an abstract pattern, which when its variables are filled with constants becomes a particular explanation. The exemplars and abstractions may be depicted diagrammatically, as they are in the cases from Mendelian and molecular genetics to be discussed. One way that a theory grows is by adding new types of exemplars to its explanatory repertoire. Model anomalies show the need for a new exemplar; they turn out to be examples of a typical, normal pattern that had not been included in the previous stage of theory development. A special-case anomaly indicates the need for a new exemplar or abstraction, but it has a small scope of applicability. Thus, our subjects here are exemplars, abstractions, diagrammatic representations, and anomalies and the roles they play in the representation of explanatory theories and in the change of such theories. Examples will be taken from Mendelian and molecular genetics.

[1] This chapter was originally published as Darden, Lindley (1995), "Exemplars, Abstractions, and Anomalies: Representations and Theory Change in Mendelian and Molecular Genetics," in James G. Lennox and Gereon Wolters (eds.), *Concepts, Theories, and Rationality in the Biological Sciences*. Pittsburgh, PA: University of Pittsburgh Press, pp. 137–158. Reprinted by permission of the University of Pittsburgh Press and the Universitaetsverlag Konstanz ©1995. I thank Joshua Lederberg for calling to my attention the model-driven aspect of the reverse transcriptase discovery during a discussion of model-driven versus data-driven discoveries in molecular biology; Pnina Abir-Am for calling my attention to the valuable book by Studer and Chubin; Alana Suskin for her able skills as a research assistant; and Sara Vollmer for comments on an earlier draft.

## 10.2  EXEMPLARS, ABSTRACTIONS, AND DIAGRAMS

Thomas Kuhn (1970, 1974) discussed the importance of "exemplars," which he characterized as concrete problem solutions in which a formalism is applied and given empirical grounding. Kuhn said that exemplars are taught by the use of problems in textbooks. The student learns to see other cases as similar to an exemplary problem and learns to use the pattern again to solve a new puzzle in a similar way. (This point was also made in Schaffner 1986b.)

Philip Kitcher (1981, 1989, 1993) analyzed theories by saying that they provide one or more abstract argument patterns that are invoked, in particular cases, by instantiating the abstract patterns, which he called "schematas." According to Kitcher, a scientific theory explains a particular case by showing that the case is an instance of one of its schematas. The domain that the theory explains consists of all the cases in which the argument pattern can be properly instantiated. Kitcher's examples included Newtonian laws and a pattern extracted from Darwin's work for explaining the evolution of a particular species. Kitcher, with his interest in philosophy of language, talked in terms of providing "arguments." The idea of an abstract schema, however, can be applied to abstract steps representing biological processes, rather than schematic premises in arguments. Other work (Darden and Cain 1989; see Chapter 8, this book) discussed an abstraction for selection type theories. The selection abstraction is not an argument pattern, as in Kitcher's analysis. Instead, the abstraction for selection type theories outlines the preconditions for and the temporal steps in a selection process. When instantiated, the abstraction is transformed into an account of an actual selection process.

Kitcher's and Kuhn's insights may be combined and expanded. Exemplars may serve as the source for constructing one or more abstract patterns, not only by a science student, as Kuhn notes, but also by a creative scientist constructing a new theory or changing an old one. Schema construction involves substituting variables for constants in the exemplar. Applying the theory then involves instantiating the schema in a new instance, that is, supplying new values for the variables. Such instantiation shows that the new instance can be explained by the abstract theory. A scientific theory is thus considered a set of abstract, problem-solving schemas (among other things). (For a similar view of explanation as schema-instantiation, see Schank 1986.)

Mendelian genetics exemplifies the role of shared examples and abstract diagrams as a means of representing a theory (Darden 1991, Ch. 12). In both Mendel's (1865) original paper and in the papers announcing the rediscovery

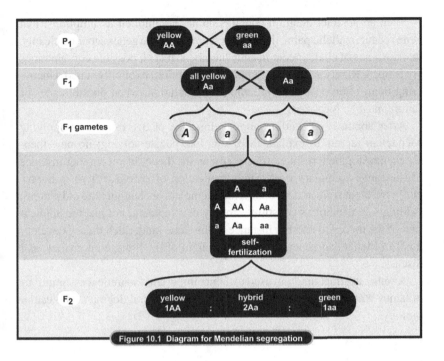

Figure 10.1 Diagram for Mendelian segregation

of his work in 1900 (e.g., de Vries 1900), the theory is represented by showing how it works in concrete exemplary cases. Providing examples of Mendel's pea crosses and showing how segregating genes explain the 3:1 ratios was a typical way of communicating Mendelian theory (e.g., Bateson 1902) and is still used in genetic textbooks (e.g., Strickberger 1985, p. 106). Such exemplars are often depicted in diagrams, such as the diagram in Figure 10.1.

It is possible to extract an abstract representation from the exemplary cases, as T. H. Morgan did in 1926. Morgan's abstract representation is shown in Figure 10.2. The diagram depicts the steps in a Mendelian breeding experiment, showing the initial conditions (a cross between two varieties), the stages in the experiment, and the data of 1:2:1 ratios. But Morgan's diagram represents not only the visible phenotypic characters, it also illustrates the theoretical assumptions, the unobservable, bead-like, paired genes that are postulated to segregate during the formation of germ cells and assort in all possible combinations during fertilization. This abstract schema can be instantiated to explain any 3:1 ratio in a Mendelian breeding experiment. It is an abstract representation of the segregation postulate of the Mendelian theory of the gene.

Such an abstract diagram contains numerous implicit assumptions: that genes occur in allelic pairs, that during segregation the genes separate cleanly, that the types of alleles occur in equal numbers, that all possible combinations are formed. Rarely are all these implicit assumptions explicitly stated. Uncovering them often occurs during anomaly resolution, when possible sites of failure must be localized.

After about 1910, finding another instance of 3:1 ratios and applying Mendelian factors to explain them was like puzzle solving; no new theoretical development resulted from applying the theory in yet another instance of the same exemplary type. Such application of an old pattern is importantly different from adding additional exemplars or changing the old pattern. Adding new exemplars constitutes theory development, not just the application of the theory to another instance of the same kind. Such theory development in Mendelian genetics occurred in light of the linkage anomalies, as I discuss.

Another abstract diagram used to depict important theoretical assumptions is James Watson's diagram for his version of the central dogma of molecular biology:

DNA –> RNA –> protein

In his retrospective account in *The Double Helix,* Watson (1968) claimed that he put a diagram of this sort above his desk in Cambridge in 1952, one year before his discovery with Francis Crick of the double helix structure of DNA. Watson does not discuss the empirical evidence that he used in constructing this abstract diagram. Despite not being explicitly abstracted from a single exemplar, Watson's version of the central dogma fits my analysis well. It is a diagrammatic representation of the steps in the process of information transfer in genetic systems. It is a general schema. It can be instantiated to account for the normal flow of genetic information in most cells. Or a piece of the process can be used to account for how an RNA virus, such as tobacco mosaic virus, transfers information just from RNA to protein, omitting the DNA step.

Crick introduced the term "central dogma" in a paper in 1958 for the more abstract idea that information flows from nucleic acid to protein but not from protein to nucleic acid. Crick explicitly said that this "dogma" was introduced in light of little empirical evidence. However, he said that attempting to build a useful theory without the hypothesis "generally ends in wilderness" (Crick 1958, p. 152). Between 1958 and 1970, evidence in favor of the central dogma mounted. In 1970, an anomaly challenged Watson's version of the central dogma, a topic to be discussed later.

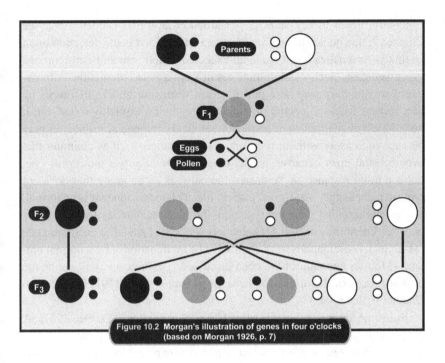

Figure 10.2 Morgan's illustration of genes in four o'clocks (based on Morgan 1926, p. 7)

So, for both Mendelian genetics and for molecular biology, important theoretical components were presented in abstract diagrammatic form, depicting entities and relations and sequential steps in biological processes. In Mendelian genetics, exemplary cases – usually Mendel's peas – provided an important way of making concrete the abstract symbols in the segregation diagrams. Instantiating the abstract schemas in additional, similar cases served to explain those cases. In molecular biology, the central dogma was generated as a theoretical assumption, without being explicitly abstracted from one concrete exemplar. Nonetheless, as in the Mendelian case, instantiations of the abstract central dogma served to explain instances of information transfer as particular genetic systems were investigated.

### 10.3   ANOMALIES AND THEORY CHANGE

Both Mendelian genetics of 1900–1910 and molecular genetics of 1952–1970 faced important challenges that resulted in new theoretical developments. Linkage anomalies and reverse transcriptase are the examples of anomalies to be examined for their role in driving theory change. I call this process "anomaly driven theory redesign" (Darden et al. 1991; Darden 1992; see Chapter 9, this book). It consists of several steps: First is the detection of an anomaly. Anomalies result when an expected pattern cannot be instantiated in a given case. Such a failure may result from a failed prediction. Alternatively, an anomaly may be a puzzling and surprising finding that turns up in a context where a specific prediction is not being explicitly tested, but it nonetheless challenges some general theoretical component. Anomalies may be explained away without requiring theory change, such as claiming that experimental error occurred or claiming that the case is a monstrous one, not a normal instance. A "monster anomaly" is an anomaly that does not present a challenge to a general theory; it is a phenomenon that is abnormal. For example, lethal gene combinations provided anomalies for 3:1 ratios, but the general abstraction for selection was not changed to accommodate them. Such combinations were claimed to be malfunctions of the normal system. (Monster anomalies have been discussed elsewhere and will not be the primary focus of the discussion here. See Darden 1990, 1991, Ch. 12; also Chapter 9, this book.)

But some anomalies are what I call "model anomalies" (Darden 1991). A model anomaly requires a change, either the alteration of a typical pattern or the addition of one or more new patterns to the explanatory repertoire of the theory. A model anomaly, after the needed theory change has been made, turns out not to have been an anomaly at all; instead, it can be viewed as a "model," or an instance, of the normal. A good example of a model anomaly is the case of linkage of genes. The model linkage anomalies required expansion of the set of exemplars in Mendelian genetics and will be discussed in more detail.

As we will see in the case of reverse transcriptase, an anomaly that provides a model for a small domain (e.g., one class of RNA viruses) may require a "special-case" exemplar to be added to a more general theory (e.g., the central dogma of molecular biology). So, in addition to monster anomalies and model anomalies, there are "special-case" anomalies. These are anomalies that indicate normal exemplars within a small class of instances but are different from the normal exemplar of a larger class.

After the anomaly is detected and is determined to require theory change, the next step in anomaly-driven theory redesign is localization. The step of localizing where the theory is failing is a diagnostic reasoning step (Darden 1990). Next, the theory must be changed, such as by changing an existing exemplary pattern or adding another one. Changing the theory is like reasoning in design tasks (Karp 1990) or, better, in redesign tasks. Details of reasoning in localization and redesign have been discussed in other work and will not be discussed further here (see Darden 1990, 1991; also Chapter 9, this book). The focus here is on the types of anomalies and the effects of their resolution on the abstract, exemplary representations of theories.

## 10.4   MENDELIAN GENETICS AND LINKAGE ANOMALIES

The model linkage anomalies required a change in the exemplars and abstract patterns of Mendelian genetics. Prior to the discovery of linkage, what are now called "Mendel's two laws" were not clearly separated. The explanation of 3:1 ratios by segregation was claimed to occur, even when two gene pairs were followed in a cross. For example, the independent assortment of yellow-green and tall-short characters in peas to produce 9:3:3:1 ratios was said to be a simple extension of the 3:1 case to a case with two different traits. But, in 1905, Bateson and his colleagues found exceptions to independent assortment, which were not exceptions to the 3:1 segregation pattern.

Bateson and his associates reported results from crosses with sweet peas (*Lathyrus odoratus*, not to be confused with Mendel's *Pisum sativum*). Flower color and pollen shape were traits that showed normal Mendelian segregation. Purple flowers crossed with red showed purple to be dominant in the $F_1$; self-fertilization produced approximately three purple to one red in the $F_2$. Similarly, long pollen grains proved dominant to round and produced approximately three long to one round. Anomalous data, however, resulted from dihybrid crosses in which the two traits were considered together. When plants with purple flowers and long pollen were crossed with plants with red flowers and round pollen, the predicted 9:3:3:1 ratios were not found. Instead, the results were 1,528 purple long, 106 purple round, 117 red long, and 381 red round (Bateson, Saunders, and Punnett 1905).

Several explanations were proposed for this anomaly. All the hypotheses localized the problem in the step of equal numbers of types of germ cells and were alternatives at different levels of organization that resulted in unequal numbers. We will not pause to consider each hypothesis here (discussed in Darden 1991, Ch. 9). The one that proved correct was Morgan's postulation of linkage of Mendelian factors on chromosomes and crossing-over between homologous chromosomes (Morgan 1910b, 1911a) (Figure 10.3).

The sweet pea anomaly could be explained by the linkage of genes for flower color and pollen shape on the same chromosome. Usually they stay together in a hybrid cross; hence, the larger-than-expected classes of purple long and red round. But, occasionally, crossing-over occurs between flower-color and pollen-shape genes, and those chromosomes give rise to the equal numbers of purple round and red long plants.

To recap: In localizing the problem of exceptions to 9:3:3:1 ratios, the exemplar for segregation explaining the 3:1 ratios was, for the first time, clearly distinguished from the exemplar for independent assortment, explaining the 9:3:3:1 ratios. Moreover, a new exemplar was added for "typical" linkage cases. Morgan, in his 1926 exposition of the theory, provided alternative examples and diagrams for linkage to accompany the usual Mendelian ones in peas that did not show linkage. The linkage anomalies were transformed into models for normal cases of linkage. This model anomaly thus resulted in an expansion of the set of exemplars. The repertoire of typical cases expanded. (For more on the linkage case, see Darden 1991, Ch. 9.)

## 10.5 MOLECULAR GENETICS AND REVERSE TRANSCRIPTASE

The discovery of the enzyme reverse transcriptase is another good case for examining the role of anomalies in theory change, in a case from molecular biology rather than Mendelian genetics.

The enzyme reverse transcriptase was discovered in 1970 by Howard Temin and David Baltimore. This discovery resulted in changes in models of viral reproduction, as well as a change in Watson's version of the central dogma of molecular biology, which had been formulated in the 1950s. Temin explained two sets of experimental findings by postulating a new model of viral reproduction. One line of evidence was genetic; another came from the study of a biochemical inhibitor of nucleic acid synthesis (Temin 1964a). (This is an example of an interfield relation in which evidence from one field is used to direct the search for evidence from another; see Darden and Maull 1977; Chapter 5, this book.) By studying the genetic mutations in Rous sarcoma viruses, Temin found that these RNA tumor viruses transformed normal cells into genetically stable cancer cells, which for generations continued to yield a small amount of the virus. This genetic work led Temin to postulate the provirus hypothesis. As he said later in his Nobel lecture: "The provirus hypothesis was [originally] a genetic hypothesis and contained no statement about the molecular nature of the provirus. However, the regular inheritance of the provirus led me to postulate that the provirus was integrated with the cell genome" (Temin 1977, p. 512).

Temin then proceeded to design new experiments, biochemical rather than genetic, to characterize and isolate the provirus. The *genetic* evidence about stable inheritance of the virus needed to be interrelated to *biochemical* findings

236

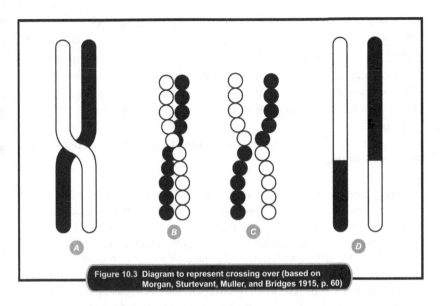

Figure 10.3 Diagram to represent crossing over (based on Morgan, Sturtevant, Muller, and Bridges 1915, p. 60)

in order to ground it molecularly. He found that viral production was inhibited in the presence of actinomycin-D (Temin 1963). Since actinomycin-D was known to selectively inhibit the synthesis of RNA on a DNA template, a functional role for DNA was implicated in viral reproduction. This was a surprising phenomenon, an anomaly for known patterns of viral reproduction. Other known RNA viruses were not so affected. Temin carried out other experiments to investigate the implications of the actinomycin-D anomaly. These experiments showed that new DNA synthesis was required for viral infection (Temin 1964a) and that new viral-specific DNA was found in infected chicken cells (Temin 1964b, 1977, p. 513).

Based on this work, in 1964 Temin postulated the provirus hypothesis; namely, that the provirus was DNA (Temin 1964c). This proposal reflected Temin's belief in the importance of models or hypotheses in science. He said:

> I (and some other people) am a strong believer that the hypothesis is the important thing, that facts without a hypothesis to give them meaning don't really exist.
>
> (Interview with Temin, 1978; quoted in Studer and Chubin 1980, p. 130)

According to the provirus hypothesis, the RNA of the virus is copied into DNA, which inserts itself into the host genome. This DNA copy is called the "provirus." Later, new RNA viruses were made by the transcription of the provirus DNA back into RNA (Temin 1977, p. 514). For the next six years, this provirus hypothesis was, as Temin put it, "essentially ignored" (Temin 1977, p. 514). Robert Gallo, reflecting on that period, went further: "This notion [the provirus hypothesis] was met with almost uniform incredulity. Chiefly because genetic information was known to go only from DNA to DNA or from DNA to RNA [Watson's version of the central dogma], some critics went further and ridiculed the experiments and the idea....." (Gallo 1991, p. 69).

The inhibitor work was especially problematic. As Baltimore put it, additional evidence was needed:

> Inhibitor experiments always have a problem, which is, if they fit into the concepts that are ongoing and you can bolster them by other kinds of things, then they make sense. Standing by themselves, everybody is itchy. You don't know enough about the inhibitor to interpret the experiment. And Howard [Temin] was broadly criticized on that ground, unfairly. But you can't take an inhibitor experiment any further. You can't show that you are right, unless you can go at it by an orthogonal new mechanism. And that's what the RT [reverse transcriptase work] did. But the problem was that up until the Spring of 1970 [when reverse transcriptase was discovered], no one had an orthogonal way to go at it. And so it [the provirus hypothesis?] just stayed around and was knocked about.
>
> (Interview with Baltimore, 1978; quoted in Studer and Chubin, p. 139)

In other words, some kind of independent biochemical evidence for the existence of the provirus, produced by some different experimental technique, was needed. The biochemical evidence from inhibitor experiments was not sufficient.

Temin persevered. He used the provirus hypothesis to predict the existence of a hitherto unknown enzyme to copy RNA into DNA. Note how the prediction arises. Temin accounted for anomalous data by constructing what might be variously termed: a picture; an abstract pattern; a view of the components in a process; or a new causal, mechanistic model (Bechtel and Richardson 1993, p. 231) of a new type of RNA virus reproduction. In order to carry out one of the steps in the process – namely, the step from RNA to its copy in provirus DNA – a hitherto unknown enzyme would be required. Thus, the existence of such an enzyme was hypothesized. Next, specific experiments were planned to detect the enzyme.

Complicated relations exist between, on the one hand, the model with its accompanying hypothesis of the existence of the enzyme, and, on the other hand, experiment planning and detailed predictions as to what would be expected to be observed in order to detect the enzyme. Analysis of this kind of reasoning is a good topic for further work; little has been done on reasoning strategies in experiment planning. (For more on reasoning in experiment planning, see Kettler and Darden 1993).

Both Temin and David Baltimore independently devised similar experiments to assay viral extracts for an enzyme with DNA polymerase activity. The assay involved measuring the incorporation of radioactively labeled thymidine triphosphate (needed for DNA but not RNA synthesis) in a cell-free system. In Baltimore's (1970) experiments, extracts from both Rauscher mouse leukemia virus and Rous sarcoma virus eventually gave positive results in the DNA polymerase assay system. Furthermore, the enzyme activity was shown to be sensitive to ribonuclease, showing the role of RNA in the synthesis of the DNA (Baltimore 1970). Temin and Mizutani found the same results in Rous sarcoma virus (Temin and Mizutani 1970). An enzyme for converting the virus's RNA into provirus DNA had been discovered and the prediction of Temin's provirus hypothesis was confirmed.

Originally called "RNA-dependent DNA polymerase," an anonymous correspondent in *Nature* (Anonymous 1970b) dubbed the new enzyme "reverse transcriptase." (Gallo says that John Tooze was responsible; see Gallo 1991, p. 69.) That catchy name stuck.

Baltimore (1977, p. 501) supplied a diagram (Figure 10.4) and explained the action of reverse transcriptase. The virion somehow allows reverse transcriptase, along with the virus RNA, to get into the infected cell's cytoplasm. Once in the cytoplasm, reverse transcriptase catalyzes the synthesis of the proviral DNA. Then, the proviral DNA integrates into the cellular DNA. It is expressed during normal transcription in two products: new virion RNA and messenger RNA for making reverse transcriptase and other viral proteins. This diagram presents the new model of viral reproduction resulting from the new discovery.

Temin (1977) in his Nobel lecture retrospectively reflected on the period in the late 1960s, just before the discovery of reverse transcriptase in 1970. By that time, other virions had been shown to carry enzymes, called "polymerases," into their host cells rather than synthesizing them after infection. Temin, sounding regretful, said: "The conclusion that RSV virions contain a DNA polymerase could have been deduced in 1967 or 1968 from the DNA provirus hypothesis and the existence of these virion polymerases, but it was not" (Temin 1977, p. 515). Actually, Temin was reasoning by analogy, not using logical deduction. Here is how the analogical reasoning goes: Because other virions carry polymerases with them, perhaps Rous sarcoma virus does also. Furthermore, because the provirus hypothesis predicts the existence of an enzyme with the function of synthesizing DNA from RNA, then devise experiments to look for the enzyme in the virion itself rather than looking for its action once the virus infects the host cell.

The existence of reverse transcriptase was a prediction of the provirus model and its discovery provided evidence for the correctness of that new model of viral reproduction. Rous sarcoma viral reproduction became the exemplar for the class of retroviruses (as they were later called). But the ramifications of this discovery were greater than providing a new model of viral reproduction. The discovery had implications for the generalization, at the heart of molecular biology, called the "central dogma." In 1970, both Watson and Crick immediately responded to the challenge to the central dogma posed by the discovery of the new enzyme.

Between 1958 and 1970, the central dogma had fared well. In the June 27, 1970, issue of *Nature*, the issue in which the reverse transcriptase discovery was announced, an anonymous "News and Views" article screamed: "Central Dogma Reversed." The first sentence began: "The central dogma, enunciated by Crick in 1958 and the keystone of molecular biology ever since, is likely to prove a considerable over-simplification" (Anonymous 1970a, p. 1198). The strategy of complicating an oversimplification in response to an anomaly is a common reasoning strategy (Darden 1991, Ch. 6).

Watson, in the 1970 second edition of his *Molecular Biology of the Gene*, said:

> The concept of a DNA provirus for an RNA virus is clearly a radical proposal. It overturns the belief that flow of genetic information always goes in the direction of DNA to RNA, and never RNA to DNA. On the other hand, it offers an even greater variety of ways for cells to exchange genetic information. Considering the enormous complexity of biological systems, it would not be surprising if this device were uniquely advantageous in some situations.
>
> (Watson 1970, pp. 621–622)

Complicating an oversimplification is thus a strategy that has a justification in nature: The complexity of biological systems makes it reasonable to postulate a more complicated general theory.

Crick (1970) also responded immediately to the challenge but in a different way than Watson responded. Crick published a paper in *Nature* in August 1970. His version of the central dogma, he contended, had not been "reversed," as the anonymous *Nature* article had claimed. Crick claimed, correctly, that in 1958 he had stated the central dogma in terms of the general transfer of

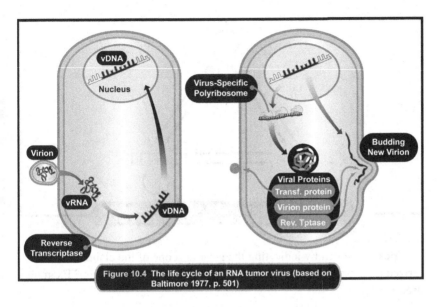

Figure 10.4 The life cycle of an RNA tumor virus (based on Baltimore 1977, p. 501)

information from nucleic acids to protein but not the reverse (Crick 1958). That abstract claim had not yet been challenged. If it were shown that information could flow from proteins to nucleic acids, he said, then such a finding would "shake the whole intellectual basis of molecular biology" (Crick 1970, p. 563). But, to use my terminology, such a model anomaly had not been found. No radically new model was needed. The finding that information could be transferred from RNA to DNA simply resulted in a small addition to the picture.

Crick (1970), as he so often did in his papers, first outlined all theoretical possibilities, then indicated the likely candidates out of all possible, and discussed ones not known and not expected. He presented a diagram in 1970, which he claimed was drawn in 1958 but which he said he was not sure was ever published. Figure 10.5 shows all the possible information transfers among DNA, RNA, and protein. Figure 10.6 shows "the situation as it seemed in 1958." Figure 10.7 shows "a tentative classification for the present day" (i.e., 1970).

Crick's reasoning often showed this jump to very general, theoretical claims, sometimes with little empirical evidence at the time. The outlining of theoretical possibilities and the narrowing to theoretical plausibilities also marked his reasoning (see, for example, his consideration of the possible explanations for the PaJaMo experiment, discussed in Olby 1970).

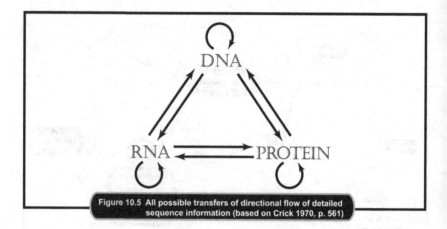

Figure 10.5 All possible transfers of directional flow of detailed sequence information (based on Crick 1970, p. 561)

Crick continued by indicating the general scope of the claims.He characterized as a "general transfer . . . one which can occur in all cells." Examples are:

DNA –>DNA

DNA –>RNA

RNA –>Protein

He continued: "A special transfer is one which does not occur in most cells but may occur in special circumstances." Possible candidates are:

RNA –>RNA

RNA –>DNA

DNA –>Protein

"At the present time the first two of these have only been shown in certain virus-infected cells" (Crick 1970, p. 562). Finally, he listed "unknown transfers." These are the three transfers which the central dogma postulates never occur:

Protein –> Protein

Protein –> DNA

Protein –> RNA

Crick was clear about the importance of the scope of the anomaly. He said: "It might indeed have 'profound implications for molecular biology' [citation to *Nature* article "Central Dogma Reversed"] if any of these special transfers could be shown to be general, or – if not in all cells – at least to be widely distributed" (Crick 1970, p. 562). But they had not. Hence, Crick's

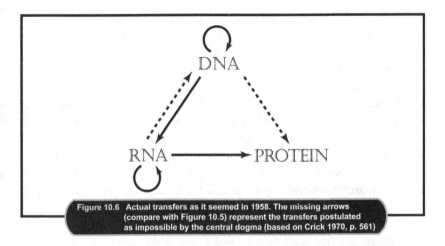

Figure 10.6 Actual transfers as it seemed in 1958. The missing arrows (compare with Figure 10.5) represent the transfers postulated as impossible by the central dogma (based on Crick 1970, p. 561)

version of the central dogma had not been reversed by the discovery of reverse transcriptase.

In my terminology, Crick claimed that the reverse transcriptase case was not a model anomaly that required a new model to show the usual flow of genetic information in most organisms; instead, it was what I call a "special-case anomaly." A special-case anomaly shows what is *normal in a few, special cases*. It is not a monster in the sense of being a *malfunction of the normal*. Nor is it a model anomaly in the sense of being a *general model representing most cases* of transfer of genetic information. The anomaly shows what is normal for a few special cases; here, for some RNA viruses.

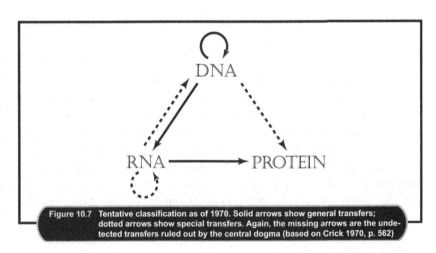

Figure 10.7 Tentative classification as of 1970. Solid arrows show general transfers; dotted arrows show special transfers. Again, the missing arrows are the undetected transfers ruled out by the central dogma (based on Crick 1970, p. 562)

The revised diagram for the central dogma now includes an arrow going backwards from RNA to DNA; however, discussion of the diagram must explain that the backwards arrow represents only special-case transfers (e.g., Berg and Singer 1992, p. 35).

Temin, however, went further than his provirus hypothesis. In contrast to Crick, and in contrast to almost everyone else, Temin interpreted reverse transcriptase as a model anomaly. He used the finding (as well as a priori reasoning about viral origins) to propose what he called the "protovirus hypothesis." In contrast to the similarly named "provirus" hypothesis, Temin's protovirus hypothesis proposed that normal cells – cells not infected with virus – had reverse transcriptase and regularly transferred information from RNA to DNA (Temin 1971). Temin (1977) proposed that "ribodeoxyviruses evolved from normal cellular components.... The normal cellular components are the endogenous ribodeoxyvirus-related genes. These genes are involved in normal DNA to RNA to DNA information transfer. This normal process of information transfer in cells could not exist only for its ability to give rise to viruses. It must exist as a result of its role in normal cellular processes, for example, cell differentiation, antibody formation and memory..." (Temin 1977, pp. 520–521; also discussed in Temin 1971). Temin was not one to shy away from bold hypotheses.

Baltimore (1977), while arguing in favor of the provirus hypothesis, expressed skepticism about Temin's new protovirus model. The protovirus model has not been confirmed. Construction of that model was an over-general response to the reverse transcriptase anomaly. Reverse transcriptase did not provide evidence about information transfer in most cells; it was not a model anomaly for a model of such general scope.

### 10.6 CONCLUSION

In summary, this chapter discussed representation of scientific theories and roles of anomalies in theory change. Theories can be represented as a set of concrete exemplars or as abstractions taken from those exemplars. The exemplars and abstractions are often presented in diagrams. Anomalies have been analyzed in light of the analysis of a theory as a repertoire of exemplary explanation patterns. Model anomalies show the need for a new exemplar; they turn out to be examples of a typical, normal pattern that had not been included in the previous stage of theory development. Special-case anomalies, in contrast, represent what is normal for a small set of cases.

New exemplars can result from the resolution of anomalies. By adding new exemplary patterns, the explanatory repertoire of the theory is expanded. Linkage and crossing-over were model anomalies in Mendelian genetics. The set of exemplars representing the normal hereditary patterns in Mendelian genetics was expanded to include not only cases of segregation and independent assortment but also linkage and crossing-over. The provirus model resulted from explaining the anomaly of the role of DNA synthesis in viral reproduction and the new model led to the prediction of the existence of the enzyme, reverse transcriptase. Rous sarcoma virus reproduction provided a new exemplar for reproduction in the class of retroviruses. Furthermore, reverse transcriptase, showing the anomaly of RNA to DNA information transfer, proved to be a special-case anomaly for the central dogma. It has not been shown to be a model for the normal transfer of genetic information in all (or most) cells. The discovery of reverse transcriptase resulted in expansion of the set of known patterns for transfer of genetic information. The abstract diagram for steps of information transfer was modified to include these special cases by adding an arrow going from RNA to DNA, but the arrow is instantiated only in special cases.

In both Mendelian and molecular genetics, new exemplars were added to previous sets and new diagrams or parts of diagrams were added to general theoretical representations. Anomaly-driven theory redesign occurred.

### REFERENCES

Anonymous (1970a), "Central Dogma Reversed," News and Views, *Nature* 226: 1198–1199.

Anonymous (1970b), "Cancer Viruses: More of the Same," *Nature* 227: 887–888.

Baltimore, David (1970), "Viral RNA-dependent DNA Polymerase," *Nature* 226: 1209–1211.

Baltimore, David (1977), "Viruses, Polymerases and Cancer," in *Nobel Lectures in Molecular Biology, 1933–1975.* [no editor] New York: Elsevier, pp. 495–508. Presented December 12, 1975.

Bateson, William (1902), *Mendel's Principles of Heredity – A Defense.* Cambridge: University Press.

Bateson, William, E. R. Saunders, and R. C. Punnett (1905), "Further Experiments on Inheritance in Sweet Peas and Stocks: Preliminary Account," *Proceedings of the Royal Society*, B, LXXVII. Reprinted in Punnett, R. C. (ed.) (1928), *Scientific Papers of William Bateson.* Cambridge: Cambridge University Press, v. 2, pp. 139–141.

Bechtel, William and Robert C. Richardson (1993), *Discovering Complexity: Decomposition and Localization as Strategies in Scientific Research.* Princeton, NJ: Princeton University Press.

Berg, Paul and Maxine Singer (1992), *Dealing with Genes: The Language of Heredity.* Mill Valley, CA: University Science Books.
Crick, Francis (1958), "On Protein Synthesis," *Symposium of the Society of Experimental Biology* 12: 138–167.
Crick, Francis (1970), "Central Dogma of Molecular Biology," *Nature* 227: 561–563.
Darden, Lindley (1990), "Diagnosing and Fixing Faults in Theories," in J. Shrager and P. Langley (eds.), *Computational Models of Scientific Discovery and Theory Formation.* San Mateo, CA: Morgan Kaufmann, pp. 319–346.
Darden, Lindley (1991), *Theory Change in Science: Strategies from Mendelian Genetics.* New York: Oxford University Press.
Darden, Lindley (1992), "Strategies for Anomaly Resolution," in R. Giere (ed.), *Cognitive Models of Science*, Minnesota Studies in the Philosophy of Science, v. 15. Minneapolis, MN: University of Minnesota Press, pp. 251–273.
Darden, Lindley and Joseph A. Cain (1989), "Selection Type Theories," *Philosophy of Science* 56: 106–129.
Darden, Lindley and Nancy Maull (1977), "Interfield Theories," *Philosophy of Science* 44: 43–64.
Darden, Lindley, Dale Moberg, Satish Nagarajan, and John Josephson (1991), "Anomaly Driven Redesign of a Scientific Theory: The TRANSGENE.2 Experiments," *Technical Report 91-LD-TRANSGENE.* Laboratory for Artificial Intelligence Research, The Ohio State University, Columbus, OH.
Gallo, Robert (1991), *Virus Hunting, Aids, Cancer, and the Human Retrovirus: A Story of Scientific Discovery.* New York: HarperCollins Publishers, Basic Books.
Karp, Peter (1990), "Hypothesis Formation as Design," in J. Shrager and P. Langley (eds.), *Computational Models of Scientific Discovery and Theory Formation.* San Mateo, CA: Morgan Kaufmann, pp. 275–317.
Kettler, Brian and Lindley Darden (1993), "Protein Sequencing Experiment Planning Using Analogy," in L. Hunter, D. Searls, and J. Shavlik (eds.), *ISMB-93, Proceedings of the First International Conference on Intelligent Systems for Molecular Biology.* Menlo Park, CA: AAAI Press, pp. 216–224.
Kitcher, Philip (1981), "Explanatory Unification," *Philosophy of Science* 48: 507–531.
Kitcher, Philip (1989), "Explanatory Unification and the Causal Structure of the World," in Philip Kitcher and Wesley Salmon (eds.), *Scientific Explanation.* Minnesota Studies in the Philosophy of Science, v. 13. Minneapolis, MN: University of Minnesota Press, pp. 410–505.
Kitcher, Philip (1993), *The Advancement of Science: Science without Legend, Objectivity without Illusions.* New York: Oxford University Press.
Kuhn, Thomas (1970), *The Structure of Scientific Revolutions.* 2nd ed. Chicago, IL: University of Chicago Press.
Kuhn, Thomas (1974), "Second Thoughts on Paradigms," in Frederick Suppe (ed.), *The Structure of Scientific Theories.* Urbana, IL: University of Illinois Press, pp. 459–482.
Mendel, Gregor ([1865] 1966), "Experiments on Plant Hybrids," in Curt Stern and Eva Sherwood (eds.), *The Origin of Genetics, A Mendel Source Book.* San Francisco: W. H. Freeman, pp. 1–48.
Morgan, Thomas H. (1910b), "Sex-Limited Inheritance in *Drosophila*," *Science* 32: 120–122.

Morgan, Thomas H. (1911a), "An Attempt to Analyze the Constitution of the Chromosomes on the Basis of Sex-Limited Inheritance in *Drosophila*," *Journal of Experimental Zoology* 11: 365–413.

Morgan, Thomas H. (1926), *The Theory of the Gene.* New Haven, CT: Yale University Press.

Morgan, Thomas H., A. H. Sturtevant, H. J. Muller, and C. B. Bridges (1915), *The Mechanism of Mendelian Heredity.* New York: Henry Holt and Company.

Olby, Robert (1970), "Francis Crick, DNA, and the Central Dogma," reprinted in G. Holton (ed.), *The Twentieth Century Sciences.* New York: W. W. Norton, pp. 227–280.

Schaffner, Kenneth (1986b), "Exemplar Reasoning About Biological Models and Diseases: A Relation Between the Philosophy of Medicine and Philosophy of Science," *The Journal of Medicine and Philosophy* 11: 63–80.

Schank, Roger C. (1986), *Explanation Patterns: Understanding Mechanically and Creatively.* Hillsdale, NJ: Lawrence Erlbaum.

Strickberger, Monroe (1985), *Genetics.* 3rd ed. New York: Macmillan.

Studer, Kenneth E. and Daryl E. Chubin (1980), *The Cancer Mission: Social Contexts of Biomedical Research.* Sage Library of Social Research, v. 103. Beverly Hills, CA: Sage Publications.

Temin, Howard M. (1963), "The Effects of Actinomycin D on Growth of Rous Sarcoma Virus in Vitro," *Virology* 20: 577–582.

Temin, Howard M. (1964a), "The Participation of DNA in Rous Sarcoma Virus Production," *Virology* 23: 486–494.

Temin, Howard M. (1964b), "Homology between RNA from Rous Sarcoma Virus and DNA from Rous Sarcoma Virus-infected Cells," *Proceedings of the National Academy of Sciences (USA)* 52: 323–329.

Temin, Howard M. (1964c), "Nature of the Provirus in Rous Sarcoma," *National Cancer Institute Monograph* 17: 557–570.

Temin, Howard M. (1971), "Guest Editorial. The Protovirus Hypothesis: Speculations on the Significance of RNA-directed DNA Synthesis for Normal Development and Carcinogenesis," *Journal of the National Cancer Institute* 46: III–VIII.

Temin, Howard M. (1977), "The DNA Provirus Hypothesis: The Establishment and Implications of RNA-directed DNA Synthesis," in *Nobel Lectures in Molecular Biology, 1933–1975.* [no editor] New York: Elsevier, pp. 509–529. Presented December 12, 1975.

Temin, Howard M. and Satoshi Mizutani (1970), "RNA-dependent DNA Polymerase in Virions of Rous Sarcoma Virus," *Nature* 226: 1211–1213.

Vries, Hugo de ([1900] 1966), "Das Spaltungsgesetz der Bastarde," *Berichte der deutschen botanischen Gesellschaft* 18: 83–90. English translation: "The Law of Segregation of Hybrids," in C. Stern and E. Sherwood (eds.), *The Origin of Genetics, A Mendel Source Book.* San Francisco: W. H. Freeman, pp. 107–117.

Watson, James D. (1968), *The Double Helix.* New York: New American Library.

Watson, James D. (1970), *Molecular Biology of the Gene.* 2nd ed. New York: W. A. Benjamin.

# 11

# Strategies for Anomaly Resolution in the Case of Adaptive Mutation[1]

## 11.1 INTRODUCTION

Anomalies often challenge biological generalizations. Among the generalizations of widest scope in biology are natural selection and the central dogma of molecular biology. Because natural selection often produces novel variants, it might be expected that wide-scope generalizations would be infrequent (other than the theory of natural selection itself). Furthermore, biological regularities are evolutionarily contingent (Beatty 1995); whatever regularity has evolved can evolve away. Nonetheless, the central dogma of molecular biology is one of the most general findings in all of biology. It provides a schema for protein synthesis that involves unidirectional information flow from nucleic acids to proteins but not back. If information does not flow back into the genetic material, then there seems to be no mechanism for the inheritance of adaptive acquired characters. Although adaptive characters might be acquired during the life of one organism, their inheritance requires a change in the genetic material passed to the next generation. Because NeoLamarckian mechanisms require such inheritance, the lack of backwards flow of information strengthens the case for NeoDarwinian natural selection as the account of adaptations against any version of NeoLamarckism. The contemporary NeoDarwinian theory of natural selection claims that mutations arise spontaneously; that is, they are produced independently of their fitness in a given environment.

[1] This work was supported by the U.S. National Science Foundation under grant SBR-9817942 and by an award from the General Research Board of the Graduate School at the University of Maryland. I thank Joshua Lederberg, David Thaler, Sri Sastry, and others in the Laboratory for Molecular Genetics and Informatics at Rockefeller University for their hospitality and help on these topics. The Center for Philosophy of Science at the University of Pittsburgh provided a hospitable environment and support for the writing. I had fruitful discussions with Douglas Allchin, Jason Baker, Kevin Elliott, David Hull, Peter Machamer, Jim Marcum, Greg Morgan, and Bob Olby.

Anomalies challenging generalizations of such wide scope get attention. In a paper in 1988, John Cairns and colleagues claimed to find a new class of mutations in bacteria, called "directed mutations" and later "adaptive mutations." ("Directed," as we will see, is more specific than "adaptive.") Cairns et al. (1988) claimed that their findings provided a challenge to one version of the central dogma of molecular biology. Further, the molecular mechanisms that they proposed to account for directed mutations provided possible mechanisms for the inheritance of adaptive acquired characteristics (Cairns et al. 1988, p. 145). Unsurprisingly, this purported anomaly has received and continues to receive much attention from biologists.

The history of this dispute from 1988 until 2003 will be briefly sketched. This case illustrates a whole range of responses to this anomaly, from questioning the adequacy of the empirical evidence for the existence of the phenomenon to proposing a radically new mechanism that would have required changes to the central dogma and the theory of natural selection. As of 2003, the controversy shows that the most radical challenges to NeoDarwinism and the central dogma have been abandoned; the phenomenon has been reinterpreted and looks as if it can be explained without postulating a radically new type of mechanism to produce directed mutations. This case provides further ground for refining reasoning strategies for anomaly resolution (Darden 1991; also see Chapters 9 and 10, this book).

First, discussion of the Cairns et al. (1988) work details the experiments, the results, and the hypothesized radical new mechanisms. Then follows a discussion of exactly what the new anomaly challenges. As we will see, alternative theoretical perspectives can make those who hold them more or less willing to embrace the implications of an anomaly. The responses to the Cairns anomaly are then divided into critical examinations: first, of the empirical evidence for the existence and reinterpretation of the phenomenon; and, second, of the hypothesized mechanisms to produce it. Concluding the chapter are implications of this case for strategies for anomaly resolution.

## 11.2 "THE ORIGIN OF MUTANTS" (CAIRNS, OVERBAUGH, AND MILLER 1988)

The model system for the Cairns et al. (1988) experiments was the favorite experimental system of molecular biology: the *lac* operon in *E. coli*. This strain of bacteria (*E. coli* K12 FC40 delta *lac pro* F' *lac Z$_{am}$*) contained a mutation in the *lac* gene, which codes for beta galactosidase, one of the enzymes needed to utilize the sugar lactose. The strain was engineered so as to have no *lac* gene on its chromosome and to have a (frameshift) mutant version on a plasmid, an extra-chromosomal bit of DNA. At the beginning of the experiment, these *lac⁻* mutant bacteria were grown on a medium so that they were in a stationary, nongrowing phase (M9 plus 0.1 percent glycerol). Mutagenesis had been assumed to require DNA synthesis in actively dividing cells (although some evidence against this conclusion had been provided by Ryan as early as 1955). Hence, the assumption was that stationary, nongrowing bacteria would not be mutating according to the usual mutagenesis mechanism of imperfect DNA replication. The surprising mutations that the Cairns group found were thus labeled "stationary phase mutants."

To the agar with the *lac⁻* stationary phase bacteria, Cairns and his colleagues plated a top layer containing lactose and afterwards noted the numbers of colonies that showed growth. The procedure selected for the Lac⁺ revertant (mutant) phenotype[2]: The revertants were able to utilize lactose to grow and thus appeared as visible colonies. The Cairns group claimed that they found two classes of mutants in which *lac⁻* gene reverted to *lac⁺*. The first class consisted of spontaneous mutants that had occurred prior to plating that showed an expected distribution. The second class of mutants showed some unexpected properties: they appeared later, several days after plating with lactose. Most surprising, in these late-growing mutations, there were more Lac⁺ mutants than would be expected, given the usual assumptions about the frequency of spontaneous mutants. As a control, they scored for mutations in another gene, valine resistance, which was located on the chromosome and which was not adaptive in the lactose environment. The control did not show increased numbers of mutants, comparable to the Lac⁺ ones on the plasmid (Cairns et al. 1988, p. 144).

In addition to the experiments selecting for *lac⁺* mutants on an engineered plasmid, Cairns and his colleagues also cited work by others in two other experimental systems that, they claimed, also showed directed mutations. One was another engineered system, Shapiro's Lac(Ara)⁺ construct that requires excision of a short segment of bacteriophage Mu in order to be functional. A second was found in normal, nonengineered *E. coli* in which cryptic genes that allowed utilization of lactose if two particular mutations occurred; these mutants appeared within about two weeks time.

In addition to the *data* from these three cases interpreted as evidence for the phenomenon of directed mutations, Cairns and his colleagues proposed *mechanisms* that could have produced this phenomenon: They claimed that "cells

---

[2] Small case italics (e.g., *lac*) designate the gene; capitalized, nonitalics (e.g., Lac) designate the phenotype.

may have mechanisms for *choosing* which mutations *will* occur" (Cairns et al. 1988, p. 142, italics added); later, they expressed this by saying: "bacteria can *choose* which mutations they *should* produce" (Cairns et al. 1988, p. 145, italics added).

Cairns and his colleagues localized their work in the step of information transfer denied by the central dogma, which they formulated as: "the central dogma of molecular biology ... denies any possible effect of a cell's experience upon the sequence of bases in its DNA" (Cairns et al. 1988, p. 145). They proposed two alternative hypothetical *mechanisms* for ways the cell's experience might affect DNA sequences, one more radical than the other. The most radical mechanism has been called the "specific reverse transcription" mechanism (Sarkar 1991): The cell, it was proposed, produced a highly variable set of mRNA molecules and then reverse transcribed the one that made the best protein back into the DNA. This mechanism required that the cell have a special "organelle" (Cairns et al. 1988, p. 145) containing the reverse transcriptase enzyme, which copied RNA into DNA.[3] Also, this hypothetical organelle was claimed to have some element for monitoring the protein products. It would determine that the messenger RNA of the best protein was reverse transcribed back into the DNA. Thus, directed changes in the DNA sequence reflected "effects of the cell's experience."

Cairns and his colleagues mentioned a second, less radical mechanism, called elsewhere the "non-specific reverse transcription" mechanism (Sarkar 1991). This second mechanism would be less efficient than the first because it had a more random component. Again, the cell was assumed to make a set of variant mRNA molecules but all were reverse transcribed; the one that allowed growth would produce more descendants.

Both proposed mechanisms involved a set of mutant mRNAs, which were produced when the cell was assumed not to be undergoing DNA replication. These mechanisms would be a new type of mutagenesis mechanism, an addition to the usual type of imperfect DNA repair during DNA replication. The new type would be a new method for mutagenesis to occur – during transcription of DNA into mRNA (known to be error-prone), prior to reverse transcription back into DNA. Cairns later reflected on the role that such newly proposed mechanisms can play:

> When novel phenomena are observed in biology, they sometimes lead to the discovery of a new, unexpected mechanism. And, from that moment on, the

---

[3] On reverse transcriptase as an anomaly for the central dogma, see Darden (1995); see Chapter 10, this book. On the use of reverse transcriptase in other mechanisms for inheritance of adaptive acquired characters, see Steele (1981); discussed in Parascandola (1995); Steele (et al. 1998).

new mechanism becomes accessible as an explanation for other phenomena in
biology. (Cairns 1997, p. 169)

To sum up the Cairns et al. (1988) work: Data provided evidence that certain
"mutations in *Escherichia coli* seem to occur at a higher frequency when they
are beneficial" (Cairns and Foster 1991, p. 695). Two hypothetical mechanisms accounted for this purported phenomenon. Each mechanism began with
error-prone RNA transcription, which produced a pool of variant messenger
RNAs. Some mRNAs became fixed by reverse transcription back in the DNA
of cells. Those cells were able to grow because they contained a revertant,
the directed mutant *lac*$^+$. The proposed mechanisms were molecular-level
mechanisms producing the inheritance of an adaptive acquired character.

## 11.3   FOR WHAT WAS DIRECTED MUTATION AN ANOMALY?

The purported phenomenon of adaptive mutation was claimed to be an
anomaly for two wide-scope generalizations in biology: the central dogma
of molecular biology and the assumption of spontaneous mutation in the
theory of natural selection. The anomaly appeared to challenge the central
dogma's claim that information was not transferred back into the DNA. The
anomaly was also viewed as providing an exception to the claim in the theory of natural selection that all mutations were spontaneous and produced
without regard to their fitness in a given environment. (For more on "characterizing an anomaly," that is, determining where the conflict occurs, see
Elliott 2004.)

It is not a trivial task to determine exactly what is challenged by a new
empirical finding. Both of these challenges need to be examined carefully.
As we have seen, Cairns and his colleagues located their work in the step
of information transfer denied by their version of the central dogma. They
said: "the central dogma of molecular biology ... denies any *possible effect
of a cell's experience* upon the sequence of bases in its DNA" (Cairns et al.
1988, p. 145, italics added). After mentioning other cases which they took
to be exceptions, including information transfer from viral RNA to DNA via
reverse transcriptase, as well as genomic instability in times of stress (the SOS
response in bacteria), they said: "The only major category of informational
transfer that has not been described is between proteins and the messenger
RNA (mRNA) molecules that made them" (Cairns et al. 1988, p. 145). Their
purposed mechanisms did so.

In order to localize what was challenged by this anomaly, it is useful
to recall the historical proposal of the central dogma. James Watson's (1965,

252

p. 298) version of the central dogma of molecular biology was a diagrammatic representation of the information flow from genes to proteins:

DNA –> RNA –> protein

Watson's version, promulgated in his successful 1965 textbook, has come to be the standard representation. However, Watson's version differed from Francis Crick's original, more abstract version. In 1958, Crick christened the "Central Dogma" and stated it verbally in terms of information flow:

> . . . the transfer of information from nucleic acid to nucleic acid, or from nucleic acid to protein may be possible, but transfer from protein to protein, or from protein to nucleic acid is impossible. Information means here the *precise* determination of sequence, either of bases in the nucleic acid or of amino acid residues in the protein.        (Crick 1958, p. 153)

When Crick formulated the central dogma in 1958, he was concerned with the stereochemical issue of whether the linear sequence of amino acids in proteins could somehow serve as a template for the sequentialization of the bases in a nucleic acid, either RNA or DNA. On stereochemical grounds, he rejected such transfer. There seemed no possibility of the many differently shaped amino acids providing a template for assembling nucleic acids (Crick 1970; 1988). Even the Cairns group's most radical hypothetical mechanism does not go that far. The sequence of the "best" protein is not itself being used as a template for constructing a nucleic acid sequence, which is the kind of information transfer explicitly excluded by Crick's 1958 version.

But the Cairns group's version was more abstract than either Watson's or Crick's versions of the central dogma. They did not require that a protein's amino acid sequence serve as a *template* to determine nucleic acid sequence. Instead, they stated the central dogma using the abstract phrase: "any possible effect of a cell's experience." This phrase could be instantiated in numerous ways other than via the amino acid sequence of the protein providing a template for a nucleic acid sequence. Their rewording opened many more possibilities for transfer of *something* back into the DNA sequence and, consequently, allowed for many more kinds of anomalies for it. One could interpret the Cairns group's specific reverse-transcription mechanism as providing a kind of information transfer from protein to DNA. A hypothetical organelle (of an unknown type) discriminated good from bad proteins and sent a signal to reverse transcribe only that best protein's messenger RNA back into the DNA.

Although Crick (1958) did not connect the two, the central dogma has often been interpreted (e.g., Jablonka and Lamb 1995; Judson 1996) as providing a molecular level denial of the inheritance of acquired characters (for an argument against this interpretation, see Sarkar 1996). For example, the proteins in the neck muscles of the giraffe do not pass information about the response to the environmental challenge that produced a longer neck into the DNA that the baby giraffe inherits. One can provide a molecular challenge to Neo-Lamarckism: by what molecular mechanism is an adaptive, acquired character passed from the phenotype to the genotype? There are a number of steps that would be necessary for such an instructive mechanism to operate at the molecular level. Some hypothetical entity or mechanism would have to detect some aspect of the environment and pass this "information" "back" into the DNA. "Information" here could be characterized in an abstract way to mean "something that constrains the space of possibilities of DNA mutations."[4] Then, the DNA would have to process this information (this instruction) to produce or preserve mutants that would be adaptive in that environment. The kind of information needed would be something from the environment that could produce an adaptive sequence change in DNA. Then, the changed DNA would have to be replicated, passed to offspring, and, subsequently, play a role in producing the adaptive, acquired character (e.g., an enzyme) in the daughter.[5]

The Cairns "specific reverse transcription" mechanism does seem to fulfill these conditions for a possible molecular-level mechanism for the inheritance of adaptive acquired characters. Such a mechanism challenged the component of the theory of natural selection in which variations were assumed to be spontaneous and produced without regard to their fitness in a given environment. If directed mutation occurred in natural populations in the wild, then the variation component of the theory of natural selection would have to be altered. Two kinds of variations would be found and the preadapted variants would, of course, fare well in the selective interaction step of the selection process (or perhaps we would say that only adaptive forms are produced as a result of the critical factor in the environment and there is no selective interaction). (For steps in the mechanism of natural selection, see Darden and Cain 1989, Chapter 8, this book.)

In addition to providing a challenge to the central dogma and to the theory of natural selection, the Cairns anomaly also provided a challenge to the

---

[4] Peter Machamer suggested this "constraint on space of possibilities" interpretation of information. A more specific molecular biological account is David Thaler's suggestion that "information" be viewed as "templating," that is, as "ordering molecules in space."

[5] For more on instructive theories, see Darden and Cain (1989; see Chapter 8, this book). For discussion of various examples of inheritance of *nonadaptive* acquired characters, see Landman (1991); Jablonka and Lamb (1995).

generalization that mutations were always formed in growing cells during DNA replication. If mutations could form during a stationary phase, with little or no DNA replication, then some other mechanism for forming mutations was needed.

## 11.4   THEORETICAL PERSPECTIVES IN RESPONSES TO ANOMALIES

Before plunging into details of specific responses to the Cairns anomaly, it is instructive to consider broader perspectives. At least some biologists involved in the dispute can be classified as having conflicting perspectives on the nature of DNA. On one side, some of the defenders of the adaptive mutation phenomenon (though not necessarily the particular Cairns 1988 mechanisms) were committed to a theoretical view that DNA and the genetic machinery were an active response system, capable of detecting aspects of the environment and mounting adaptive responses. For example, Eva Jablonka and Marion Lamb (1995, p. vii) said: "The genome is described as 'flexible', 'dynamic', or 'clever', to emphasize its role as an active response system as well as a passive information carrier." James Shapiro (1995, p. 374), who studied transposons extensively, said: "The discovery that cells use biochemical systems to change their DNA in response to physiological inputs moves mutation beyond the realm of 'blind' stochastic events and provides a mechanistic basis for understanding how biological requirements can feed back onto the genome structure." David Thaler (1994, p. 225) discussed the role of the environment in influencing the "genes and physiology of DNA metabolism" and speculated "if environmental influences affecting the generation of variation can become coherent with those affecting subsequent selection, then the conditions are ripe for the evolutionary bootstrapping of genetic intelligence."

This view of DNA as an active response system was in sharp contrast to the alternative theoretical framework in which DNA was viewed as a passive information carrier. Such passive DNA merely undergoes occasional spontaneous mutations due to physical or chemical mutagens and imperfect DNA replication. Passive DNA has been the traditional view, a molecular perspective underpinned by the central dogma and compatible with NeoDarwinian natural selection. Such general perspectives may affect whether an anomaly is welcomed as evidence for a given perspective or as evidence against a rival. Given its potential implications for supporting alternative theoretical perspectives, as well as for such widespread generalizations as the central dogma and the theory of natural selection, it is not surprising that the Cairns work received much attention.

Challenges to adaptive mutation can be divided into at least two groups. First are challenges to the adequacy and scope of the *empirical evidence* for the existence of the phenomenon. Second are challenges to and postulation of hypothetical *mechanisms* claimed to produce the phenomenon of adaptive mutation.

## 11.5  ADEQUACY OF THE EMPIRICAL EVIDENCE
### FOR ADAPTIVE MUTATION

There are a number of ways that an anomaly can be dissolved by challenging the empirical evidence. One way is to replicate the experiments to see if the same result appears. Attempts at replicating the experiments leveled the charge that Cairns and his colleagues had not provided sufficient experimental details in the 1988 paper so that others could be sure about the experimental conditions (Mittler and Lenski 1990a, 1990b).

Once that problem was resolved, subsequent replication and extension with the *lac⁻* frameshift mutants on the plasmid established two classes of mutations, with different kinds of DNA sequence changes. In Lac⁺ revertants in normal, rapidly growing cells, many different sequence changes restored the Lac⁺ phenotype; however, in the slow-growing, late-appearing Lac⁺ revertants, only a single base change occurred to correct a frameshift mutation, from CCCC to CCC (Rosenberg et al. 1994; Foster and Trimarchi 1994). These sequence differences provided additional, molecular-level evidence for the Cairns group's claim that there were two different classes of mutants in this experimental system. "None of the earlier results have been as pervasive as the sequence data" (Culotta 1994).

A key empirical challenge focused on the adequacy of the controls (Bridges 1995; response in Cairns 1995). The 1988 data seemed to show that late-arising mutations were occurring in the *lac⁻* gene in the plasmid but not in a control gene on the chromosome that was not adaptive in the lactose environment. Additional experiments were designed to provide better controls (Hall 1997). Putting the control gene on the plasmid with *lac⁻* showed that the control gene also showed a higher rate of mutation (Foster 1997). Consequently, the specificity of "directed" mutations was challenged.

Another question was raised as to whether the engineered laboratory strains in the plasmid system (and the Lac(Ara)⁺ Mu excision system) were adequate as model systems for normal *E. coli* plasmid genes. One could accept that the phenomenon of adaptive mutation occurred in a plasmid in an engineered laboratory strain, but one could then employ the strategy of monster barring

(Lakatos 1976) to limit the impact of this anomaly. The cases could be claimed to be monsters, not within the scope of applicability of mutagenesis mechanisms in normal *E. coli* plasmid genes. Thus, the cases could be barred from requiring changes to the accepted mechanism for mutagenesis in normal, nonengineered cells.

Others accepted the empirical evidence that adaptive mutations were occurring in the experimental system of *E. coli* with *lac⁻* mutants on a plasmid and might apply to nonengineered genes on plasmids but argued that the phenomenon was "trivial" (Bridges 1995). Even in *E. coli*, most genes are on the chromosome, not in plasmids. Thus, even if genes on plasmids have the ability to adaptively mutate – for example, based on a mechanism involving plasmid recombination (Harris, Longerich, and Rosenberg 1994; Rosenberg 1994) – this might not be a mutational phenomenon occurring in most genes in *E. coli* or, more generally, other organisms. Thus, this case would become a "special-case anomaly" (Darden 1995; Chapter 10, this book), having very small scope. However, others (Foster and Trimarchi 1994) countered this charge of triviality by saying that bacteria could mobilize chromosomal genes to plasmids to allow adaptive mutations in them.

The scope of the purported phenomenon was thus a matter of debate. Cairns and his colleagues in the 1988 paper had attempted to respond to this kind of challenge by discussing mutants in cryptic genes in *natural* populations of *E. coli*. Additional work by, for example, Galitski and Roth (1996), entitled "A Search for a General Phenomenon of Adaptive Mutability," investigated thirty different chromosomal *lacZ* mutants in another bacterium, *Salmonella*. They concluded that some of the "excess late revertants" arise from "general adaptive mutability available to any chromosome gene" (Galitski and Roth 1996, p. 645). Work by Bjedov and colleagues tested 787 *E. coli* isolates from diverse natural environments and found evidence for what they termed "stress-induced mutagenesis" (Bjedov et al. 2003; discussed in Rosenberg and Hastings 2003). Additional work extended the scope of the phenomenon to other bacterial species and to yeast. However, no instances of adaptive mutations in nonmicroorganisms had been found as of 2000 (Foster 2000).

These challenges and responses illustrate the importance of determining the scope of the anomalous phenomenon. An anomaly whose scope was only a laboratory-engineered strain could be barred as a monster; that is, it would require no change in any theoretical generalization about nonengineered strains. An anomaly that provided evidence for adaptive mutation of genes on only bacterial plasmids might provide some evidence about normal plasmid mutations but would be more of a special-case anomaly than one that included more numerous chromosomal genes in many species of microorganisms.[6] An anomaly found in wild strains of many species of microorganisms would have wider ramifications.

There is still some dispute about the adequacy of the empirical evidence, its interpretation, and its scope. Nonetheless, there is now good evidence that the phenomenon exists: A second class of late-appearing mutations occurs in stationary bacteria and some other microorganisms under conditions of nonlethal selection. However, these mutations are no longer labeled "directed." One refined account of "adaptive" mutation is the following: "a process that during nonlethal selection produces mutations that relieve the selective pressure, whether or not other, nonselected mutations are also produced" (Foster 2000).

In addition to the controversy about the adequacy of the empirical evidence, even more controversy has centered on the hypothesized mechanisms to produce these adaptive mutations.

## 11.6   PROPOSED MECHANISMS FOR ADAPTIVE MUTATION

The space of "how possibly" mechanisms to account for adaptive mutation was, at the outset, a large one and various portions of this hypothesis space were explored. They may be grouped into types.

The characterization of the phenomenon as either directed or not demanded different kinds of mechanisms to produce it. Directed mutations appeared to require some radically new kind of mutagenesis mechanism, one evolved to produce specific mutants in response to an environmental challenge. The radically new hypothesized mechanisms proposed by the Cairns group involving reverse transcriptase were merely sketched, with important components never fully specified. Never filled was the gray box representing a mechanism component whose functional role was to monitor a successful enzyme and direct the reverse transcribing of its mRNA back into the DNA.

Sadly or fortunately (depending on one's perspective), further work led Cairns (Foster and Cairns 1992) to abandon the reverse transcriptase mechanisms. Reverse transcriptase was not found in *E. coli* K12, the strain used in their experiments. Furthermore, adaptive mutations appeared even when there was no link between the altered protein and its DNA sequence.[7] Abandonment of directed mutation removed one of the challenges to the nature of mutation

---

[6] Compare Beatty (1997) on the issue of "relative significance," Darden (1996) on the issue of scope of generalizations, and Chapter 10, this book, on the scope of special-case anomalies.

[7] For discussion of anomaly resolution for a research protocol designed to test the hypothesis of mutagenesis during transcription, see Darden and Cook (1994).

in NeoDarwinism. Such abandonment also removed the challenge to the central dogma. Large sections of the possible hypothesis space had been pruned away. The anomaly appeared to be localized in error-prone DNA replication. This localization required abandoning the previous assumption that bacteria in stationary phase, not undergoing cell division, did not replicate their DNA. Those cells appeared to be undergoing small amounts of DNA replication. This finding allowed the construction of hypotheses for adaptive mutation as occurring via the previously characterized mechanism of imperfect DNA replications. As mutations accumulate, one that allows the cell to grow was designated as "adaptive."

In addition to the category of new type versus non-new type of mechanism, another division between types of hypotheses is between a mechanism evolved to produce adaptive mutation versus a mechanism not specifically evolved to produce adaptive mutations (Brisson 2003). The evolved mechanism might be an adaptive response to a stressful environmental challenge. Various types of alternative hypotheses accounted for apparent adaptive mutations as, for example, some sort of side effect or a breakdown in the functioning of a normal mechanism (e.g., the failure of mismatch repair mechanisms, an early idea discussed in Stahl 1988).

Roth et al. (2003b) discussed three types of hypotheses for adaptive mutation. The directed mutation model (DMM) suggested that stress-induced mutagenesis focused on the relevant target, the *lac*⁻ gene (Cairns et al. 1988). This type of hypothesis has evidence against it, as discussed previously, and has been abandoned. The hypermutable state model (HSM) suggested that stress induced genome-wide mutagenesis in a subset of the cells; this generated some mutations that were adaptive but killed the rest of the mutagenized cells (Hall 1990). According to HSM, mutation appeared directed because only Lac⁺ revertants survived mutagenesis. Roth (et al. 2003a) argued that the mutation rate was insufficient to generate the number of detected Lac⁺ revertants, but this conclusion was disputed (Cairns and Foster 2003).

Yet a third type, advocated by Roth and his colleagues, was the amplification mutagenesis model (AMM), in which growth under selection increased revertant number with no required change in the rate or target specificity of mutation (Andersson et al. 1998). The AMM model proposed that rare preexisting cells with a *lac* duplication (two copies of the *lac*⁻ gene) grew slowly when placed under selection and improved their growth by further *lac* amplification within each developing colony. It had been assumed that *lac*⁻ was dysfunctional, but Roth discussed evidence that some enzyme was produced in the *lac*⁻ cells. As copies of *lac*⁻ (within colonies) were amplified, some reverted to *lac*⁺ without any increase in the underlying (per base pair)

| Table 11.1 Types of Mechanisms for Adaptive Mutation | | |
|---|---|---|
| | Evolved via selection for adaptive mutation | Not evolved via selection for adaptive mutation |
| **Directed** | Reverse transcribe mRNA | No possible mechanism here? |
| **Not directed adaptive capture** | Mutations due to increased Pol IV | Amplification and reversion |

mutation rate (Andersson et al. 1998; Hendrickson et al. 2002). Selection appeared to direct mutations to *lac⁻* because this gene was amplified during growth under selection and because only the *lac⁺* revertant allele was maintained during subsequent selected loss of the many (now deleterious) copies of the mutant *lac⁻* allele (Roth et al. 2003b, p. 2320).

The controversy continued as, contra Roth et al., Foster and her colleagues proposed a refined version of the hypermutation hypothesis that again viewed the mechanism for increasing mutations as one evolved via selection for an adaptive response to stress (Layton and Foster 2003). Construction of this hypothesis was aided by the discovery of an error-prone DNA polymerase, Pol IV. This polymerase was found to be induced during stress, as part of the previously characterized SOS response. The SOS system had been found to consist of a group of coordinately regulated genes that were activated by DNA damage, which occurred when bacteria were under stress (Friedberg 1997, Ch. 6). Pol IV was found to be up-regulated by a stress-response SOS factor and to produce increased errors during DNA replication. Layton and Foster speculated about the origin of this increase in mutation rate: "The fact that Pol IV is induced in late stationary phase under control of the general stress-response sigma factor suggests that the ability to transiently increase genetic diversity is one of the functions that have evolved to promote survival during stress" (Layton and Foster 2003, p. 50). They also noted that the Pol IV mechanism seemed not to be sufficient to account for all the adaptive mutations, so they speculated that additional mechanisms would also be discovered.

If adaptive mutation is shown to be a result of normal amplification/reversion, then this anomaly will be resolved without viewing it as a result of a mechanism that has been selected for adaptive mutation. The SOS response is a response to stress, so subsumption of adaptive mutation under this rubric would possibly include it as a phenomenon evolved to aid bacteria in coping with a challenging environment. (Presumably, it might be just part of general decay and not actively selected for?)

Table 11.1 categorizes some of the types of mechanisms proposed to account for adaptive mutation. The horizontal dimension is evolved versus nonevolved via selection for some type of adaptive mutation. The vertical is whether the phenomenon is directed and thereby requires a new, previously unknown type of mechanism or whether the phenomenon is not directed. The Cairns reverse transcriptase mechanisms were directed/evolved. If the SOS response produced error-prone DNA synthesis via Pol IV in response to stress, then the anomaly would be resolved by subsuming it under a known type of (presumably) evolved mechanism. Finally, if reversion without increased mutagenesis and then subsequent amplification produced the adaptive mutations, then usual chance mutations would be shown to be sufficient. No new type of evolved mechanism would be required.

Invoking a type of hypothesis is a method for discovering mechanisms (Darden 2002; Chapter 12, this book). Hence, it is useful to delineate types in historical episodes so as to outline types within the hypothesis space and to make these types available for future hypothesis construction episodes, as in Table 11.1.

## 11.7  IMPLICATIONS FOR STRATEGIES FOR ANOMALY RESOLUTION

The many diverse responses to this anomaly make this a rich case for examining strategies for anomaly resolution. They are summarized in Table 11.2. The first step indicates the importance of reproducing the anomalous data and carefully analyzing the controls. If the data are not reproducible, then the issue is closed. Raw data must be interpreted as evidence for the existence of a phenomenon. James Bogen and James Woodward (1988) discussed the way in which raw data provide evidence for phenomena. This distinction was particularly important in the directed/adaptive mutation phenomenon because the existence of the phenomenon depended on analyzing data to show a numerical bias in the types of mutations produced. Reproduction of the data with the particular engineered plasmid strain and reanalysis of the data occurred in this case.

Critique of the adequacy of the control proved crucial in establishing the existence of the phenomenon. When the control gene was moved from the chromosome to the plasmid, then that gene, too, was shown to be undergoing increased mutation. Later, the phenomenon was reinterpreted: directed mutation became selective capture of adaptive mutations.

The scope of the anomaly was and continues to be thoroughly explored. Because the phenomenon originally was found in a highly engineered system, some argued that it should be barred as a monster, something not occurring in nonengineered organisms and having no implications for generalizations

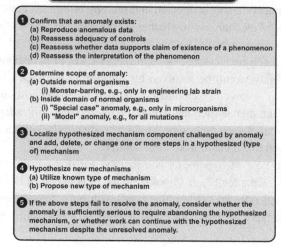

**Table 11.2 Strategies for Anomaly Resolution**

**1** Confirm that an anomaly exists:
(a) Reproduce anomalous data
(b) Reassess adequacy of controls
(c) Reassess whether data supports claim of existence of a phenomenon
(d) Reassess the interpretation of the phenomenon

**2** Determine scope of anomaly:
(a) Outside normal organisms
(i) Monster-barring, e.g., only in engineering lab strain
(b) Inside domain of normal organisms
(i) "Special case" anomaly, e.g., only in microorganisms
(ii) "Model" anomaly, e.g., for all mutations

**3** Localize hypothesized mechanism component challenged by anomaly and add, delete, or change one or more steps in a hypothesized (type of) mechanism

**4** Hypothesize new mechanisms
(a) Utilize known type of mechanism
(b) Propose new type of mechanism

**5** If the above steps fail to resolve the anomaly, consider whether the anomaly is sufficiently serious to require abandoning the hypothesized mechanism, or whether work can continue with the hypothesized mechanism despite the unresolved anomaly.

about nonengineered mechanisms in organisms. However, subsequent work showed that indeed the phenomenon occurred widely: in chromosomal genes and plasmid ones, in laboratory strains of *E. coli* and in strains captured in the wild, and in other species of bacteria and yeast. This additional empirical work showed that the anomaly could not just be barred as a monster. A phenomenon in microorganisms in need of explanation had been established.

An anomaly of wide scope may be a "model" for all or most normal cases. For example, Cairns (1995, p. 293) speculated: "For all we know, most sequence changes in prokaryotes and eukaryotes (and, in particular, in the somatic cells of animals) may be adaptive events occurring in nonmultiplying cells." Recall that all mutations had previously been assumed to occur during DNA replication in multiplying cells. If Cairns's view here had proved to be correct, then whatever mechanism(s) accounts for adaptive mutations would be the "model" for mechanism(s) for normal mutagenesis in "most" cases, eucaryotes, as well as bacteria. However, current evidence limits the scope of the anomaly to microorganisms and only to some mutations in microorganisms, so adaptive mutation appears to be a "special-case anomaly" for general claims about mechanisms of mutagenesis. But within this limited scope, it appears to be a typical response to nonlethal selection of stationary, nonmultiplying cells.

In sum: In different responses, adaptive mutations were claimed to be monstrous, special-case, or model anomalies.

In addition to debates about the scope of the anomaly, this case also illustrates alternative ways of generating a new hypothesis to account for an anomaly. These include utilizing known mechanisms (amplification/reversion), hypothesizing an altered step in a known mechanism (mismatch repair slowed down in DNA replication), or postulating a new type of mechanism (reverse transcription of variant mRNAs). Utilizing or tweaking hypotheses about previously known types of mechanisms is a more conservative method than hypothesizing a new type of mechanism.

When anomaly resolution requires hypothesizing a new mechanism, one method is to invoke a type of mechanism. The history of science provides "compiled hindsight" (Darden 1987) about types of theories and their successes and failures in past problem-solving episodes. Two such types are instructive versus selective type mechanisms to produce adaptations. An instructive type mechanism, such as the Cairns specific reverse transcription mechanism, involves receiving an instruction from the environment and making an adaptive change in response to it. This contrasts with selective type mechanisms in which variations are produced spontaneously with a subsequent selective interaction step to choose the adapted ones (Darden and Cain 1989; Chapter 8, this book).

It is more radical to propose a type of mechanism that has not been shown to operate in other relevantly similar episodes, such as proposing an instructive type mechanism for mutagenesis. More radical hypotheses will appropriately be subjected to greater scrutiny before they are accepted into the explanatory repertoire of types of mechanisms within biology. It is an important question whether past successes should lead one to have a strong preference for selective as opposed to instructive types of mechanisms. The biologist Joshua Lederberg clearly stated his preference for selective mechanisms, based on their past record:

> Any heuristic can be treacherous, but a Darwinian explanation is the first I would seek in explaining a biological enigma. I do not insist that it will always last, but it has had enormous power in bringing us to our present understanding.
>
> (Lederberg 1989, p. 398)

Preference for a type of mechanism may be influenced, as we have seen, by general theoretical commitments, such as viewing DNA as active or passive.

The philosopher Gilbert Harman (1986), in *Change in View,* advocated the strategy of making the least change in one's theoretical framework when faced with an anomaly. This is a very conservative strategy. Scientists do not always follow such a conservative path.

Proposing a radical response to an anomaly can play the pragmatic role of generating attention and additional work to confirm and resolve the anomaly. Certainly, Cairns and his colleagues' radical proposals, with far-reaching consequences for the central dogma and NeoDarwinism, generated much additional work to test their claims. On the other hand, seeing every unexpected data point as a potential crisis anomaly, in Thomas Kuhn's (1962) sense of an anomaly requiring a paradigm change, is not desirable either. Some middle ground seems preferable.

Once the existence of an anomaly has been empirically established, the space of possible mechanisms needs to be constructed and explored, either by a single researcher or, as this case shows, by different researchers in the scientific community. This space should include more radical and more conservative hypotheses; then, experiments to prune the space should be devised.

11.8   CONCLUSION

The wide-scope generalizations of the central dogma and the theory of natural selection have thus far survived the challenge of directed mutation. The diverse responses to this anomaly aided in developing strategies for anomaly resolution. The anomalous phenomenon needs to be characterized and the space of possible mechanisms that could produce it developed. More general theoretical perspectives may influence whether an anomaly is welcomed or disparaged and may provide the incentive for additional work. Radical challenges to previously known types of mechanisms and to wide-scope biological generalizations provoke much work to prove or disprove them. The work to find the mechanisms producing adaptive mutations in microorganisms continues.

REFERENCES

Andersson, D. I., E. S. Slechta, and J. R. Roth (1998), "Evidence that Gene Amplification Underlies Adaptive Mutability of the Bacterial *lac* Operon," *Science* 282: 1133–1135.

Beatty, John (1995), "The Evolutionary Contingency Thesis," in James G. Lennox and Gereon Wolters (eds.), *Concepts, Theories, and Rationality in the Biological Sciences*. Pittsburgh, PA: University of Pittsburgh Press, pp. 45–81.

Beatty, John (1997), "Why Do Biologists Argue Like They Do?" *Philosophy of Science 64 (Proceedings)*: S432–S443.

Bjedov, Ivana, Olivier Tenaillon, Benedicte Gerard, Valeria Souze, Erick Denamur, Miraslav Radman, Francois Taddei, and Ivan Matic (2003), "Stress-Induced Mutagenesis in Bacteria," *Science* 300: 1404–1409.

Bogen, James and James Woodward (1988), "Saving the Phenomena," *Philosophical Review* 97: 303–352.

Bridges, Bryn A. (1995), "Sexual Potency and Adaptive Mutation in Bacteria," *Trends in Microbiology* 3: 291–292.

Brisson, Dustin (2003), "The Directed Mutation Controversy in an Evolutionary Context," *Critical Reviews in Microbiology* 29: 25–35.

Cairns, John (1995), "Response [to Bridges 1995]," *Trends in Microbiology* 3: 293.

Cairns, John (1997), *Matters of Life and Death: Perspectives on Public Health, Molecular Biology, Cancer, and the Prospects for the Human Race*. Princeton, NJ: Princeton University Press.

Cairns, John (1998), "Mutation and Cancer: The Antecedents to Our Studies of Adaptive Mutation," *Genetics* 148: 1433–1440.

Cairns, John and Patricia L. Foster (1991), "Adaptive Reversion of a Frameshift Mutation in *Escherichia coli*," *Genetics* 128: 695–701.

Cairns, John and Patricia L. Foster (2003), "Letter to the Editor: The Risk of Lethals for Hypermutating Bacteria in Stationary Phase," *Genetics* 165: 2317–2318.

Cairns, John, Julie Overbaugh, and S. Miller (1988), "The Origin of Mutants," *Nature* 335: 142–145.

Crick, Francis (1958), "On Protein Synthesis," *Symposium of the Society of Experimental Biology* 12: 138–163.

Crick, Francis (1970), "Central Dogma of Molecular Biology," *Nature* 227: 561–563.

Crick, Francis (1988), *What Mad Pursuit: A Personal View of Scientific Discovery*. New York: Basic Books.

Culotta, Elizabeth (1994), "A Boost for 'Adaptive' Mutation," *Science* 265: 318–319.

Darden, Lindley (1987), "Viewing the History of Science as Compiled Hindsight," *AI Magazine* 8(2): 33–41.

Darden, Lindley (1990), "Diagnosing and Fixing Faults in Theories," in J. Shrager and P. Langley (eds.), *Computational Models of Scientific Discovery and Theory Formation*. San Mateo, CA: Morgan Kaufmann, pp. 319–346.

Darden, Lindley (1991), *Theory Change in Science: Strategies from Mendelian Genetics*. New York: Oxford University Press.

Darden, Lindley (1992), "Strategies for Anomaly Resolution," in R. Giere (ed.), *Cognitive Models of Science*, Minnesota Studies in the Philosophy of Science, v. 15. Minneapolis, MN: University of Minnesota Press, pp. 251–273.

Darden, Lindley (1995), "Exemplars, Abstractions, and Anomalies: Representations and Theory Change in Mendelian and Molecular Genetics," in James G. Lennox and Gereon Wolters (eds.), *Concepts, Theories, and Rationality in the Biological Sciences*. Pittsburgh, PA: University of Pittsburgh Press, pp. 137–158.

Darden, Lindley (1996), "Generalizations in Biology," Essay Review of K. Schaffner's *Discovery and Explanation in Biology and Medicine. Studies in History and Philosophy of Science* 27: 409–419.

Darden, Lindley (2002), "Strategies for Discovering Mechanisms: Schema Instantiation, Modular Subassembly, Forward/Backward Chaining," *Philosophy of Science (Supplement)* 69: S354–S365.

Darden, Lindley and Joseph A. Cain (1989), "Selection Type Theories," *Philosophy of Science* 56: 106–129.

Darden, Lindley and Michael Cook (1994), "Reasoning Strategies in Molecular Biology: Abstractions, Scans and Anomalies," in D. Hull, M. Forbes, and R. M. Burian (eds.), *PSA 1994*, v. 2. East Lansing, MI: Philosophy of Science Association, pp. 179–191.

Elliott, Kevin (2004), "Error as Means to Discovery," *Philosophy of Science* 71: 174–197.

Foster, Patricia L. (1997), "Nonadaptive Mutations Occur on the F' Episome During Adaptive Mutation Conditions in *Escherichia coli*," *Journal of Bacteriology* 179: 1550–1554.

Foster, Patricia L. (2000), "Adaptive Mutation: Implications for Evolution," *BioEssays* 22: 1067–1074.

Foster, Patricia L. and John Cairns (1992), "Mechanisms of Directed Mutation," *Genetics* 131: 783–789.

Foster, Patricia L. and Jeffrey M. Trimarchi (1994), "Adaptive Reversion of a Frameshift Mutation in *Escherichia coli* by Simple Base Deletions in Homopolymeric Runs," *Science* 265: 407–409.

Friedberg, Errol C. (1997), *Correcting the Blueprint of Life: An Historical Account of the Discovery of DNA Repair Mechanisms*. Plainview, NY: Cold Spring Harbor Laboratory Press.

Galitski, T. and John R. Roth (1996), "A Search for a General Phenomenon of Adaptive Mutability," *Genetics* 143: 645–659.

Hall, Barry G. (1990), "Spontaneous Point Mutations That Occur More Often When Advantageous Then When Neutral," *Genetics* 126: 5–16.

Hall, Barry G. (1997), "On the Specificity of Adaptive Mutations," *Genetics* 145:39–44.

Harman, Gilbert (1986), *Change in View: Principles of Reasoning*. Cambridge, MA: MIT Press.

Harris, Reuben S., Simonne Longerich, and Susan M. Rosenberg (1994), "Recombination in Adaptive Mutation," *Science* 264: 258–260.

Hendrickson, Heather, E. Susan Slechta, Ulfar Bergthorsson, Dan I. Andersson, and John R. Roth (2002), "Amplification Mutagenesis: Evidence that 'Directed' Adaptive Mutation and General Hypermutability Result from Growth With a Selected Gene Amplification," *Proceedings of the National Academy of Sciences* 99: 2164–2169.

Jablonka, Eva and Marion J. Lamb (1995), *Epigenetic Inheritance and Evolution: The Lamarckian Dimension*. New York: Oxford University Press.

Judson, Horace F. (1996), *The Eighth Day of Creation: The Makers of the Revolution in Biology*. Expanded Edition. Cold Spring Harbor, NY: Cold Spring Harbor Laboratory Press.

Kuhn, Thomas (1962), *The Structure of Scientific Revolutions*. Chicago, IL: University of Chicago Press.

Lakatos, Imre (1976), *Proofs and Refutations: The Logic of Mathematical Discovery*. J. Worrall and E. Zahar (eds.). Cambridge: Cambridge University Press.

Landman, Otto E. (1991), "The Inheritance of Acquired Characteristics," *Annual Review of Genetics* 25: 1–20.

Layton, Jill C. and Patricia L. Foster (2003), "Error-prone DNA Polymerase IV is Controlled by the Stress-response Sigma Factor, RpoS, in *Escherichia coli*," *Molecular Microbiology* 50: 549–561.

Lederberg, Joshua (1989), "Replica Plating and Indirect Selection of Bacterial Mutants: Isolation of Preadaptive Mutants in Bacteria by Sib Selection," *Genetics* 121: 395–399.

Mittler, John E. and Richard E. Lenski (1990a), "New Data on Excisions of Mu from *E. coli* MCS2 Cast Doubt on Directed Mutation Hypothesis," *Nature* 344: 173–175.

Mittler, John E. and Richard E. Lenski (1990b), "Mittler and Lenski Reply," *Nature* 345: 213.

Parascandola, Mark (1995), "Philosophy in the Laboratory: The Debate Over Evidence for E. J. Steele's Lamarckian Hypothesis," *Studies in History and Philosophy of Science* 26: 469–492.

Rosenberg, Susan M. (1994), "In Pursuit of a Molecular Mechanism for Adaptive Mutation," *Genome* 37: 893–899.

Rosenberg, Susan M. and P. J. Hastings (2003), "Modulating Mutation Rates in the Wild," *Science* 300: 1382–1384.

Rosenberg, Susan M., Simonne Longerich, Pauline Gee, and Reuben S. Harris (1994), "Adaptive Mutation by Deletions in Small Mononucleotide Repeats," *Science* 265: 405–407.

Roth, John R., Eric Kofoid, Frederick P. Roth, Otto G. Berg, Jon Seger, and Dan I. Andersson (2003a), "Regulating General Mutation Rates: Examination of the Hypermutable State Model for Cairnsian Adaptive Mutation," *Genetics* 163: 1483–1496.

Roth, John R., Eric Kofoid, Frederick P. Roth, Otto G. Berg, Jon Seger, and Dan I. Andersson (2003b), "Letter to the Editor: Adaptive Mutation Requires No Mutagenesis – Only Growth Under Selection: A Response," *Genetics* 165: 2319–2321.

Ryan, F. J. (1955), "Spontaneous Mutation in Non-dividing Bacteria," *Genetics* 40: 726–738.

Sarkar, Sahotra (1991), "Lamarck *Contre* Darwin, Reduction *Versus* Statistics: Conceptual Issues in the Controversy Over Directed Mutagenesis in Bacteria," in Alfred I. Tauber (ed.), *Organism and the Origins of Self*. Dordrecht: Kluwer, pp. 235–271.

Sarkar, Sahotra (1996), "Biological Information: A Skeptical Look at Some Central Dogmas of Molecular Biology," in S. Sarkar (ed.) *The Philosophy and History of Molecular Biology: New Perspectives*. Dordrecht: Kluwer, pp. 187–231.

Shapiro, James A. (1995), "Adaptive Mutation: Who's Really in the Garden?" *Science* 268: 373–374.

Stahl, F. W. (1988), "A Unicorn in the Garden," *Nature* 335: 112–113.

Steele, E. J. (1981), *Somatic Selection and Adaptive Evolution: On the Inheritance of Acquired Characters*. 2nd ed. Chicago, IL: University of Chicago Press.

Steele, Edward J., Robyn A. Lindley, and Robert B. Blanden (1998), *Lamarck's Signature*. Reading, MA: Perseus Books.

Thaler, David S. (1994), "The Evolution of Genetic Intelligence," *Science* 264: 224–225.

Watson, James D. (1965), *Molecular Biology of the Gene*. New York: W. A. Benjamin.

# III

## Discovering Mechanisms

### Construction, Evaluation, Revision

# 12

## Strategies for Discovering Mechanisms

### Construction, Evaluation, Revision[1]

#### 12.1  INTRODUCTION

The key ideas in the chapters in Part I on mechanisms and those in Part II on reasoning strategies now need to be integrated and extended. The chapters in Part I provide a characterization of mechanisms, based on an analysis of paradigm cases from molecular and neurobiology. Those chapters also begin a search for reasoning strategies for discovering mechanisms and for analyzing the role of mechanisms in understanding the relations among fields in biology. The chapters in Part II discuss reasoning strategies in scientific change, including the use of analogies, interfield theories, and abstract types of theories, as well as strategies for anomaly resolution.

Biologists often seek to discover mechanisms, as we have seen in previous chapters. The task now is to summarize and extend previous work on reasoning strategies for discovering mechanisms. When the discovery task is construed in an unconstrained way to find some sort of hypothesis, little guidance is available. It is not surprising that some philosophers relegated discovery to psychology, to the study of presumed "a-ha" discovery moments (Popper 1965). They had little to say about how to discover hypotheses or to revise them in light of anomalies. However, when the task is specified as the discovery of a mechanism, much can be said about how to proceed. The product guides the process of discovery, as stressed in Chapters 2 and 3. Guidance can be given as to what counts as an adequate description of a mechanism; which

[1] This work was supported by a grant from the General Research Board of the Graduate School of the University of Maryland. I thank Jason Baker, Jim Bogen, Carl Craver, Greg Morgan, Rob Skipper, and members of the DC History and Philosophy of Biology Discussion Group for helpful comments on earlier drafts of this chapter.

271

features need to be specified; and which reasoning strategies aid construction, evaluation, and revision of hypotheses about mechanisms.

The methodology to find such reasoning strategies is to produce compiled hindsight about strategies exemplified in cases from the history of biology. The strategies are stated more explicitly than scientists are likely to do. They are philosophical hypotheses about methods that are exemplified in historical changes that did occur. The strategies are "advisory," not descriptive or prescriptive (Nickles 1987; Darden 1991, pp. 15–17). In a future discovery episode, the philosopher may be able to provide advice that one or more of these strategies may prove useful. Further, the strategies are good candidates for items to be taught in science education.

As argued in earlier chapters, discovery should be viewed as an extended, piecemeal process with hypotheses undergoing iterative refinement. Construction, evaluation, and revision are tightly connected in ways that philosophers have often not recognized, given their neglect of hypothesis construction and revision. For example, imposing more constraints during construction satisfies, at the outset, some evaluative criteria and obviates the need to apply those criteria subsequently. More specifically, using a strategy to systematically generate rival hypotheses serves to satisfy the evaluative demand that plausible rivals be investigated. Alternatively, weaker construction strategies are a useful starting point if one has good evaluative and revision strategies. As we have seen in Chapters 9, 10, and 11, revision guided by an anomaly provides a positive role for purportedly falsifying instances. Instead of eliminating a hypothesis when an anomaly arises (Popper 1965), one may be able to diagnose the failure by localizing the anomaly and, then, use the anomaly to guide the redesign of the hypothesis. Instead of viewing science as a series of unconstrained conjectures followed by refutations (Popper 1968) or as irrational paradigm changes (Kuhn 1970), science is viewed as an error-correcting process, with iterative refinement via cycles of construction, evaluation, and revision.

The goal in this chapter is to summarize and expand strategies for the piecemeal discovery of mechanisms. Features of mechanisms guide insights into discovery strategies. First, this chapter presents an expanded characterization of mechanisms and responds to some of the criticisms raised about our earlier work. Then follows discussion of strategies for constructing, evaluating, and revising hypotheses about specific mechanisms, more abstract mechanism schemas, or incomplete sketches. The strategies are illustrated with brief examples of mechanisms from molecular biology, biochemistry, immunology, neuroscience, and evolutionary biology, many discussed in more detail in previous chapters.

## 12.2   CHARACTERIZATION OF MECHANISMS

A mechanism is sought to explain how a *phenomenon* is produced (Machamer, Darden, Craver 2000; Chapter 1, this book) or how some *task* is carried out (Bechtel and Richardson 1993) or how the mechanism as a whole *behaves* (Glennan 1996). In Chapter 1, mechanisms were characterized in the following way:

> Mechanisms are entities and activities organized such that they are productive
> of regular changes from start or set-up to finish or termination conditions.
> (Machamer, Darden, Craver 2000, p. 3; Chapter 1, this book)

This characterization that Peter Machamer, Carl Craver, and I (hereafter MDC) proposed in 2000 (reproduced in Chapter 1) has been subjected to refinements and critiques. The characterization was not presented as a definition of "mechanism" to provide necessary and sufficient conditions for the usage of the term in all cases. This characterization was informed by our detailed examination of cases from molecular biology and neurobiology, as well as reflection on requirements for productive changes.

The MDC characterization and my refinements of it have several key components in need of further discussion. These include the perspectival nature of this characterization, the nature of working entities and of activities, critiques of the claim of regularity, the importance of production and productive continuity, and discussion of the organizational aspects of a mechanism.

When biologists identify mechanisms, there is an inherent perspectival aspect as to what is picked out of interest from all the goings on in the world (cf. Kauffman 1971). First, the choice of phenomenon is relative to the scientist's interests. This is the first step in an investigation of a mechanism. How phenomena are individuated and characterized are open to revision in the light of empirical inquiry. There are reciprocal relations between identifying a phenomenon and finding the mechanism(s) that produces it. If, for example, two mechanisms are found instead of one, the phenomenon may be more appropriately divided into two phenomena (or maybe not, if one mechanism is a backup that operates if the first fails to produce the same phenomenon).

As we discussed in Chapters 1 and 2, to some extent and in some cases, the choice of beginning, ending, topping off, and bottoming out points may also be related to the interests of the investigator. In an ongoing series or cycles of mechanisms, with nested levels of mechanisms, exactly where investigators choose to focus may well be influenced by their interests, as well as their available model organisms, model experimental systems, techniques, and tools.

However, various factors mitigate rampant perspectivalism, that is, completely arbitrary choices in individuating phenomena and mechanisms. Some phenomena come more naturally packaged than others. One example in MDC was the mechanism of DNA replication, which has a natural beginning point of one double helix and a natural ending point of two helices. Some bottoming out and topping off points are not arbitrary. In order to operate, some mechanisms require entities of a given shape and size (and other properties) as the working entities of that mechanism. Chapter 4 supplied the example of chromosomes as the working entities in mechanisms of assortment of linkage groups of genes. Some activities require entities having certain properties. Hydrogen bonding, for example, requires polarized molecules exhibiting slight asymmetric charges. If hydrogen bonding is the operative activity, then entities smaller than molecules cannot be recruited as working entities to carry out that activity. Neutral molecules in the cellular milieu will not participate.

Often biologists engage in much investigative work to discover the level at which certain mechanisms operate. Geneticists worked to find the operative level for genetic linkage, ruling out the coupling of paired alleles and the reduplication of germ cells, and ruling in chromosomal mechanisms (discussed in Darden 1980; Chapter 6 this book; Darden 1991, Ch. 2). In immunology, the working entities in clonal selection were at first hypothesized to be self-replicating protein molecules but were later found to be self-reproducing immune cells (Darden and Cain 1989; Chapter 8, this book). These two examples show that biologists do not always discover working entities by going to a smaller size level; sometimes the operative units are larger than at first hypothesized.

Another factor constraining choice of perspective is that some mechanisms operate only within natural boundaries, such as the nuclear or cellular membrane. These boundaries protect the mechanism's components from buffeting and dissipation. Biologists do not arbitrarily choose such locations and boundaries of mechanisms; instead, they find in their investigations that being inside or outside such boundaries makes a difference.

As discussed in Chapter 4, various general features of entities (whether working or not) aid in identifying the working entities of a mechanism. The goal is not to define an entity by giving necessary and sufficient conditions. The goal here, as in other analyses throughout this book, is to provide strategies that have been found to be useful; hence, these features of entities should be seen as guides for finding entities. An entity may have a spatio-temporal location. An entity may have a clear boundary, such as a membrane bounding it. An entity may be composed of chemically bonded subparts that are not similarly bonded to the parts of other entities. It may be composed of specific chemicals that differ from chemicals in the surroundings. It may be stable over some period of time, as are chromosomes, or it may be rapidly synthesized and degraded, as are some messenger RNAs. It may have a developmental history; that is, it may be formed during embryological development. It may have an evolutionary history; that is, it may be a descendant in a lineage.

In addition to these general features of entities, working entities in mechanisms have additional features. A working entity *acts* in a mechanism. It may move from one place to another. It may have one or more *localized* active sites. For example, the centromeres of chromosomes are active sites that attach to the spindles during the mechanisms of meiosis. Similarly, enzymes have localized active sites that bind to substrates. Alternatively, the active sites may be *distributed* throughout the entity, as are the slightly charged bases along the entire double helix, which serve as the active sites in DNA replication. An entity may have activity-enabling properties or it may bear activity signatures that indicate an activity operated on it previously (discussed in more detail in Chapter 3).

Working entities in a given mechanism may be and often are different sizes. For example, ions, macromolecules, and cell organelles may all be working entities in the same mechanism, such as the mechanism of protein synthesis. Because working entities in a given mechanism may be of different sizes, mechanism levels may not correspond tidily to size levels, as emphasized in Chapter 2. Of course, all biological entities are composed of smaller parts; however, most subcomponents do not change during the activities of the working entities of which they are parts. For example, atomic nuclei are parts of working entities but merely stable subcomponents in most biological mechanisms. Atomic nuclei are not *working* entities or active sites in the DNA replication mechanism. They are parts of the structure, providing stability, but buried away behind electrons from active sites. In other conditions, nuclei of atoms can become working entities (e.g., in nuclear fission mechanisms when atoms are split). But such an activity during DNA replication would completely disrupt that biological mechanism.

The linear and three-dimensional structures of biological entities often hold clues to the activities in which they can engage. The double helical structure of DNA is a famous example, immediately suggesting a copying mechanism. The concept of an activity has traditionally received less attention by philosophers than the concept of an entity and is sometimes viewed as more problematic than the concept of an entity. Breaking with the more usual entity, property, interaction ontology, MDC stressed that mechanisms are composed of both *entities* (with their properties) and *activities*. Activities are the "doings" or the "producings" in mechanisms. That is, activities are

275

the producers of change; they are constitutive of the transformations that yield new states of affairs. As Machamer (2004) stressed, activities are often referred to by verbs or verb forms (e.g., participles, gerunds). Molecules *bond*, helices *unwind*, ion channels *open*, chromosomes *pair* and *separate*.

In MDC, we named four types of activities that we claimed operated in molecular and neurobiological mechanisms: geometrico-mechanical, electro-chemical, energetic, and electromagnetic. Those names are in need of refinement. "Electro-chemical" should be called merely "chemical bonding." ("Electrochemistry" historically referred to the relations between electricity and living things.) "Energetic" may not be the best term for the activity of diffusion, such as in osmosis.

An important feature of activities is that they come in types that have been discovered as science has changed. Over the centuries, scientists discovered new types of activities and their ways of operating. They then became part of the "store" or "library" of mechanism parts available for new discovery episodes. Modules may be put together in novel ways to form new hypotheses about possible mechanisms (discussed in Chapter 2 and in Darden 2001).

Marcel Weber (2005, Ch. 2) correctly endorsed the importance of mechanisms in biology. He discussed MDC's example of the mechanism of action potentials in neurotransmission. He identified entities, such as $Na^+$ channels (see Figure 1.2, this book), and curiously separated activities into "active" and "passive" categories. The $Na^+$ channel, he claimed, after changing conformation, "passively" allows $Na^+$ ions to pass through it. Perhaps there is a useful distinction to be made between active and passive transport but, as MDC noted, the $Na^+$ channel protein has numerous "active" activities, such as attracting, repelling, opening, closing, and rotating. Calling these "passive" is at best curious; however, his discussion does point to the need for further analysis of the roles of passive diffusion and active transport as activities in molecular biological mechanisms.

After identifying various other biological entities and their activities, Weber (2005, p. 32) puzzlingly found osmosis problematic because he could not identify a single kind of entity with that activity. However, he too readily retreated to the need for physical laws (see Bogen 2005 for a critique) and equated MDC's view of activities with Nancy Cartwright's (1989) view of capacities (as Weber perhaps concedes in footnote 17, p. 299). MDC's view does not require that each activity be associated with a single entity as a single activity-enabling property (a capacity); that view is more entity-property centered than our view. MDC explicitly discussed osmosis as a kind of activity.

Machamer said more about activities and their relations to entities:

> Entities, most often, are the things that act. This may be taken to imply that there
> is no activity without an entity. However, this is not to say that activities belong
> to entities in the same way structural properties belong to entities: running
> does not belong to Lisa in the same way that her nose does. It is to say that
> activities are how entities express themselves. . . . some philosophers want to
> treat activities as dispositions or propensities. . . . it is not clear that all activities
> are necessarily the activity of some entity, or less strongly, that one always can
> or needs to identify an entity to which an activity belongs. It is unclear to me
> that forces, fields, or energy are entities . . . or that the process or activities of
> equilibrating or reaching stasis need entities in order to be understood.
>
> (Machamer 2004, p. 30)

Activities can sometimes be identified independently of the entities that
engage in them. For example, the melting temperature of the DNA helix
indicated that it contained weak hydrogen bonds, even before the specific
bases exhibiting those bonds had been identified. More generally, activities
may sometimes be investigated to find their order, rate, and duration more or
less independently of the entities that engage in those activities.

James Tabery (2004) contrasted MDC's activities and Stuart Glennan's
(1996, 2002) interactions in our different characterizations of mechanisms.
Tabery suggested that the two concepts should be synthesized to emphasize
"interactivity." This concept, he said, "draws on the property changes that
occur between entities of a mechanism emphasized in Glennan's analysis."
Furthermore, "it also takes advantage of the fact that the production of these
property changes is a dynamic process," as MDC emphasized (Tabery 2004,
p. 12). Tabery helpfully pointed out the dynamic aspect of activities. But
activities can be carried out by one entity, by two entities interacting, or by a
more generalized process not easily attributed to the capacity of or change in
the property of a single entity (e.g., osmosis). Interaction between two entities
is a kind of activity, conceived in a more entity-centered way and missing the
*productive* nature of activities in general.

Machamer replied to Tabery's talk of *inter*action by saying:

> Here, and in MDC (2000), it may have been unclear that activity is meant to
> include activities that are mutually effective and affected. There is no dispute
> about interaction if the "action" part is taken to refer to activities (so they'll be
> interactivities), and not as is usually done to refer to relations that exist among
> static states. (Machamer 2004, footnote 4)

In addition to talk of "phenomena," "entities," and "activities," the MDC characterization of mechanism included the phrase "regular changes." Jim Bogen (2005) suggested omitting "regular." He pointed to instances of irregularly operating and stochastic mechanisms. A mechanism, he claimed, might operate just once. Machamer (2004) embraced Bogen's suggestion. However, even if it is not constitutive of what it is to be a mechanism, most of the biological mechanisms discussed in this book are *regular*. They usually work in the same way under the same conditions. The regularity is exhibited in the typical way that the mechanism runs from start to finish. As Bogen conceded, regularity is important to biologists' ability to investigate mechanisms. The following discussions on experimental strategies, as well as the generality and scope of mechanism schemas, show the importance of finding regularly operating mechanisms.

Bogen (2005) correctly stressed that the most important aspect of mechanisms is their *productivity*. An activity produces a change. I imagine a world with no productive activities being a completely static one.[2] Others have called for a definition of production (see, e.g., Woodward 2002; Psillos 2004). Finding necessary and sufficient conditions for recognizing the many diverse kinds of production is difficult. As Machamer suggested in MDC, human beings directly experience many kinds of activities, such as collision, pushing, pulling, and rotating (the favorite activities in seventeenth-century mechanisms). Scientists have since discovered many kinds of activities not directly detectable by human senses, such as attraction and repulsion, hydrophobicity and hydrophilicity, and movements across membranes to achieve equilibrium. Science students must be trained to understand how these activities work so that, with education, they can "see" how mechanisms employing them operate. Understanding how an activity (even one far removed from sense experience) works makes a proposed mechanism employing that activity intelligible.

Skipper and Millstein (2005) joined Bogen in criticizing (among other things) MDC's claim of "regular changes." Rather than stochastic mechanisms in neuroscience, which were Bogen's examples, Skipper and Millstein examined the mechanism of natural selection. They are correct that there is an inherently probabilistic component in the way that the mechanism of selection operates. However, there must be a bias in the way that the adapted property variant interacts with the critical environmental factor (the P and F in the schema for natural selection discussed in Chapter 8). As Joe Cain and I argued (Cain and Darden 1988; Darden and Cain 1989; see Chapter 8,

---

[2] In my imagination this takes the form of a frozen world but, then, ice couldn't even form without productive activities. I think I just can't image a world without activities.

this book), the key difference-making step of selection is that P-F interaction, which serves to benefit the organism with property P in contrast to the lack of benefit for other organisms without P. That bias produces the phenomenon of the increase of Ps in subsequent generations, the outcome of the operation of the mechanism of natural selection. And, one expects that, given the same set of variant properties and the same environmental challenge, the same property P would be the adaptive one, producing (in the absence of disrupting factors) the same outcome. So, one might interpret the "regularity" of the way the mechanism of selection operates as something between deterministic and completely random, producing (yes, probabilistically producing) a bias in favor of increase of property P.[3]

In the original MDC characterization, we did not perhaps sufficiently stress that mechanisms have *productive continuity* between stages. Carl Craver and I (see Chapters 2 and 3) placed more emphasis on this aspect of mechanisms: The entities and activities of each stage give rise to, allow, drive, or make the next. There are no gaps from the set-up to the termination conditions. Furthermore, Craver and I stressed the aspects of the *organization* of mechanisms crucial to this productive continuity. For those such as Glennan (1996), who characterize a mechanism as a "complex system," some sort of organization is already assumed. MDC did not use the term "system" because we suspected that characterizing what was in and what was out of the system was just as difficult (if not more so) than characterizing a mechanism itself. More recently, William Bechtel and Adele Abrahamsen (2005) have stressed the spatial and temporal organization of the parts of a mechanism and their orchestrated effects.

As Craver and I argued, organization of the mechanism has a number of features. The general features of a mechanism are listed in Table 12.1 (which is a modified version of Table 2.1, where features were called "constraints"). An adequate description of a mechanism includes these features; hence, the search for them guides the discovery of mechanisms. The first feature is "phenomenon" because, as discussed previously, the first step in the search for a mechanism is to identify a phenomenon of interest. Next are componency features. The mechanism's component entities and activities, sometimes further organized into modules, are sought. These components have spatial and temporal organization. The components of mechanisms have *spatial* organization, including location, internal structure, orientation, connectivity, and

---

[3] No doubt Skipper and Millstein (2005) are correct that more work needs to be done to characterize natural selection as a mechanism. I have not addressed all their critiques here. Exactly what changes are needed in the MDC characterization of mechanism remains to be seen.

**Table 12.1  Features of Mechanisms**

Phenomenon

Components
  Entities and activities
  Modules

Spatial arrangement of components
  Localization
  Structural
  Orientation
  Connectivity
  Compartmentalization

Temporal aspects of components
  Order
  Rate
  Duration
  Frequency

Contextual locations
  Location within hierarchy
  Location within series

sometimes compartmentalization. Furthermore, mechanisms have *temporal* organization. The stages of the mechanism occur in a particular *order* and they take certain amounts of *time* (*duration*). Some stages occur at a certain *rate* or repeat with a given *frequency*.

In addition to the componency, spatial, and temporal features of a specific mechanism, that mechanism may be situated in wider contexts – both in a hierarchy of mechanism levels (discussed in Chapter 2) and in a temporal series of mechanisms (discussed in Chapter 4). Biological mechanisms often can be situated in a hierarchy of mechanisms with lower level ones playing roles in higher level ones. For example, mechanisms of spatial memory, as discussed in Chapter 2, may be characterized at the membrane, synapse (intercellular), brain region, and organismal behavior levels. As discussed in Chapter 4, a temporal series of mechanisms spanning generations is the wider context for the mechanisms of heredity.

All of these features of mechanisms (listed in Table 12.1) can play roles in the search for mechanisms, and then they become parts of an adequate description of a mechanism. This detailed characterization of mechanisms supplies constraints and guidance for reasoning in their discovery. As emphasized in Chapter 2, the product shapes the process of discovery.

In providing a characterization of something, it is useful to have examples of things that fail to satisfy that characterization. The MDC characterization of mechanism points to its *operation*. Although someone (perhaps Glennan 1996?) might call a stopped clock, for example, a mechanism, I would not. It is a machine, not a mechanism. The MDC characterization views mechanisms

280

as inherently active. In the stopped clock, the entities are in place but not operating, not engaging in time-keeping activities. When appropriate set-up conditions obtain (e.g., winding a spring, installing a battery), then the clock mechanism may operate. The stopped clock as a mechanism is a poor analogy for many biological mechanisms that make and destroy entities as they operate. For example, in the mechanism of protein synthesis (as in any synthesis reaction), some of the entities of the mechanism are not intact and in place prior to the initiation of the start conditions. Some are made on the fly and rapidly degraded after they play their role (e.g., some messenger RNAs). Thus, the analogy to a system with stable parts that either operate or do not fails for some changing components in this mechanism.

Another example of something that fails to satisfy the MDC characterization of a mechanism is a crystallized form of a molecule. No productive change is occurring in a crystal, even though chemical bonds are holding it together. One might regard stabilizing forces as kinds of activities, resisting dissipation. From such a perspective (suggested to me by Joshua Lederberg, see his 1965 paper), a crystal might be viewed as composed of a kind of degenerate mechanism, an extreme in a continuum from mechanisms producing stability to mechanisms producing orderly changes (the kinds of interest here) to random motions producing nothing in particular.

Now let's turn from discussion of the nature of mechanisms in the world to scientists' representations of mechanisms. As discussed in earlier chapters, a *mechanism schema* is a truncated abstract description of a mechanism that can be filled with more specific descriptions of component entities and activities. An example is James Watson's (1965) diagram of his version of the central dogma of molecular biology:

DNA –>RNA –>protein

This is a schematic representation (with a high degree of abstraction) of the mechanism of protein synthesis.[4]

A less schematic description of a mechanism shows how the mechanism operates to produce the phenomenon in a productively continuous way and satisfies the componency, spatial, temporal, and contextual constraints. A goal is to find a description of a mechanism that produces the phenomenon,

---

[4] To describe abstractions, I now use the term "degree" of abstraction, not "level." The term "level" is used in Craver's (2001) discussion of mechanisms to point to part-whole relations among nested mechanisms. "Level" is often used by others to refer to entities of different sizes. The "degree" of abstraction of a schema may be "specified" by adding more detail and finally "instantiated" for a particular fully specified instance.

and for which there is empirical evidence for its various componency and organizational features. A mechanism schema can be instantiated to yield such an adequate description.

In contrast, a mechanism sketch cannot (yet) be instantiated. Components are (as yet) unknown. Sketches may have black boxes for missing components whose function is not yet known. They may also have gray boxes, whose functional role (Craver 2001) is known or conjectured; however, which specific entities and activities carry out that function in the mechanism are (as yet) unknown. The goal in mechanism discovery is to transform black boxes (components and their functions unknown) to gray boxes (component functions specified) to glass boxes (components supported by good evidence), to use Hanson's (1963) metaphor. A schema consists of glass boxes; one can look inside and see all the parts. Well-supported theories in biology are often represented by schemas and schematic diagrams. An instantiated schema shows details of how the mechanism operates in a specific instance to produce the phenomenon. Hence, mechanistic theories explain the phenomena in their domains.

Glennan (2005) criticized the schema/sketch distinction, saying that articulation of a mechanistic model is a continuous process. For a period of time, during discovery, that is right. However, biologists come to believe that they have sufficiently identified the components of mechanism schemas after some period of work and amassing of good evidence so as to claim that they know how actually the mechanism operates. Schemas then become textbook knowledge and are no longer the source for research projects; incomplete sketches indicate where fruitful work may be directed to produce new discoveries.

In passing, it is worth noting that biologists and philosophers use the term "model" in many ways. When the model is a model of a mechanism, "model" may refer to a mechanism sketch, a schema at any degree of abstraction, or an instantiation of a schema. Sometimes the terms "model" and "theory" are used synonymously, in which case a mechanism schema is appropriate. Sometimes a model is said to be an instance of a theory, showing how the abstract theory is to be applied in a particular case, in which case an instantiation of a mechanism schema is appropriate. Sometimes biologists say, "I want to propose a model" (as opposed to saying "I want to propose a mechanism") when their hypothesis is a mechanism sketch with many missing parts and/or little empirical support. (Curiously, I rarely find biologists using the term "mechanistic model." "Model of a mechanism" occasionally turns up, but I haven't done a thorough survey.) Mathematical and computer simulation models of mechanisms usually have equations or functions which produce state transitions while omitting representations of structures and the activities that produce the transitions. Sometimes a scientific model is not a model of a mechanism. Given the many uses of "model," I avoid it here. This discussion makes explicit the differences among a sketch, a schema, and an instantiation; hence, those terms are more informative than "model."

As MDC mentioned and Bechtel and Abrahamsen (2005) have discussed in more detail, diagrams are especially propitious for representing many mechanisms. They show more or less structural details of the entities and overall organization of the parts. Activities are more difficult to represent in static drawings. Sometimes they are symbolized by arrows, although arrows are also often used to represent mere movement or to show time slices. Cognitive psychologists have studied how humans manipulate visual representations in order to run "mental simulations" of mechanisms (on mental models, see Nersessian 1993). This enables the person to "see" how some mechanisms work and to use the representation to make predictions (discussed in Bechtel and Abrahamsen 2005). But, in more complex cases, humans may need aids, such as computer simulations, to represent and run a simulation to make a prediction.

Also in passing it is worth noting that this discussion of mechanisms has as yet found little need for a general philosophical analysis of the problematic concept of cause. As noted in MDC, and as further developed by Bogen (2004, 2005), cause is a schema term. To talk of A causing B seems to be a rather impoverished way of talking compared to describing a mechanism. "Cause" may point to a mechanism connecting A with B, or it may point to an appropriately bottomed-out *specific* activity that *produces* a successor stage B from stage A, by, for example, pushing, pulling, bonding, or breaking. Woodward's (2003) interventionist view of causal explanation furnishes a useful method for investigating the components of mechanisms, given appropriate experimental setups. However, the analysis of mechanisms here conveniently sidesteps the philosophical enterprise of finding a single analysis of the concept of cause (for a critique of the search for a single analysis, see Cartwright 2004).

In sum: Reasoning to find any particular mechanism is guided by the description of the phenomenon of interest, aided by the characterization of what a mechanism is, and elaborated by specifying the features that an adequate description of a mechanism should satisfy. Discoverers need no longer flounder in an unconstrained space of finding vaguely characterized theories. If the goal is to discover a mechanism, much can now be said about reasoning strategies to aid that task.

Reasoning strategies may guide construction of hypotheses about mechanisms. Guidance may come from schema instantiation, modular subassembly, or reasoning forward or backward about the entities and activities themselves. Schema instantiation utilizes an abstract type of mechanism, often found via analogy, that may be further specified and instantiated for a particular case. In modular subassembly, one searches not for an entire schema but rather for types of modules to assemble into a hypothesized mechanism. A strategy operating at an even finer grain is to reason stage by stage about how gaps in what is known about the productive continuity of a mechanism are to be filled, either forward chaining from a convenient starting point or backward from a later stage.

### 12.3.1   Schema Instantiation

Schema instantiation begins with a highly abstract framework for a mechanism. The mechanism schema is then rendered less abstract by instantiating it. Instantiation is usually characterized as supplying values for the variables in a schema, as in Kitcher's (e.g., 1989) discussion of the instantiation of an argument schema. This view of a two-place relation – a variable and its value – is too restrictive. Schemas may be stated with varying degrees of abstraction, that is, with more or fewer details. One may specify details to make a mechanism schema less abstract before one gets all the way down to a description of a particular mechanism. For example, DNA –>template –>protein is more abstract than DNA –>RNA –>protein, which is more abstract than an instantiated schema for protein synthesis. Furthermore, the parts of a mechanism schema are not premises and conclusions in an argument pattern, as in Kitcher's analysis of a schema. They are, instead, placeholders for ordered steps in an operating mechanism. (For more discussion of this point, see Skipper 1999.)

There are many types of mechanisms: transport mechanisms, control mechanisms, repair mechanisms. Consider the example of selection mechanisms. To find a mechanism to carry out the task of producing adaptations, one might consider a selection schema (Darden and Cain 1989; Chapter 8, this book). At a high degree of abstraction, a selection schema may be characterized as having the following modules: first comes a stage of variant production, then a selective interaction that poses a challenge to the variants, followed by differential benefit for some of the variants. Specifying this abstraction, that is, rendering it less abstract, can produce natural selection in evolutionary biology. Supplying further details yields an instantiation, that is, a description of a particular mechanism for producing stout beaks on Galapagos finches for cracking hard seeds during drought conditions, for example. Alternatively, a different specification yields the clonal selection theory in immunology. A set of lymphocyte cells have variant reactive sites on their surface; they react differently with invading antigens; those reacting with the antigen are activated; more cells with such reactive sites are cloned to deactivate the invading antigen.

There are several sources of schemas, as Chapter 8 discussed. One is the history of science. Analogous theories may be grouped and an abstract schema constructed by dropping the specific details (Holyoak and Thagard 1995). Dropping details of genic, organismic, and group selection yields a natural selection schema; dropping further details yields an even more abstract selection schema that may be instantiated for other selection mechanisms, such as clonal selection in immunology. In addition to finding analogs in the history of science, scientists often use "local analogies" to similar mechanisms in their own field and "regional analogies" to mechanisms in other, neighboring fields, as the cognitive psychologist Kevin Dunbar (1995) found in his studies of reasoning strategies in molecular biology laboratories (Dunbar 1995). Thus, closely related areas of contemporary science are also a source for mechanism schemas.

Another method for schema construction is to sketch hypothetical roles that components of the mechanism being sought are expected to carry out and work to specify them. In 1952, Watson sketched the protein synthesis mechanism that began with DNA, had some as-yet unknown stage involving RNA, and ended in the synthesis of the protein (Watson 1968). Further work converted Watson's sketch into a schema with three different RNA components playing their various roles in the mechanism (see Chapter 3, this book). Crick (1988) later analogized the role of the ribosome to a reading head of a tape recorder, which moves along the tape-like messenger RNA and reads the genetic code. However, no evidence exists that this distant analogy played a role in the discovery of the role of the ribosome or furnished a source for constructing a schema in this case. This historical evidence is consistent with Dunbar's work. He found that in contemporary molecular biological laboratories, such "long-distance" analogies were not used in research but were instead used to bring home a point or to educate new staff members (Dunbar 1995).

Once a schema is hypothesized or sketched in a discovery episode, then the task is to find the entities and activities, or modular groups of them, that play the roles outlined in the abstract schema. Black and gray boxes may be filled piecemeal, as empirical evidence is found for the various components. The lack of an entity or activity or module to fill a place or a role in a sketch points to the need for further work. Continued failure to find role fillers may lead to abandonment of the schema/sketch in favor of another one. The goal is to produce a mechanism schema in which the parts are glass boxes – one can see what is inside for any given instantiation.

By about 1970, the details of the protein synthesis mechanism had been worked out. By then, the schema DNA –>RNA –>protein became textbook knowledge that could be instantiated whenever a protein synthesis mechanism was needed. The protein synthesis mechanism came to serve as a module in other mechanisms.

## 12.3.2   Modular Subassembly

The use of analogues in discovery has been much discussed (e.g., Harré 1970; Darden 1980; see Chapter 6 this book; Holyoak and Thagard 1995). Gentner (1983) argued that sophisticated analogy users transfer a tightly interconnected system of causal relations from analogue to subject. However, such analogical reasoning is conservative: One expects to find the entire abstract framework, the entire schema, elsewhere. More creative is putting together parts, modules, in new ways.

Thus, another strategy to guide mechanism discovery, in addition to schema instantiation, is modular subassembly. This strategy involves reasoning about groups of mechanism components. One hypothesizes that a mechanism consists of (perhaps known) modules or types of modules. One cobbles together different modules to construct a hypothesized mechanism.

Evolution itself often works by copy and edit: Copies of genes can be found in mutated form, playing similar or different roles in the same or related organisms. Finding these recurrent motifs has been a powerful tool in discovery in biology. There are various types of receptors, types of neurotransmitters, types of enzymes, and types of gene regulatory components (e.g., inducers, repressors).

Somewhat larger modules are recurrent components in developmental mechanisms. The developmental biologist Scott Gilbert and his colleagues note:

> ... in the Wnt signaling case, every element in the insect pathway has a homologue in the vertebrate embryo.... The genes and the protein interactions are the same: only the "readout" (the target genes) is changed between tissues and species.... The Wnt pathways in vertebrates, nematodes, and arthropods comprise proteins arranged in similar fashions.... The Wnt pathway... then is

a homologous cassette of information, a homologous module of gene-protein
interactions. . . .                                    (Gilbert and Bolker 2001, p. 4)

Knowledge of such types of modules may be useful in constructing a
plausible candidate mechanism in a particular discovery episode.

Finding a new type of module opens a new hypothesis space of possible
mechanisms. As discussed in Chapter 10, the 1970 discovery of the enzyme
reverse transcriptase, which copies RNA back into DNA, is such a module.

RNA – via reverse transcriptase –>DNA

The discovery opened up a space of possible mechanisms with feedback
into DNA from elsewhere. After this reverse transcriptase module was found
to function in the mechanism by which retroviruses copy their RNA back
into the host DNA, it then became a module for hypotheses about nonviral
mechanisms. Controversial proposals of possible mechanisms for directed
mutation in bacteria (Cairns et al. 1988) and for feedback from the soma to
germ line cells in the immune system in mammals (Steele et al. 1998) are
examples of instructive mechanisms utilizing reverse transcriptase (Chapter
11, this book).

### 12.3.3   Forward/Backward Chaining

A schema provides the overall framework of the mechanism. Modular subassembly provides working subcomponents of a schema. Finally, at a finer
grain, one can construct a hypothesized mechanism by reasoning about the
entities or activities themselves. Forward and backward chaining are reciprocal strategies for reasoning about one part of a mechanism on the basis of what
is known or conjectured about other parts in the mechanism. Forward chaining uses the early stages of a mechanism to reason about the types of entities
and activities that are likely to be found in later stages. Backward chaining
reasons from the entities and activities in later stages in a mechanism to find
entities and activities appearing earlier.

Forward chaining is illustrated by Watson and Crick's suggestion about
DNA replication. As soon as the double helix structure of DNA was proposed,
properties of the double helix indicated how it might be copied. Watson and
Crick's famous line in their first 1953 paper showed their ability to reason to
the next stage: "It has not escaped our attention that the specific [base] pairing
we have postulated immediately suggests a possible copying mechanism for
the genetic material" (Watson and Crick 1953a, p. 737). DNA has polarly
charged bases that hold the two helices together with their complementary
hydrogen bonds. These entities could obviously play a role in the first stage of a

copying mechanism. The double helix could open and allow complementary bases to line up along it. The polar charges and their spatial arrangements are *activity-enabling properties*. They suggested to Watson and Crick what happened in the next stage of the mechanism. Continuing to forward chain, one then could see how two identical helices would result.

Chapter 3 discussed backward chaining as illustrated by Zamecnik and Hoagland's work on protein synthesis in the 1950s and 1960s. Biochemists knew that the endpoint of the protein synthesis mechanism was a string of amino acids held together by strong covalent bonds. They thus reasoned back toward free amino acids. Since energy was required to form such strong bonds, that activity required a high-energy intermediate in the immediately preceding step. They isolated such a high-energy intermediate. Surprisingly, the activated amino acid was associated with RNA. The biochemical reaction schema had no role for RNA to fill, and reasoning backward from protein to free amino acids did not suggest an RNA intermediate.

Meanwhile, the molecular biologists were reasoning forward from the DNA double helix to the next stage in the protein synthesis mechanism. Molecular biologists suggested that RNA played the functional role of template. The order of the bases in DNA would be transcribed into similarly ordered bases in RNA, which would then be translated, thereby serving as the template to order the amino acids in the protein during protein synthesis. These tandem strategies served to fill gaps in the productive continuity of the proposed mechanism. The molecular biologists reasoned forward from the DNA, while the biochemists reasoned backward from the finished protein; their work met in the middle of the mechanism, with the discovery of the various types of RNAs and their roles in the middle of the mechanism.

Even if one cannot find a possible overall schema or familiar modules, one may be able to reason forward from the beginning or backward from the end of the mechanism in a search for its productive continuity from start to finish. The nature of the entities and activities at each stage guides the discovery of the prior and subsequent stages, based on activity-enabling properties, activity consequences, as well as entity and activity signatures (Darden and Craver 2002; Chapter 3, this book). With cyclic (e.g., feedback) mechanisms, some separable stage can serve as a relative starting point for reasoning about earlier or later ones. Thus, the strategy of forward/backward chaining seems likely to be available when anything is known or can be conjectured about entities and activities anywhere in the hypothesized mechanism.

Construction of one or more hypothesized mechanisms (or schemas or incomplete sketches) is but the first step in discovery. Experimental testing and other evaluation strategies are needed to assess adequacy.

## 12.4   STRATEGIES FOR EVALUATION: HOW POSSIBLY
## TO HOW PLAUSIBLY TO HOW ACTUALLY

Richard Westfall criticized the mechanical philosophy in seventeenth-century chemistry:

> Since there were no criteria by which to judge the superiority of one imagined mechanism over another, the mechanical philosophy itself dissolved into as many versions as there were chemists.          (Westfall 1977, p. 81)

Clearly, finding methods for evaluating proposed mechanisms is an important task. Happily, contemporary biologists have much more guidance than was available to seventeenth-century chemists.[5] Molecular biological mechanisms have many components that are supplied by modern chemistry's well-supported theories of molecular structure and types of bonding. Furthermore, modern biology itself has developed good strategies for evaluating hypothesized mechanisms. Some of those strategies are the subject of this section.

As we saw in the previous section on construction strategies, the nature of the desired product – an adequate description of a mechanism – guides the process of construction. Similarly, much guidance is provided for how to evaluate a proposed mechanism. Let's consider in more detail evaluative strategies tailored to evaluating mechanistic hypotheses.

As we have seen, the search for a mechanism begins not with a conjectured generalization but rather with the characterization of a puzzling phenomenon.[6] Then one seeks a well-supported hypothesis about the mechanism that produces the phenomenon, perhaps constructed using one or more of the strategies for construction discussed in the previous section. Hence, an important step in evaluation is to consider if, indeed, the phenomenon exists and

---

[5] The history of the usage of the concept of mechanism from the seventeenth century to molecular biology has yet to be written. For a brief overview, see Machamer, Darden, Craver (2000; see Chapter 1, this book, Section 1.5.2) and Craver and Darden (2005).

[6] Abduction is supposedly reasoning that begins with a puzzling phenomenon and then involves an inference to the best explanation for that phenomenon. However, accounts of abduction do not provide any guidance as to how to identify a puzzling phenomenon (on this account, one for which there is no known mechanism to produce it). Nor do such accounts say what kind of inference can allow finding an explanation (here, using strategies for constructing a how possibly mechanism). Nor do such accounts provide ways of generating rivals or of knowing how thorough the search for such rivals has been (a point emphasized in Josephson and Josephson 1994). Nor do such accounts provide evaluative criteria for choosing the best explanation (here, strategies for evaluating hypothesized mechanisms). So, talk of abduction either is vacuous or else the reasoning strategies in this chapter for construction, evaluation, and revision of mechanistic hypotheses are needed to supplement the too abstract schema for abduction. See further discussion in Darden 1991 (pp. 10–11, 267–269).

how it is best characterized. Is there good data (reproducible, with adequate controls) to provide evidence that a phenomenon is produced? (On the relation of data and phenomena, see Bogen and Woodward 1988.) Is there a single phenomenon or can two or more be distinguished, calling for two or more different mechanisms? What clues does the phenomenon itself provide about the kind of mechanism and its location? (These queries flow from the "phenomenal constraints" discussed by Bechtel and Richardson 1993, pp. 235–239.)

Once a phenomenon is at least provisionally characterized and credentialed and the goal of finding a mechanism that produces it is chosen, then evaluation may commence with one of several beginning points. The starting point for evaluation depends on which of the several construction strategies was used. One need not begin with a thoroughly articulated mechanistic hypothesis. One might begin with a very incomplete sketch, full of black boxes. Or, evaluation might commence with a sketch with gray boxes, that is with functional specifications as to which roles the as-yet-unknown components play. Given that any proposed mechanism must be shown to be better than rivals, one might begin with not one but two or more abstract types of possible mechanisms, that is, abstract points on two different branches in an abstract space of possible mechanisms. Or, one might propose one well-articulated mechanistic hypothesis, constructed via such methods as schema instantiation or modular subassembly.

One might be able to identify the type of schema used in hypothesis construction and find specific testing strategies for that type of theory. For example, in reflecting on his own reasoning strategies, Crick enunciated the importance of such considerations: "One should ask: What is the essence of the type of theory I have constructed, and how can that be tested?" (Crick 1988, pp. 141–142).

The goal of evaluation (more fully, the goal of iterative construction-evaluation-revision) is to produce an adequate description of a mechanism with all the features of Table 12.1 specified and with evidence for each of them. How possibly needs to be converted to how plausibly and finally to how actually the mechanism works. Some questions to ask as part of this evaluation process are the following[7]: What are the component entities and activities? How are they spatially and temporally organized? Is the hypothesized mechanism an incomplete sketch in need of being fully articulated so as to describe a productively continuous mechanism? Have rival hypotheses

---

[7] Stuart Glennan (2005, p. 457) provides a similar checklist of questions to ask when evaluating a mechanistic model, based on his own characterization of a mechanism.

(or rival types) been systematically generated, considered, and ruled out? Does the proposed mechanism fit into the wider hierarchical and serial contexts around it? As more of the evaluative criteria are met, the status of the hypothesis changes from incomplete to how possibly to how plausibly to how actually (i.e., how most plausibly). Let's consider each of these evaluative strategies in more detail.

### 12.4.1   Evaluating and Improving Sketches

Evaluation can begin with an incomplete sketch of components of the mechanism with black and gray boxes. The requirement of productive continuity demands evaluation to see if each stage can give rise to the next. Evaluating a sketch requires locating missing pieces to pinpoint what is incomplete. The location and nature of the missing pieces in a sketch provide guidance as to what is to be sought: something that can play the appropriate role in the mechanism, with appropriate connections to the stages before and after it.

In addition to employing the construction strategies discussed in the previous section, one method for removing incompleteness is exploratory investigation of an experimental setup. The setup may be an instance in which the mechanism runs to produce the phenomenon, or it may be only a working module of the mechanism. As Bechtel and Richardson (1993, p. 239) noted, a smoothly working mechanism may not immediately provide clues to its working parts. Tools for manipulation are needed so that the experimental setup can be poked and prodded and decomposed in various ways to find its components. One commonly used method for discovering unknown components in molecular biological mechanisms is to centrifuge or otherwise physically decompose the natural system and then attempt to isolate the working entities and characterize their activities and role in the mechanism (see examples in Bechtel and Richardson 1993; Rheinberger 1997; Darden and Craver 2002; Chapter 3, this book). More will be said about types of manipulations in the following discussion on experimental strategies.

This procedure shows the tight connection that may exist between articulating sketches and exploratory experimentation. Poking and prodding an experimental system may reveal the presence (or absence) of component entities and activities that then will need to find a place in a more articulated sketch. If a hypothesized mechanism component is missing, then a particular sketch can be ruled out (e.g., no reverse transcriptase in normal *E. coli*, as discussed in Chapter 11). If one finds an unexpected component, then one may need to revise the sketch to give it a functional role (e.g., RNA attached to activated amino acids, which proved to be transfer RNA; discussed in Chapter 3).

Successive refinement occurs as an incomplete sketch is converted to a how possibly/plausibly mechanism. Then that proposed mechanism(s) is available for other types of evaluation.

## 12.4.2 Evaluation Strategies for Assessing Possibility/Plausibility

Rather than begin with an incomplete sketch, one may instead conjecture how possibly the mechanism could work. Any answer to "how possibly" is a description of a productively continuous mechanism that could produce the phenomenon. If one understands how the types of activities in the how possibly mechanism work, one may be able to evaluate whether the operative mechanism could produce the phenomenon. This is a conceptual evaluation task. Bechtel and Abrahamsen (2005) refer to this as running a "mental animation" of the possible mechanism. A failure of such an animation may even support an argument that the presumed how possibly mechanism is impossible. We will discuss an example of an argument for impossibility, based on insufficient degrees of freedom. In a complex case, human cognitive capacities for envisioning a mechanism may need to be supplemented with computer (or physical) simulations in order to assess possibility.

One consideration in assessing the plausibility of a proposed mechanism is to consider the construction strategies used. More or fewer constraints may be used during construction. One can brainstorm with few constraints to get, perhaps, quite implausible how possibly hypotheses and then weed them out by introducing additional constraints. Alternatively, more constraints (if available) can be imposed to produce a smaller search space, thereby yielding a more plausible hypothesis at the outset. As more empirical evidence is found, and other evaluation strategies are successfully employed, the how possibly hypothesized mechanism becomes how plausibly.

Using reasoning by analogy in construction provides a measure of plausibility at the outset (Harré 1970; Darden 1980; Chapter 6 this book; Darden 1991). If such a schema has been successfully instantiated elsewhere, nature is claimed to act in such a way. Similarly, using known kinds of modules provides a measure of plausibility because such modules are found elsewhere. However, analogies confer only a weak measure of plausibility; appeal to evolutionary homology provides more. When there is evidence that the similar modules are evolutionarily homologous, then one is justified in expecting more similarity. A similar point is made by Weber (2005, p. 181), who discusses making inferences about mechanisms that are "phylogenetically conserved." Nonetheless, a type of mechanism never before encountered might produce the phenomenon. Consequently, appeal to analogies or presumed homologies cannot obviate the need for more direct empirical evidence about the specific mechanism and its components.

Again, consider the example of a working part of the protein synthesis mechanism discussed in Chapter 3. In 1954, Watson conjectured that the RNA that carried the template from DNA to order amino acids in protein synthesis might be an RNA helix. The grooves or holes in the turns of the helix, he conjectured, would have different shapes so as to dock in geometrical lock and key ways with the twenty differently shaped amino acids. Watson's conjecture for how this module of the protein synthesis mechanism worked gained plausibility because it was based on the analogy with the recently discovered helical DNA. A chemically similar nucleic acid RNA might also form a helix; the four bases, in various arrangements in the turns of the helix, might provide three-dimensional holes where amino acids could dock. One could possibly imagine elaborating a sketch with twenty differently shaped holes, in various orders, serving to order amino acids in proteins.

Watson worked with Alex Rich to try to get empirical evidence about the shape of this hypothesized RNA to support this "holes in the RNA" hypothesis. However, that evidence was inconclusive (Rich and Watson 1954a; 1954b). Meanwhile, Crick considered how possibly to build a working model. He recalled: ". . . by assuming that there are twenty different cavities and trying to build a structure that *had* twenty different cavities. And as soon as you put it that way, you saw that it was almost impossible to *do*" (quoted in Judson 1996, p. 283). The shapes of holes formed by the four different nucleic acid bases seemed not to have sufficient possible shapes to discriminate between twenty different amino acids. Making the task even more difficult was that some of the twenty different amino acids were known to have very similar shapes.

Hence, a three dimensional template RNA at first seemed plausible. Plausibility resulted from a presumed close analogy to the three-dimensional structure of the other nucleic acid, DNA, as well as intelligibility of a possible mechanism based on differently shaped holes. However, attempts at constrained model building (the method that has been so successful with DNA) pointed to the difficulties of finding a structure with appropriate activity enabling properties. Crick introduced additional constraints, based on the three-dimensional structures of the molecules. As a result, he was unable to build a possible structural model that could play the required template role in the protein synthesis mechanism. Even more strongly, he could argue for likely impossibility, based on insufficient degrees of freedom in the structures. Furthermore, the empirical evidence for a helical RNA was not found. The "holes in the RNA" hypothesis was originally plausible because of an analogy. However, it failed as model building and empirical constraints were introduced. As more empirical tests are passed, the plausibility increases that the proposed possible mechanism is operating in the particular case.

### 12.4.3   Consideration of Rivals as an Evaluative Strategy

Because a given hypothesis should be evaluated against its rivals, evaluation and systematic generation are closely tied (Josephson and Josephson 1994). An effective construction procedure should be sought to find rivals. Ideally, some sort of exhaustive procedure (given certain constraints) is used to generate a hypothesis space, to search among the alternatives, and to rule out all but the most plausible. For example, as we saw in Chapters 8 and 11, selective and instructive theories are the types of mechanisms that have been used to account for adaptive changes. So, an evaluative query can take the form: Have alternative types of mechanisms that can produce the type of phenomenon been considered and ruled out? As we have discussed, the history of science yields compiled hindsight about types of hypotheses that may be plausibly considered to generate rivals in some cases. Actually, in contrast to the ideal, scientists usually come up with only a few (types of) hypotheses. Perhaps efforts to systematically generate rivals should be undertaken more often.

Joshua Lederberg (e.g., 1965) proposed one heuristic procedure for constructing a hypothesis space. Find one solution and then ask: Of what set(s) is this solution a member? Consider what component might become a variable or what dimensions (the axes of the hypothesis space) might vary. Then systematically scan the alternatives. Ask if some other member of the set or some other point in the space is more plausible or at least equally worthy of consideration. Crucial experiments may be designed (and other evaluative strategies used) to choose among the rival hypotheses.

Thus, we are adding to our list of strategies for moving a proposed mechanism from the status of how possibly to how plausibly. A plausible hypothesized mechanism is one that can be seen to possibly produce the phenomenon; one that may have analogies or homologies to other known mechanisms (or their components); one that does not violate any known empirical constraints; and one that has been generated by a systematic process that uncovers plausible rivals. Plausible hypotheses passing such tests nonetheless need to be subjected to direct empirical testing to provide evidence for how, actually, the mechanism operates.

## 12.4.4  Experimental Strategies for Discovering How Actually the Mechanism Operates

Given one or more fairly well-specified possible/plausible hypothesized mechanisms, direct empirical evidence (Schaffner 1993, Ch. 4) is needed to evaluate them. Experimental investigations of mechanisms supply such direct evidence.

Chapter 2 (Craver and Darden 2001) introduced and then Craver (2002b) discussed in more detail experimental strategies for testing a hypothesized mechanism. Such experiments have three basic elements: (i) an experimental setup in which the mechanism (or a part of it) is running, (ii) an intervention technique, and (iii) a detection technique.

Biologists use many kinds of experimental setups. Intact organisms have many mechanisms running; the challenge with intact organisms is to find ways of individuating single mechanisms and ruling out confounding factors. As discussed in Chapter 3, in vitro preparations solve some problems encountered with in vivo preparations. The challenge is to find the appropriate components and make them work in vitro. In devising the in vitro protein synthesis system, Zamecnik and his colleagues originally ground up and centrifuged fractions from rat livers. The task then became to "drain the biochemical bog" (to use a metaphor from Rheinberger 1997) to find the entities and activities in the various fractions. Such a physical decomposition method is particularly useful when the starting point is an incomplete sketch, with many unknown components. A goal is to be able to isolate and identify working components or to use off-the-shelf components or synthesize them to engineering specifications. The ability literally to construct such an in vitro setup shows that all the components are known.

Given an experimental setup still in need of investigation to find the mechanism's components, a usual method employs the oft-discussed strategies – an intervention and detection of its effects. Several different kinds of intervention strategies have been used historically. First are activation strategies in which the mechanism is activated and then some downstream effect is detected. Craver (2002b) provided the example of putting a rat into a maze and detecting activity in its brain cells with a recording device. Whether such effects illuminate earlier stages in the mechanism varies with experimental setup. A common biochemical intervention is to put in a tracer, such as a radioactive element, activate the normal mechanism, and detect the tracer as it runs through the mechanism. Good tracers do not significantly alter the running of the activated mechanism; they merely allow observation of its workings.

Second are experimental strategies that involve not merely activating but modifying the normal working of the mechanism. As Wimsatt (personal communication), Bechtel and Richardson (1993), and Glennan (2005) discussed, a way to learn about a mechanism is to break a part of it and diagnose the failure. A fruitful way to learn about the action of a gene is to knock it out and note the effects in the organism (Craver and Darden 2001; Chapter 2, this book). As with the notorious ablation experiments in physiology in the nineteenth century, the problem with gene knockout techniques in intact animals is that such a missing part may have multiple effects that are difficult to disentangle, given the often-complex reactions between genotype and phenotype (Culp 1997).

Another kind of modification strategy is what Craver (2002b) called an "additive strategy," which is similar to what Bechtel and Richardson (1993, p. 20) called "excitatory studies." Some component in the mechanism is augmented or overstimulated, then effects are detected downstream. Craver's (2002b) example was of engineered mice with more of a specific kind of neural receptor. Those mice learned faster and retained what they learned longer, thereby providing evidence for the role of such receptors in learning and memory.

Craver (2002b) suggested using all three types of strategies: activation, ablation, addition. Consistent results strengthen the evidence for the hypothesized mechanism. Each helps to compensate for the weaknesses of the others to yield a robust (Wimsatt 1981) conclusion, namely, a conclusion supported by a variety of types of evidence (Lloyd 1987).

In addition to manipulating an intact, operating mechanism, one may seek evidence for the existence and nature of hypothesized entities, activities, and/or modules separately. For example, an ion channel protein may be isolated and its structure investigated to find its role in a neuronal mechanism. The melting point of the DNA double helix structure may be measured to detect the presence of hydrogen bonds. The results of such investigation of parts of a mechanism is an example of what Lloyd (1987) called "independent support for aspects of the model," to distinguish it from the "outcome of the model" (the latter is what I call "testing a prediction of the operation of the hypothesized mechanism as a whole").

Strategies for credentialing experimental evidence, in general, are, of course, important for assessing the evidence for mechanisms. These include use of adequate controls, reproducibility of results, and demonstration of the adequacy of instruments, to name only a few. (For a more extensive list, see Skipper 2000.) How these credentialing issues may be specifically informed by this mechanistic approach is a topic in need of further investigation.

Finally, in the section on empirical evidence, it should be noted that evidence from two or more fields further strengthens the claim that the conclusion is robust. As discussed in Chapters 2 and 3, different fields supplied evidence for the hypothesized mechanisms of spatial memory and protein synthesis. Sometimes a single researcher uses techniques from two different fields, as did Temin in his search for evidence for the provirus, using genetic and biochemical techniques (discussed in Chapter 10). Sometimes researchers from different fields provide evidence for different modules of the mechanism, as did the biochemists and molecular biologists for the mechanism of protein synthesis and the working of the genetic code (discussed in Chapter 3).

## 12.4.5 Strategies for Finding Scope

Once a hypothesized mechanism has good evidence that it is the actual mechanism operating to produce a specific phenomenon, it becomes available for possible generalization. Details may be dropped to produce a more abstract schema. To determine its generality, the scope of the domain of the abstract schema needs to be investigated. As Chapter 1 noted, abstraction and generalization should be distinguished. Abstraction is the dropping of details; generalization is the application to a domain of wide scope. A more abstract schema may be a more general one; however, an abstract schema might have no domain of applicability and not be generally applicable at all.

Determining the scope of a schema is an important part of theory evaluation (Beatty 1995; Skipper 2002). One needs to know the scope of the domain where such a type of mechanism is operating so that the schema can be confidently instantiated. One strategy for determining scope is to conjecture and examine likely sites of failure, such as at the extreme boundaries of the possible domain. The heuristic of "look at extremes" was used fruitfully in a simulation of mathematical discovery (Lenat 1976). Furthermore, a classification scheme that partitions the domain provides places to test scope, such as eucaryotes versus procaryotes, animals versus plants, invertebrates versus vertebrates, or heart cells versus liver cells. Variation in one part of the domain but not in others points to special-case or model anomalies (as discussed in Chapters 9, 10, 11, and later in this chapter).

As Bechtel and Abrahamsen insightfully noted, after a mechanism is found in a model system, the task is to determine how widely it occurs. They said: "As research proceeds, scientists find variants of what initially might seem to be the same mechanism, for example, the mechanisms responsible for oxidative phosphorylation in liver versus heart cells in cows" (Bechtel and Abrahamsen 2005, p. 437). They continued: "Examining the counterpart mechanism in other organs and species, any differences can be identified and their importance assessed" (Bechtel and Abrahamsen 2005, p. 439). Papers reporting such findings are a regular part of the scientific literature and textbooks often note such variations in mechanisms.

Given that evolution operates by producing variants and selecting the adapted ones, it is not surprising that biological mechanisms exhibit much variability. What is perhaps more surprising is the amount of similarity exhibited by molecular biological mechanisms and their homologous modules. As noted in Chapter 3, the protein synthesis mechanism schema, at a high degree of abstraction, has a very wide scope; it applies to almost all instances of protein synthesis in all living things on earth.

This issue of determining the scope of mechanism schemas is further discussed later in Section 12.5 in relation to anomaly detection and resolution. When a mechanism schema fails to apply in what had been thought to be its domain, then anomaly resolution is required. Again, we see the close relationships among construction, evaluation, and revision of mechanistic hypotheses – in this case, in relation to questions about unity and generality versus variability of biological mechanisms.

### 12.4.6   Strategies for Evaluating Context

Finally, a newly discovered mechanism needs to be evaluated as to how well it can be situated hierarchically and in a series of the mechanisms that are known to come before and after it (see "Contextual locations" in Table 12.1). Such contextual evaluation serves to evaluate a proposed mechanism to assess whether it fits into the currently accepted biological knowledge.[8] Ideally, a newly discovered mechanism can be tightly integrated with mechanisms coming before and after it and with those (if any) above it in a hierarchy. Such integration serves to put it into the context of the matrix of biological knowledge (Morowitz 1985; Morowitz and Smith 1987). Such a fit with other well-established mechanisms provides indirect empirical evidence for the newly proposed one. This corresponds to the strategy that Newton-Smith (1981) called "intertheory support," and that I previously called "relations to other accepted theories" (Darden 1991, p. 265).

---

[8] Carl Craver (2001) uses "contextual description" to refer only to situating a mechanism hierarchically, that is, the role the mechanism plays in the next higher level of a nested, part-whole hierarchy of mechanisms. My usage here is broader and includes, I think, what Craver calls an "etiological aspect," namely, what comes before a mechanism in the recent causal nexus in its past. My usage also includes noting which mechanisms follow it, if any. See Didion (2003) on the role that the end product of one mechanism may play in a subsequent one.

## 12.4.7   Comparison to Previous Work on Strategies
## for Theory Evaluation

To sum up this discussion of strategies for evaluating mechanism sketches and schemas: Finding a well-supported hypothesis about a mechanism progresses piecemeal. Incomplete sketches have their black boxes converted to gray boxes whose functional roles are conjectured, and gray boxes converted to glass boxes, whose entities and activities are specified. Evaluation of a proposed mechanism proceeds by moving through successive stages of how possibly to how plausibly to how actually a mechanism works to produce the phenomenon. A systematic scan of alternative hypotheses and good methods for choosing among them helps to ensure that plausible rivals have been considered and ruled out. Experimental results provide direct evidence for specific mechanism components, as well as how actually the mechanism as a whole operates. The scope of an abstract schema is investigated to determine how generally the mechanism occurs and what variants are found. The proposed mechanism is assessed as to how well it fits into a wider context of other mechanisms – up, down, before, and after.

These evaluation strategies are similar to but in some cases different from strategies for theory assessment discussed in Chapter 9 (see Table 9.2). Theory assessment strategies have been divided into categories (Darden 1991, Ch. 15; Skipper 2000, Ch. 4). The relevant categories here are conceptual, empirical, relations to other theories, and contextual.

Previously discussed conceptual evaluation strategies rely on such criteria as internally consistent and nontautologous (i.e., potentially testable), and clarity (Darden 1991, 15). These criteria are more relevant to theories with components represented as statements. (On different kinds of representations of theories, see Darden 1991, Ch. 11 and 12; Darden 1995; Chapter 10, this book). More important for mechanistic hypotheses is the conceptual issue of whether the description of the mechanism is incomplete and whether the components can be seen to operate to produce the phenomenon (perhaps via mental simulation or a computer simulation model). For assessing how possibly the mechanism operates, these conceptual evaluation strategies are much more specific and provide much more guidance than the vaguer conceptual theory evaluation strategies, such as demanding "clarity." If the hypothesized mechanism can be demonstrated to possibly produce the phenomenon, presumably it is not inconsistent. Furthermore, it can be used to make predictions and (given adequate techniques and an experimental setup) is potentially testable, hence, nontautologous.

A previously discussed empirical evaluative strategy for theories is explanatory adequacy. For mechanistic hypotheses, the proposed mechanism should be shown to produce the phenomenon; furthermore, there should be good empirical evidence for its various componency and organizational features. When that has been done, then a mechanistic explanation for the phenomenon has been given. Hence, the mechanistic view comes with an analysis of explanation. The features in an adequate description of a mechanism (see Table 12.1) supply some of the adequacy criteria for how to provide a mechanistic explanation.

Another theory evaluation strategy is predictive adequacy. The corresponding mechanistic strategy is to demonstrate that the proposed mechanism schema can be instantiated to correctly predict the occurrence of the phenomenon in a specific experimental setup. One may also be able to predict what will be found when the mechanism is decomposed and internal parts of the mechanism are investigated separately. As noted previously, separate evidence for mechanism parts provides independent evidence to supplement testing predictions of the mechanism as a whole. Different kinds of experimental strategies and experimental setups may be used to investigate the mechanism and its component parts. Such prediction testing shows that the mechanistic hypothesis is robust and supported by a variety of evidence.

In addition to the requirements to pass conceptual and empirical tests, theories should be evaluated in relation to other theories, both rivals and nonrivals. As discussed previously, systematically generating and ruling out rival mechanistic hypotheses is an important evaluation strategy. The additional guidance for construction of a mechanistic hypothesis (and variants of it) facilitates the satisfaction of this requirement. Suppose in theory evaluation one demands finding rivals. Further suppose that all one can say is: find a generalization that allows deduction of an observation statement, given statements of initial conditions. One has almost no guidance as to how to construct a rival hypothesis. But when one knows that a mechanism is to be found that can operate to produce the puzzling phenomenon, then much more guidance for finding rivals is available.

Theories need to be assessed as to their relations with nonrivals, that is, other accepted theories. Again, this vaguely stated criterion becomes much more specific when one is attempting to situate a mechanism into its hierarchical and serial contexts. Much more guidance is provided by the nature of the other mechanisms – those before and after, up and down from the one in question. The ability to fit into the matrix of biological knowledge provides indirect empirical evidence for the proposed mechanism. Such evidence comes from relations to other mechanisms (perhaps investigated in other fields), which have their own direct empirical evidence.

Theories are assessed as to how generally applicable they are and how much guidance they provide for future work. Mechanism sketches point to problems to be solved to remove their black and gray boxes. Mechanistic schemas are in need of assessment as to the scope of the domain in which they can be instantiated, that is, determine their extendibility. Anomalies guide judgments about how extendable a given schema is to new instances. The need to situate a newly discovered mechanism in its hierarchical and serial context also poses fruitful avenues to pursue.

The use of evaluative strategies may yield hypotheses about mechanisms in need of revision. We now turn to revision strategies, namely, strategies for anomaly resolution.

### 12.5   STRATEGIES FOR REVISION

A theme of this book is that discovery occurs piecemeal and incrementally, with iterative refinement via construction, evaluation, and revision. A theme throughout this chapter is that when what is to be discovered is a mechanism, much guidance is provided by the features of mechanisms (see Table 12.1). The theme of this section of this chapter is that features of mechanisms provide guidance for revising hypotheses about mechanisms. As noted previously, one can have weak construction strategies if one has good strategies for evaluation and revision. As discussed in the previous section on evaluation, two kinds of failure are incompleteness and incorrectness. The previous section suggested some strategies to remove incompleteness in mechanism sketches, namely how to revise a sketch to transform black boxes (components and their functions unknown) to gray boxes (component functions specified) to glass boxes (components supported by good evidence). This section summarizes and extends work on revision to remove incorrectness in mechanistic hypotheses when an evaluation strategy detects an anomaly (see Chapters 9, 10, and 11, this book).

An anomaly is an empirical finding that (purportedly) provides evidence against a hypothesized mechanism. Sometimes anomalies are generated by failed predictions. Alternatively, an anomaly may arise as an empirical finding in another investigation, which is then seen to have implications for the mechanism of interest. (On "characterizing an anomaly" as determining where the conflict lies, see Elliott 2004.)

Past work on anomalies often viewed a theory as a set of explanatory generalizations represented by a set of sentences (e.g., Darden 1991, Ch. 14). When an inconsistency between an empirical finding and a generalization arose, such a representation of a theory provided little guidance as to how to resolve the anomaly. However, when what is to be revised is a mechanistic hypothesis, much more guidance is available. As discussed in Chapters 9, 10, and 11, anomaly resolution is a diagnostic and redesign reasoning process. The location of the failure is sought. Then, depending on the site of localization, a redesign process may be needed to improve the hypothesized mechanism. The hypothesized mechanism or mechanism schema aids both diagnosis to localize the failure and, if required, redesign to supply an improved module.

As a first step in the anomaly resolution process, the anomalous result must be credentialed to ensure that it is not the result of an observational or experimental error. Experiments revealing an anomaly may be reproduced, using careful controls, or may be investigated using other credentialing strategies for experimental results (Franklin 1989; Rudge 1996). Error-probing in experimental results, as well as ways of classifying error types, has been discussed by several philosophers of science (Mayo 1996; Allchin 1992, 1997, 1999, 2002; Elliott 2004).

Once the anomalous result is confirmed, the location of the failure needs to be diagnosed. Chapters 9, 10, and 11 categorized kinds of anomalies, based on the extent of revision required: monster, special-case, and model anomalies. If the anomaly can be localized outside the domain of the mechanism schema, then no revision is required. Another possibility is that the anomaly might result from a disease or other abnormality. Such "monster" anomalies can be barred from requiring a change in the normal mechanism schema. The example in Chapter 9 showed that monster-barring occurred when lethal gene combinations produced anomalous genetic ratios; normally, the combination of two genetic alleles does not lead to the death of the embryo. No revision in claims about normal genetic mechanisms was required with the discovery of lethals.

Sometimes, the anomaly requires a splitting of the domain in which the mechanism is claimed to operate. If the anomaly only occurs in a small part of the domain, the anomaly is a special-case anomaly. The case discussed in Chapter 10 was an example. For the small domain consisting only of retroviruses, a RNA –>DNA step was added to the usual mechanism schema for protein synthesis.

In contrast to monster and special-case anomalies, model anomalies indicate what is normal for a domain of wide scope. Thus, the anomaly is a model in the sense of an exemplar (see Chapter 10). There may be no sharp divide between special-case and model anomalies as domains are split to accommodate variations in the ways mechanisms operate. The boundary between special-case anomalies and model anomalies is not sharp. Biologists frequently debate the "relative significance" of a hypothesis (Beatty 1995); that is, the scope of the domain of a newly revised mechanism schema needs to be determined. It may be very narrow, as in the case of mechanisms employing reverse transcriptase, or very wide, as in the case of the very abstractly formulated central dogma (DNA –>RNA –>protein).

In sum, an anomaly may be explained away as a result of experimental error, or it may be a monster, special-case, or model anomaly. One final category needs to be added to this list. A falsifying anomaly indicates that the entire proposed mechanism (and possibly its type) needs to be abandoned. For example, if a thorough search has shown that no reverse transcriptase can be found in an organism, then proposed mechanisms employing it need to be abandoned. Popper (1965) discussed falsifying instances for theories and Kuhn (1970) stressed the role of crisis anomalies in larger scale paradigm change. (For related categorizations of anomalies, see Brewer and Chinn 1994; Burian 1996b.)

Guidance in anomaly resolution has several sources. The difference between the prediction and the anomalous empirical finding needs to be examined. The nature of that difference may furnish guidance as to what failed (Karp 1990). Some guidance to aid in distinguishing monster, special-case, and model anomalies comes from the scope of the domain in which the anomaly occurs. As we saw in Chapter 11, biologists investigated the scope of the adaptive mutation anomaly. They found that it was not just a monstrous result, found only on engineered plasmids; genes in natural populations showed the same kind of late-occurring mutations. Adaptive mutations appear to be a special-case anomaly, found only in microorganisms. In contrast, the anomalies for the ribosome-as-template in the mechanism of protein synthesis (see Chapter 3) proved to be a model anomaly. In all cases of protein synthesis, both procaryotes and eucaryotes, it is messenger RNA – not the nonspecific, homologously similar ribosomes – that serves as a template.

Once the anomaly is judged to require revision of a mechanistic hypotheses, further guidance results from a diagrammatic representation of a mechanism or other means of locating its modules. As discussed in Chapters 1 and 3, in the mid-1950s, the ribosome was hypothesized to be the template for transferring the order of the bases in the DNA to the order of the amino acids in a protein:

DNA –>template RNA –>protein
DNA –>ribosomal template –>protein

Anomalies began to accumulate for the ribosomal template hypothesis. As Douglas Allchin noted (personal communication), multiple anomalies localized in the same site of a hypothesis strengthens the confidence that revision is required. (My previous work had treated each anomaly individually; see Darden 1991, Ch. 15.) Attempting to resolve the anomaly in which the base ratios of DNA and ribosomal RNA did not correspond, Crick (1959) at first systematically generated alternative hypotheses to save the "ribosome as template" hypothesis. This anomaly indicated a problem about the "DNA –> RNA" step in the proposed mechanism. He proposed alternatives localized in this module of the mechanism. Let's consider more detail than discussed in Chapter 1 or 3 so as to see how each component of this module served as a location for generating how possibly redesign hypotheses. Perhaps, Crick suggested, only part of the DNA carried coded information for proteins. Perhaps the DNA-to-RNA transcription mechanism varied somehow; Crick did not indicate how this type of hypothesis was to be specified. Focusing on the RNA step in the mechanism, Crick said that three fractions of RNA were known at the time: soluble RNA (now "transfer RNA") plus large and small RNA components of the ribosome itself. He suggested that perhaps one but not the other of the ribosomal components had base ratios similar to those of the DNA. This uses the reasoning strategy of delineate and specialize: What had been thought to be a property of the whole is instead claimed to be a property of only one of the parts. (For more discussion of this strategy, see Bechtel and Richardson 1993; Darden 1991, pp. 103, 130.) Thus, one of the mechanism's components would be separated into two components, with the functional role of template assigned to only one of the parts of the ribosome.

Conservatively, the set of alternatives Crick discussed in 1959 did not include the postulation of an as-yet undiscovered type of RNA having a base composition like DNA. This is the idea of a separate messenger RNA, different from the known types of RNA. The discovery of such a messenger RNA was how the anomaly was soon resolved (Jacob and Monod 1961). Crick employed the conservative strategy of tweaking properties of known entities rather than, more radically, postulating an as-yet-undetected entity. Reflecting on this reasoning, Crick (1988) said that there were plenty of RNAs available (e.g., the two components of the ribosome), so there seemed to be no need to postulate an undetected type of RNA. Activation, tracer experiments supplied direct evidence for the existence of mRNAs (Brenner et al. 1961; Gros et al. 1961). The functional requirement of a template, at that stage of the mechanism, with appropriate relations to the stages before and after it, acquired a role filler, namely messenger RNA.

In sum, tight relations exist among construction, evaluation, and revision of mechanistic hypotheses. Diagrams and other representations of the modules of mechanisms furnish guidance for localization and redesign. When an anomaly is localized to a stage, then redesign may need to be done by adding something before or after the stage or changing component entities and/or activities within the stage itself. Furthermore, the entities and activities of a stage must give rise to the next, thus imposing constraints on the components of a subsequent stage, based on what the prior one can produce. Also, the modules of the mechanism not implicated by the anomaly must be shown to continue to function. The constraint of having a productively continuous mechanism thus aids redesign. Finally, the newly revised mechanism must be subjected to strategies for evaluation again. When the revised hypothesis no longer has the failure and satisfies the other evaluative criteria, the anomaly has been resolved.

Will Bridewell surveyed the literature in artificial intelligence (AI) on computational models for belief revision. He listed the steps that such computational models traverse while reasoning in belief revision: (1) anomaly detection, (2) fault localization, (3) revision generation, (4) revision assessment, (5) revision application, and (6) expectation evaluation (Bridewell 2004, p. 9). Clearly, these are similar reasoning steps to those discussed previously. However, while belief revision models have similarities to reasoning in anomaly resolution in some scientific cases, these models typically do not represent mechanistic hypotheses. They struggle with what, in a set of sentences, to revise when an inconsistency is detected. Such representations (as earlier ones by philosophers of science) do not benefit from the guidance provided by the goal of discovering a mechanism and the guidance that a representation of a mechanism's components and stages affords.

Bridewell (2004) also discussed the issue raised in Chapter 11, namely the advisability of proposing conservative versus radical changes during anomaly resolution. Harman (1986) and Shapere (1980) advised conservative changes, in order to preserve as much as possible of previous beliefs. On the other hand, Popper (1968) suggested making bold conjectures (at the outset and after a falsifying instance necessitated rejecting a previous generalization). Popper's justification was that bold conjectures are more easily falsified. Work in AI suggests constructing a search space of possible revisions that includes both conservative fixes and more radical changes. As we saw in Chapter 11, hypotheses about mechanisms to account for adaptive mutation ranged from very conservative to very radical. Empirical work continues to evaluate the fairly wide range of alternatives that have been proposed. Systematic generation of rivals may be a good strategy during revision, so as to satisfy that demand in evaluation. Again, this proposal shows the close relations between construction, evaluation, and revision.

## 12.6 CONCLUSION

In sum, the goal of discovering a mechanism that produces a phenomenon provides much guidance about how to proceed in its discovery. The discovery of mechanisms occurs piecemeal via iterative refinement, which may be guided by reasoning strategies for construction, evaluation, and revision. The incompleteness of mechanism sketches is removed as black boxes for components are converted to gray boxes (with functional roles), which are converted to glass boxes (with visible contents). The discovery of how possibly, how plausibly, and how actually a mechanism operates is guided by a characterization of mechanisms, by the features of an adequate description of a mechanism, and by numerous reasoning strategies. Having good strategies for evaluation and revision obviates the need for powerful construction strategies. During anomaly resolution, representations of the hypothesized mechanism's modules aid the diagnosis of the location and nature of a failure and ways of doing redesign. Strategies for construction, evaluation, and anomaly resolution discussed in this chapter are listed in Table 12.2 and their roles are diagrammed in Figure 12.1.

This analysis uncovered a number of issues requiring more analysis. Is this characterization of mechanism adequate for capturing what is important for all biological mechanisms and for mechanisms in other fields? Do, for example, mechanisms with stochastic components require that the characterization (with entities, activities, productive continuity) be altered? To what extent is the search for mechanisms a (the?) goal in sciences other than biology? How does the inability to conceive a how possibly mechanism affect the acceptance of empirical findings? Future work needs to be done to find additional constraints and strategies for mechanism discovery. No claim is made that the ones discussed here are the only ones.

The extent to which these strategies may provide good advice to scientists and to science students needs to be investigated. Also, whether these strategies can aid researchers in artificial intelligence in designing better computational discovery models could be further explored.

Philosophers of science need to continue to analyze the implications of the importance of mechanisms in science for traditional philosophical problems. These include the structure of biological and other theories, the nature of explanations, the nature and role of laws, the relations between the analysis

Table 12.2   Strategies for Discovering Mechanisms

Goal: *Discover mechanism with features in Table 12.1*

Strategies for Construction
Schema Instantiation
Modular Subassembly
Forward/Backward Chaining

Strategies for Evaluation: How Possibly, How Plausibly, How Actually
Evaluate phenomenon
characterize the phenomenon
hypothesized mechanism produces phenomenon
Remove incompleteness (sketch to schema)
Experimental manipulation
intervene upstream, detect downstream
activate, modify via ablation or addition
Internal module testing (structure of entities; signatures of activities)
Determine scope
examine extremes of the domain
find variant mechanisms in other domains
Systematically generate and eliminate rivals
Evaluate context in matrix of biological knowledge
hierarchically
temporal series
relations to non-rival theories in other fields

Strategies for Anomaly Resolution
Credential anomalous experimental results
Localize anomaly
monster: monster barring
special case: split domain and redesign one but not other
model: requires schema redesign
Diagnosis for schema redesign: localize failure
use diagrams to locate modules
do experiments upstream or downstream
Redesign modules
use strategies for construction
constrain by retaining non-implicated modules
Reevaluate revised schema

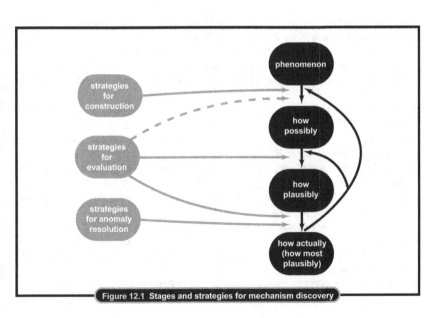

Figure 12.1 Stages and strategies for mechanism discovery

of mechanisms and the analysis of causality, as well as the role that proposed mechanisms can play in integrating results from different scientific fields.

There is much yet to be done in this research program to understand the reasoning to discover and find the roles of mechanisms in science.

### REFERENCES

Allchin, Douglas (1992), "How Do You Falsify a Question?: Crucial Tests versus Crucial Demonstrations," in D. Hull, M. Forbes, and K. Okruhlik (eds.), *PSA 1992*, v. 1. East Lansing, MI: Philosophy of Science Association, pp. 74–88.

Allchin, Douglas (1997), "A Twentieth-Century Phlogiston: Constructing Error and Differentiating Domains," *Perspectives on Science* 5: 81–127.

Allchin, Douglas (1999), "Negative Results as Positive Knowledge, and Zeroing in on Significant Problems," *Marine Ecology Progress Series* 191: 301–305.

Allchin, Douglas (2002), "Error Types," *Perspectives on Science* 9: 38–58.

Beatty, John (1995), "The Evolutionary Contingency Thesis," in James G. Lennox and Gereon Wolters (eds.), *Concepts, Theories, and Rationality in the Biological Sciences*. Pittsburgh, PA: University of Pittsburgh Press, pp. 45–81.

Bechtel, William and Adele Abrahamsen (2005), "Explanation: A Mechanist Alternative," in Carl F. Craver and Lindley Darden (eds.), Special Issue: "Mechanisms in Biology," *Studies in History and Philosophy of Biological and Biomedical Sciences* 36: 421–441.

Bechtel, William and Robert C. Richardson (1993), *Discovering Complexity: Decomposition and Localization as Strategies in Scientific Research*. Princeton, NJ: Princeton University Press.

Bogen, James (2004), "Analysing Causality: The Opposite of Counterfactual is Factual," *International Studies in the Philosophy of Science* 18: 3–26.

Bogen, James (2005), "Regularities and Causality; Generalizations and Causal Explanations," in Carl F. Craver and Lindley Darden (eds.), Special Issue: "Mechanisms in Biology," *Studies in History and Philosophy of Biological and Biomedical Sciences* 36: 397–420.

Bogen, James and James Woodward (1988), "Saving the Phenomena," *Philosophical Review* 97: 303–352.

Brenner, S., F. Jacob, and M. Meselson (1961), "An Unstable Intermediate Carrying Information from Genes to Ribosomes for Protein Synthesis," *Nature* 190: 576–581.

Brewer, William F. and Clark A. Chinn (1994), "Scientists' Responses to Anomalous Data: Evidence from Psychology, History, and Philosophy of Science," in D. Hull, M. Forbes, and R. M. Burian (eds.), *PSA 1994*, v. 1. East Lansing, MI: Philosophy of Science Association, pp. 304–313.

Bridewell, Will (2004), *Science as an Anomaly-Driven Enterprise: A Computational Approach to Generating Acceptable Theory Revisions in the Face of Anomalous Data*. Ph.D. Dissertation, Department of Computer Science, University of Pittsburgh, Pittsburgh, PA.

Burian, Richard M. (1996b), "Some Epistemological Reflections on Polistes as a Model Organism," in S. Turillazzi and M. J. West-Eberhard (eds.), *Natural History and*

*Evolution of an Animal Society: The Paper Wasp Case.* Oxford: Oxford University Press, pp. 318–337.

Cain, Joseph A. and Lindley Darden (1988), "Hull and Selection," *Biology and Philosophy* 3: 165–171.

Cairns, John, Julie Overbaugh, and S. Miller (1988), "The Origin of Mutants," *Nature* 335: 142–145.

Cartwright, Nancy (1989), *Nature's Capacities and Their Measurement.* Oxford: Oxford University Press.

Cartwright, Nancy (2004), "Causation: One Word, Many Things," *Philosophy of Science* 71: 805–819.

Craver, Carl F. (2001), "Role Functions, Mechanisms, and Hierarchy," *Philosophy of Science* 68: 53–74.

Craver, Carl F. (2002b), "Interlevel Experiments, Multilevel Mechanisms in the Neuroscience of Memory," *Philosophy of Science (Supplement)* 69: S83–S97.

Craver, Carl F. and Lindley Darden (2001), "Discovering Mechanisms in Neurobiology: The Case of Spatial Memory," in Peter Machamer, R. Grush, and P. McLaughlin (eds.), *Theory and Method in the Neurosciences.* Pittsburgh, PA: University of Pittsburgh Press, pp. 112–137.

Craver, Carl F. and Lindley Darden (2005), "Introduction: Mechanisms Then and Now," in Special Issue, "Mechanisms in Biology," *Studies in History and Philosophy of Biological and Biomedical Sciences* 36: 233–244.

Crick, Francis (1959), "The Present Position of the Coding Problem," *Structure and Function of Genetic Elements: Brookhaven Symposia in Biology* 12: 35–39.

Crick, Francis (1988), *What Mad Pursuit: A Personal View of Scientific Discovery.* New York: Basic Books.

Culp, Sylvia (1997), "Establishing Genotype / Phenotype Relationships: GeneTargeting as an Experimental Approach," *Philosophy of Science (Proceedings)* 64: S268–S278.

Darden, Lindley (1980), "Theory Construction in Genetics," in Thomas Nickles (ed.), *Scientific Discovery: Case Studies.* Dordrecht: Reidel, pp. 151–170.

Darden, Lindley (1991), *Theory Change in Science: Strategies from Mendelian Genetics.* New York: Oxford University Press.

Darden, Lindley (1995), "Exemplars, Abstractions, and Anomalies: Representations and Theory Change in Mendelian and Molecular Genetics," in James G. Lennox and Gereon Wolters (eds.), *Concepts, Theories, and Rationality in the Biological Sciences.* Pittsburgh, PA: University of Pittsburgh Press, pp. 137–158.

Darden, Lindley (2001), "Discovering Mechanisms: A Computational Philosophy of Science Perspective," in Klaus P. Jantke and Ayumi Shinohara (eds.), *Discovery Science* (Proceedings of the 4th International Conference, DS2001). New York: Springer-Verlag, pp. 3–15.

Darden, Lindley and Joseph A. Cain (1989), "Selection Type Theories," *Philosophy of Science* 56: 106–129.

Darden, Lindley and Carl F. Craver (2002), "Strategies in the Interfield Discovery of the Mechanism of Protein Synthesis," *Studies in History and Philosophy of Biological and Biomedical Sciences* 33: 1–28.

Didion, David (2003), "Relevant Bounds on Hierarchical Levels in the Description of Mechanisms," *History and Philosophy of the Life Sciences* 25: 5–25.

Dunbar, Kevin (1995), "How Scientists Really Reason: Scientific Reasoning in Real-World Laboratories," in R. J. Sternberg and J. E. Davidson (eds.), *The Nature of Insight.* Cambridge, MA: MIT Press, pp. 365–395.

Elliott, Kevin (2004), "Error as Means to Discovery," *Philosophy of Science* 71: 174–197.

Franklin, Allan (1989), "The Epistemology of Experiment," in David Gooding, Trevor Pinch, and Simon Schaffer (eds.), *The Uses of Experiment: Studies in the Natural Sciences.* New York: Cambridge University Press, pp. 437–460.

Gentner, Dedre (1983), "Structure Mapping – A Theoretical Framework for Analogy," *Cognitive Science* 7: 155–170.

Gilbert, Scott F. and Jessica A. Bolker (2001), "Homologies of Process and Modular Elements of Embryonic Construction," *Journal of Experimental Zoology* 291: 1–12.

Glennan, Stuart S. (1996), "Mechanisms and The Nature of Causation," *Erkenntnis* 44: 49–71.

Glennan, Stuart S. (2002), "Rethinking Mechanistic Explanation," *Philosophy of Science (Supplement)* 69: S342–S353.

Glennan, Stuart S. (2005), "Modeling Mechanisms," in Carl F. Craver and Lindley Darden (eds.), Special Issue: "Mechanisms in Biology," *Studies in History and Philosophy of Biological and Biomedical Sciences* 36: 443–464.

Gros, Francois, Howard Hiatt, Walter Gilbert, Chuck G. Kurland, R. W. Risebrough, and James D. Watson (1961), "Unstable Ribonucleic Acid Revealed by Pulse Labeling of *E. coli,*" *Nature* 190: 581–585.

Hanson, Norwood Russell (1963), *The Concept of the Positron: A Philosophical Analysis.* Cambridge, UK: Cambridge University Press.

Harman, Gilbert (1986), *Change in View: Principles of Reasoning.* Cambridge, MA: MIT Press.

Harré, Rom (1970), *The Principles of Scientific Thinking.* Chicago, IL: University of Chicago Press.

Holyoak, Keith J. and Paul Thagard (1995), *Mental Leaps: Analogy in Creative Thought.* Cambridge, MA: MIT Press.

Jacob, Francois and Jacques Monod (1961), "Genetic Regulatory Mechanisms in the Synthesis of Proteins," *Journal of Molecular Biology* 3: 318–356.

Josephson, John R. and Susan G. Josephson (eds.) (1994), *Abductive Inference: Computation, Philosophy, Technology.* New York: Cambridge University Press.

Judson, Horace F. (1996), *The Eighth Day of Creation: The Makers of the Revolution in Biology.* Expanded Edition. Cold Spring Harbor, NY: Cold Spring Harbor Laboratory Press.

Karp, Peter (1990), "Hypothesis Formation as Design," in J. Shrager and P. Langley (eds.), *Computational Models of Scientific Discovery and Theory Formation.* San Mateo, CA: Morgan Kaufmann, pp. 275–317.

Kauffman, Stuart A. (1971), "Articulation of Parts Explanation in Biology and the Rational Search for Them," in Roger C. Buck and Robert S. Cohen (eds.), *PSA 1970.* Boston Studies in the Philosophy of Science, v. 8. Dordrecht: Reidel, pp. 257–272.

Kitcher, Philip (1989), "Explanatory Unification and the Causal Structure of the World," in Philip Kitcher and Wesley Salmon (eds.), *Scientific Explanation.* Minnesota Studies in the Philosophy of Science, v. 13. Minneapolis, MN: University of Minnesota Press, pp. 410–505.

Kuhn, Thomas (1970), *The Structure of Scientific Revolutions*. 2nd ed. Chicago, IL: The University of Chicago Press.

Lederberg, Joshua (1965), "Signs of Life: Criterion-System of Exobiology," *Nature* 207: 9–13.

Lenat, Douglas (1976), *AM: An Artificial Intelligence Approach to Discovery in Mathematics as Heuristic Search*. Ph.D. Dissertation. Stanford University, Stanford, CA.

Lloyd, Elizabeth (1987), "Confirmation of Ecological and Evolutionary Models," *Biology and Philosophy* 2: 277–293.

Machamer, Peter (2004), "Activities and Causation: The Metaphysics and Epistemology of Mechanisms," *International Studies in the Philosophy of Science* 18: 27–39.

Machamer, Peter, Lindley Darden, and Carl Craver (2000), "Thinking About Mechanisms," *Philosophy of Science* 67: 1–25.

Mayo, Deborah G. (1996), *Error and the Growth of Experimental Knowledge*. Chicago, IL: University of Chicago Press.

Morowitz, Harold (1985), "Models for Biomedical Research: A New Perspective," Report of the Committee on Models for Biomedical Research. Washington, DC: National Academy Press.

Morowitz, Harold and Temple Smith (1987), "Report of the Matrix of Biological Knowledge Workshop, July 13–August 14, 1987," Santa Fe, NM: Santa Fe Institute.

Nersessian, Nancy J. (1993), "In the Theoretician's Laboratory: Thought Experimenting as Mental Modeling," in D. Hull, M. Forbes, and K. Okruhlik (eds.), *PSA 1992*, v. 2. East Lansing, MI: Philosophy of Science Association, pp. 291–301.

Newton-Smith, W. H. (1981), *The Rationality of Science*. Boston, MA: Routledge & Kegan Paul.

Nickles, Thomas (1987), "Methodology, Heuristics, and Rationality," in J. C. Pitt and M. Pera (eds.), *Rational Changes in Science*. Dordrecht: Reidel, pp. 103–132.

Popper, Karl R. (1965), *The Logic of Scientific Discovery*. New York: Harper Torchbooks.

Popper, Karl (1968), *Conjectures and Refutations: The Growth of Scientific Knowledge*. New York: Harper Torchbooks.

Psillos, Stathis (2004), "A Glimpse of the *Secret Connexion*: Harmonizing Mechanisms with Counterfactuals," *Perspectives on Science* 12: 288–391.

Rheinberger, Hans-Jörg (1997), *Toward a History of Epistemic Things: Synthesizing Proteins in the Test Tube*. Stanford, CA: Stanford University Press.

Rich, Alexander and James D. Watson (1954a), "Physical Studies on Ribonucleic Acid," *Nature* 173: 995–996.

Rich, Alexander and James D. Watson (1954b), "Some Relations Between DNA and RNA," *Proceedings of the National Academy of Sciences* 40: 759–764.

Rudge, David W. (1996), *A Philosophical Analysis of the Role of Natural Selection Experiments in Evolutionary Biology*. Ph.D. Dissertation, Department of History and Philosophy of Science, University of Pittsburgh, Pittsburgh, PA.

Schaffner, Kenneth (1993), *Discovery and Explanation in Biology and Medicine*. Chicago, IL: University of Chicago Press.

Shapere, Dudley (1980), "The Character of Scientific Change," in Thomas Nickles (ed.), *Scientific Discovery, Logic and Rationality*. Dordrecht: Reidel, pp. 61–101.

311

Skipper, Robert A., Jr. (1999), "Selection and the Extent of Explanatory Unification," *Philosophy of Science* 66 (Proceedings): S196-S209.

Skipper, Robert A., Jr. (2000), *The R.A. Fisher-Sewall Wright Controversy in Philosophical Focus: Theory Evaluation in Population Genetics*. Ph.D. Dissertation, Department of Philosophy, University of Maryland, College Park, MD.

Skipper, Robert A., Jr. (2002), "The Persistence of the R. A. Fisher-Sewall Wright Controversy," *Biology and Philosophy* 17: 341–367.

Skipper, Robert A., Jr. and Roberta L. Millstein (2005), "Thinking about Evolutionary Mechanisms: Natural Selection," in Carl F. Craver and Lindley Darden (eds.), Special Issue: "Mechanisms in Biology," *Studies in History and Philosophy of Biological and Biomedical Sciences* 36: 327–347.

Steele, Edward J., Robyn A. Lindley, and Robert B. Blanden (1998), *Lamarck's Signature*. Reading, MA: Perseus Books.

Tabery, James G. (2004), "Synthesizing Activities and Interactions in the Concept of a Mechanism," *Philosophy of Science* 71: 1–15.

Watson, James D. (1965), *Molecular Biology of the Gene*. New York: W. A. Benjamin.

Watson, James D. (1968), *The Double Helix*. New York: New American Library.

Watson, James D. and Francis Crick (1953a), "A Structure for Deoxyribose Nucleic Acid," *Nature* 171: 737–738.

Weber, Marcel (2005), *Philosophy of Experimental Biology*. New York: Cambridge University Press.

Westfall, Richard S. (1977), *The Construction of Modern Science: Mechanisms and Mechanics*. New York: Cambridge University Press.

Wimsatt, William C. (1981) "Robustness, Reliability, and Overdetermination," in M. Brewer and B. Collins (eds.), *Scientific Inquiry and the Social Sciences*. San Francisco, CA: Jossey-Bass, pp. 124–163.

Woodward, James (2002), "What Is a Mechanism? A Counterfactual Account," *Philosophy of Science (Supplement)* 69: S366-S377.

Woodward, James (2003), *Making Things Happen: A Theory of Causal Explanation*. New York: Oxford University Press.

# Bibliography

Ada, G. L. and N. Gustav (1987), "The Clonal Selection Theory," *Scientific American* 257: 62–69.

Alberts, Bruce, Dennis Bray, Julian Lewis, Martin Raff, Keith Roberts, and James D. Watson (1983), *Molecular Biology of the Cell*. New York: Garland.

Allchin, Douglas (1992), "How Do You Falsify a Question?: Crucial Tests versus Crucial Demonstrations," in D. Hull, M. Forbes, and K. Okruhlik (eds.), *PSA 1992*, v. 1. East Lansing, MI: Philosophy of Science Association, pp. 74–88.

Allchin, Douglas (1997), "A Twentieth-Century Phlogiston: Constructing Error and Differentiating Domains," *Perspectives on Science* 5: 81–127.

Allchin, Douglas (1999), "Negative Results as Positive Knowledge, and Zeroing in on Significant Problems," *Marine Ecology Progress Series* 191: 301–305.

Allchin, Douglas (2002), "Error Types," *Perspectives on Science* 9: 38–58.

Allemang, Dean (1990), *Understanding Programs as Devices*. Ph.D. Dissertation, Department of Computer and Information Sciences, The Ohio State University, Columbus, OH.

Allen, Garland (1978), *Thomas Hunt Morgan*. Princeton, NJ: Princeton University Press.

Andersson, D. I., E. S. Slechta, and J. R. Roth (1998), "Evidence that Gene Amplification Underlies Adaptive Mutability of the Bacterial *lac* Operon," *Science* 282: 1133–1135.

Anonymous (1970a), "Central Dogma Reversed," News and Views, *Nature* 226: 1198–1199.

Anonymous (1970b), "Cancer Viruses: More of the Same," *Nature* 227: 887–888.

Anscombe, Gertrude Elizabeth Margaret ([1971] 1981), "Causality and Determination," in *Metaphysics and the Philosophy of Mind, The Collected Philosophical Papers of G. E. M. Anscombe*, v. 2. Minneapolis, MN: University of Minnesota Press, pp. 133–147.

Arnold, A. J. and K. Fristrup (1982), "The Theory of Evolution by Natural Selection: A Hierarchical Expansion," *Paleobiology* 8: 113–129.

Ayala, F. J. and T. Dobzhansky (eds.) (1974), *Studies in the Philosophy of Biology*. Berkeley, CA: University of California Press.

Baker, Jason M. (2005), "Adaptive Speciation: The Role of Natural Selection in Mechanisms of Geographic and Non-geographic Speciation," in Carl F. Craver and Lindley Darden (eds.), Special Issue: "Mechanisms in Biology," *Studies in History and Philosophy of Biological and Biomedical Sciences* 36: 303–326.

313

Baltimore, David (1970), "Viral RNA-dependent DNA Polymerase," *Nature* 226: 1209–1211.

Baltimore, David (1977), "Viruses, Polymerases and Cancer," in *Nobel Lectures in Molecular Biology, 1933–1975.* [no editor] New York: Elsevier, pp. 495–508. Presented December 12, 1975.

Barinaga, M. (1999), "New Clues to How Neurons Strengthen Their Connections," *Science* 284: 1755–1757.

Bateson, Beatrice (1928), *William Bateson, Naturalist.* London: Cambridge University Press.

Bateson, William (1902), *Mendel's Principles of Heredity – A Defense.* Cambridge: Cambridge University Press.

Bateson, William (1913), *Problems of Genetics.* New Haven, CT: Yale University Press.

Bateson, William, E. R. Saunders, and R. C. Punnett (1905), "Further Experiments on Inheritance in Sweet Peas and Stocks: Preliminary Account," *Proceedings of the Royal Society* 77, Reprinted in R. C. Punnett (1928), *Scientific Papers of William Bateson,* v. 2. Cambridge: Cambridge University Press, pp. 139–141.

Bateson, William, E. R. Saunders, and R. C. Punnett (1906), "Experimental Studies in the Physiology of Heredity," *Reports to the Evolution Committee of the Royal Society III.* Reprinted in R. C. Punnett (1928), *Scientific Papers of William Bateson,* v. 2. Cambridge: Cambridge University Press, pp. 152–161.

Bateson, William and R. C. Punnett (1911), "On Gametic Series Involving Reduplication of Certain Terms," *Journal of Genetics* 1. Reprinted in R. C. Punnett (1928), *Scientific Papers of William Bateson,* v. 2. Cambridge: Cambridge University Press, pp. 206–215.

Beatty, John (1980), "What's Wrong with the Received View of Evolutionary Theory?" in Peter D. Asquith and Ronald N. Giere (eds.), *PSA 1980,* v. 2. East Lansing, MI: Philosophy of Science Association, pp. 397–426.

Beatty, John (1995), "The Evolutionary Contingency Thesis," in James G. Lennox and Gereon Wolters (eds.), *Concepts, Theories, and Rationality in the Biological Sciences.* Pittsburgh, PA: University of Pittsburgh Press, pp. 45–81.

Beatty, John (1997), "Why Do Biologists Argue Like They Do?" *Philosophy of Science* 64 (*Proceedings*): S432–S443.

Bechtel, William (1984), "Reconceptualizations and Interfield Connections: The Discovery of the Link Between Vitamins and Coenzymes," *Philosophy of Science* 51: 265–292.

Bechtel, William (1986), "Introduction: The Nature of Scientific Integration," in W. Bechtel (ed.), *Integrating Scientific Disciplines.* Dordrecht: Nijhoff, pp. 3–52.

Bechtel, William (1988), *Philosophy of Science: An Overview for Cognitive Science.* Hillsdale, NJ: Lawrence Erlbaum.

Bechtel, William and Adele Abrahamsen (2005), "Explanation: A Mechanist Alternative," in Carl F. Craver and Lindley Darden (eds.), Special Issue: "Mechanisms in Biology," *Studies in History and Philosophy of Biological and Biomedical Sciences* 36: 421–441.

Bechtel, William and Robert C. Richardson (1993), *Discovering Complexity: Decomposition and Localization as Strategies in Scientific Research.* Princeton, NJ: Princeton University Press.

Belozersky, Andrei N. and Alexander S. Spirin (1958), "A Correlation between the Compositions of Deoxyribonucleic and Ribonucleic Acids," *Nature* 182: 111–112.

Berg, Paul and Maxine Singer (1992), *Dealing with Genes: The Language of Heredity.* Mill Valley, CA: University Science Books.

Blackwell, R. J. (1969), *Discovery in the Physical Science.* Notre Dame, IN: University of Notre Dame Press.

Bjedov, Ivana, Olivier Tenaillon, Benedicte Gerard, Valeria Souze, Erick Denamur, Miraslav Radman, Francois Taddei, and Ivan Matic (2003), "Stress-Induced Mutagenesis in Bacteria," *Science* 300: 1404–1409.

Bogen, James (2004), "Analysing Causality: The Opposite of Counterfactual is Factual," *International Studies in the Philosophy of Science* 18: 3–26.

Bogen, James (2005), "Regularities and Causality; Generalizations and Causal Explanations," in Carl F. Craver and Lindley Darden (eds.), Special Issue: "Mechanisms in Biology," *Studies in History and Philosophy of Biological and Biomedical Sciences* 36: 397–420.

Bogen, James and James Woodward (1988), "Saving the Phenomena," *Philosophical Review* 97: 303–352.

Bourgeois, S., M. Cohen, and L. Orgel (1965), "Suppression of and Complementation among Mutants of the Regulatory Gene of the Lactose Operon of *Escherichia coli*," *Journal of Molecular Biology* 14: 300–302.

Boveri, Theodor ([1902] 1964), "On Multipolar Mitosis as a Means of Analysis of the Cell Nucleus," in B. H. Willier and J. Oppenheimer (eds.), *Foundations of Experimental Embryology.* Englewood Cliffs, NJ: Prentice-Hall, pp. 75–97.

Boveri, Theodor (1904), *Ergebnisse über die Konstitution der chromatischen Substanz des Zellkerns.* Jena: G. Fischer.

Brandon, Robert N. (1980), "A Structural Description of Evolutionary Theory," in Peter D. Asquith and Ronald N. Giere (eds.), *PSA 1980.* v. 2. East Lansing, MI: Philosophy of Science Association, pp. 427–439.

Brandon, Robert N. (1985), "Grene on Mechanism and Reductionism: More Than Just a Side Issue," in Peter D. Asquith and Philip Kitcher (eds.), *PSA 1984*, v. 2. East Lansing, MI: Philosophy of Science Association, pp. 345–353.

Brandon, Robert N. (1990), *Adaptation and Environment.* Princeton, NJ: Princeton University Press.

Brenner, Sydney, Francois Jacob, and M. Meselson (1961), "An Unstable Intermediate Carrying Information from Genes to Ribosomes for Protein Synthesis," *Nature* 190: 576–581.

Brewer, William F. and Clark A. Chinn (1994), "Scientists' Responses to Anomalous Data: Evidence from Psychology, History, and Philosophy of Science," in D. Hull, M. Forbes, and R. M. Burian (eds.), *PSA 1994*, v. 1. East Lansing, MI: Philosophy of Science Association, pp. 304–313.

Bridewell, Will (2004), *Science as an Anomaly-Driven Enterprise: A Computational Approach to Generating Acceptable Theory Revisions in the Face of Anomalous Data.* Ph.D. Dissertation, Department of Computer Science, University of Pittsburgh, Pittsburgh, PA.

Bridges, Bryn A. (1995), "Sexual Potency and Adaptive Mutation in Bacteria," *Trends in Microbiology* 3: 291–292.

315

Bridges, Calvin B. (1914), "Direct Proof Through Non-disjunction that the Sex-linked Genes of *Drosophila* are Borne by the X-chromosome," *Science,* N. S., 40: 107–109.

Bridges, Calvin B. (1916), "Non-disjunction as Proof of the Chromosome Theory of Heredity," *Genetics* 1: 1–52, 107–163.

Brisson, Dustin (2003), "The Directed Mutation Controversy in an Evolutionary Context," *Critical Reviews in Microbiology* 29: 25–35.

Brush, Stephen (1978a), "Nettie M. Stevens and the Discovery of Sex Determination by Chromosomes," *Isis* 69: 163–172.

Brush, Stephen (1978b), "A Geologist Among Astronomers: The Rise and Fall of the Chamberlin–Moulton Cosmogony," *Journal for the History of Astronomy* 9: 1–41.

Buchanan, Bruce (1982), "Mechanizing the Search for Explanatory Hypotheses," in Peter Asquith and Thomas Nickles (eds.), *PSA 1982,* v. 2. East Lansing, MI: Philosophy of Science Association, pp. 129–146.

Buchanan, Bruce (1985), "Steps Toward Mechanizing Discovery," in K. Schaffner (ed.), *Logic of Discovery and Diagnosis in Medicine.* Berkeley, CA: University of California Press, pp. 94–114.

Burian, Richard M. (1983), "Adaptation," in Marjorie Grene (ed.), *Dimensions of Darwinism.* Cambridge: Cambridge University Press, pp. 287–314.

Burian, Richard M. (1996a), "Underappreciated Pathways Toward Molecular Genetics as Illustrated by Jean Brachet's Cytochemical Embryology," in Sahotra Sarkar (ed.), *The Philosophy and History of Molecular Biology: New Perspectives.* Dordrecht: Kluwer, pp. 67–85.

Burian, Richard M. (1996b), "Some Epistemological Reflections on Polistes as a Model Organism," in S. Turillazzi and M. J. West-Eberhard (eds.), *Natural History and Evolution of an Animal Society: The Paper Wasp Case.* Oxford: Oxford University Press, pp. 318–337.

Burnet, F. M. (1957), "A Modification of Jerne's Theory of Antibody Production Using the Concept of Clonal Selection," *The Australian Journal of Science* 20: 67–69.

Bylander, T. and B. Chandrasekaran (1987), "Generic Tasks for Knowledge-Based Reasoning: The 'Right' Level of Abstraction for Knowledge Acquisition," *International Journal of Man-Machine Studies* 28: 231–243.

Bylander, T. and Sanjay Mittal (1986), "CSRL: A Language for Classificatory Problem Solving and Uncertainty Handling," *AI Magazine* 7(3): 66–77.

Cain, Joseph A. and Lindley Darden (1988), "Hull and Selection," *Biology and Philosophy* 3: 165–171.

Cairns, John (1995), "Response [to Bridges 1995] Cairns," *Trends in Microbiology* 3: 293.

Cairns, John (1997), *Matters of Life and Death: Perspectives on Public Health, Molecular Biology, Cancer, and the Prospects for the Human Race.* Princeton, NJ: Princeton University Press.

Cairns, John (1998), "Mutation and Cancer: The Antecedents to Our Studies of Adaptive Mutation," *Genetics* 148: 1433–1440.

Cairns, John and Patricia L. Foster (1991), "Adaptive Reversion of a Frameshift Mutation in *Escherichia coli,*" *Genetics* 128: 695–701.

Cairns, John and Patricia L. Foster (2003), "Letter to the Editor: The Risk of Lethals for Hypermutating Bacteria in Stationary Phase," *Genetics* 165: 2317–2318.

Cairns, John, Julie Overbaugh, and S. Miller (1988), "The Origin of Mutants," *Nature* 335: 142–145.

Carlson, E. A. (1971), "An Unacknowledged Founding of Molecular Biology: H. J. Muller's Contributions to Gene Theory, 1910–1936," *Journal of the History of Biology* 4: 149–170.

Carothers, E. E. (1913), "The Mendelian Ratio in Relation to Certain Orthopteran Chromosomes," *The Journal of Morphology* 24: 487–509.

Cartwright, Nancy (1989), *Nature's Capacities and Their Measurement.* Oxford: Oxford University Press.

Cartwright, Nancy (2004), "Causation: One Word, Many Things," *Philosophy of Science* 71: 805–819.

Castle, W. E. and C. C. Little (1910), "On a Modified Mendelian Ratio among Yellow Mice," *Science* 32: 868–870.

Cech, T. R. (2000), "The Ribosome is a Ribozyme," *Science* 289: 878.

Chadarevian, Soraya de (1996), "Sequences, Conformation, Information: Biochemists and Molecular Biologists in the 1950s," *Journal of the History of Biology* 29: 361–386.

Chadarevian, Soraya de (2002), *Designs for Life: Molecular Biology after World War II.* New York: Cambridge University Press.

Chadarevian, Soraya de and Jean-Paul Gaudilliere (1996), "The Tools of the Discipline: Biochemists and Molecular Biologists," Introduction to Special Issue, *Journal of the History of Biology* 29: 327–330.

Chandrasekaran, B., John Josephson, and Anne Keuneke (1986), "Functional Representation as a Basis for Generating Explanations," *Proceedings of the IEEE Conference on Systems, Man, and Cybernetics*, Atlanta, GA, pp. 726–731.

Chandrasekaran, B., John Josephson, Anne Keuneke, and David Herman (1989), "Building Routine Planning Systems and Explaining Their Behaviour," *International Journal of Man-Machine Studies* 30: 377–398.

Changeux, J. P. (1985), *Neuronal Man: The Biology of Mind.* New York: Pantheon, Random House.

Chapuis, N., M. Durup, and C. Thinus-Blanc (1987), "The Role of Exploratory Experience in a Shortcut in Golden Hamsters *(Mesocricetus auratus),*" *Animal Learning and Behavior* 15: 174–178.

Charniak, Eugene and Drew McDermott (1985), *Introduction to Artificial Intelligence.* Reading, MA: Addison-Wesley.

Coleman, William (1965), "Cell, Nucleus, and Inheritance: A Historical Study," *Proceedings of the American Philosophical Society* 109: 124–158.

Coleman, William (1970), "Bateson and Chromosomes: Conservative Thought in Science," *Centaurus* 15: 228–314.

Conrad, M. (1976), "Complementary Molecular Models of Learning and Memory," *BioSystems* 8: 119–138.

Correns, Carl ([1900] 1966), "G. Mendel's Law Concerning the Behavior of Progeny of Varietal Hybrids." Translated from German and reprinted in C. Stern and E. Sherwood (eds.), *The Origin of Genetics, A Mendel Source Book.* San Francisco, CA: W. H. Freeman, pp. 119–132.

Craver, Carl F. (1998), *Neural Mechanisms: On the Structure, Function, and Development of Theories in Neurobiology.* Ph.D. Dissertation, University of Pittsburgh, Pittsburgh, PA.

317

# Bibliography

Craver, Carl F. (2001), "Role Functions, Mechanisms, and Hierarchy," *Philosophy of Science* 68: 53–74.

Craver, Carl F. (2002a), "Structures of Scientific Theories," in Peter K. Machamer and M. Silberstein (eds.), *Blackwell Guide to the Philosophy of Science*. Oxford: Blackwell, pp. 55–79.

Craver, Carl F. (2002b), "Interlevel Experiments, Multilevel Mechanisms in the Neuroscience of Memory," *Philosophy of Science (Supplement)* 69: S83-S97.

Craver, Carl F. (2003), "The Making of a Memory Mechanism," *Journal of the History of Biology* 36: 153–195.

Craver, Carl F. (2005), "Beyond Reduction: Mechanisms, Multifield Integration, and the Unity of Neuroscience," in Carl F. Craver and Lindley Darden (eds.), Special Issue: "Mechanisms in Biology," *Studies in History and Philosophy of Biological and Biomedical Sciences* 36: 373–397.

Craver, Carl F. and Lindley Darden (2001), "Discovering Mechanisms in Neurobiology: The Case of Spatial Memory," in Peter Machamer, R. Grush, and P. McLaughlin (eds.), *Theory and Method in the Neurosciences*. Pittsburgh, PA: University of Pittsburgh Press, pp. 112–137.

Craver, Carl F. and Lindley Darden (2005), "Introduction: Mechanisms Then and Now," in Special Issue, "Mechanisms in Biology," *Studies in History and Philosophy of Biological and Biomedical Sciences* 36: 233–244.

Crick, Francis (unpublished MS of 1955), "On Degenerate Templates and the Adaptor Hypothesis: A Note for the RNA Tie Club."

Crick, Francis (1957), "Discussion Note," in E. M. Crook (ed.), *The Structure of Nucleic Acids and Their Role in Protein Synthesis: Biochemical Society Symposium* 14 (February 18, 1956). London: Cambridge University Press, pp. 25–26.

Crick, Francis (1958), "On Protein Synthesis," *Symposium of the Society of Experimental Biology* 12: 138–163.

Crick, Francis (1959), "The Present Position of the Coding Problem," *Structure and Function of Genetic Elements: Brookhaven Symposia in Biology* 12: 35–39.

Crick, Francis (1970), "Central Dogma of Molecular Biology," *Nature* 227: 561–563.

Crick, Francis (1988), *What Mad Pursuit: A Personal View of Scientific Discovery*. New York: Basic Books.

Crick, Francis (1996), "The Impact of Linus Pauling on Molecular Biology," in Ramesh S. Krishnamurthy (ed.), *The Pauling Symposium: A Discourse on the Art of Biography*. Corvallis, OR: Oregon State University Libraries Special Collections, pp. 3–18.

Crick, Francis, Leslie Barnett, Sydney Brenner, and R. J. Watts-Tobin (1961), "General Nature of the Genetic Code for Proteins," *Nature* 192: 1227–1232.

Crick, Francis and James D. Watson (1956), "Structure of Small Viruses," *Nature* 177: 473–475.

Crick, Francis and James D. Watson (1957), "Virus Structure: General Principles," in *Ciba Foundation Symposium on The Nature of Viruses*. London: J. & A. Churchill, pp. 5–13.

Cuénot, Lucien (1905), "Les Races Pures et Leurs Combinaisons Chez Les Souris," *Archives de Zoologie Expérimentale et Générale* 4 Serie, T. 111: 123–132.

Culotta, Elizabeth (1994), "A Boost for 'Adaptive' Mutation," *Science* 265: 318–319.

Culp, Sylvia (1997), "Establishing Genotype/Phenotype Relationships: Gene Targeting as an Experimental Approach," *Philosophy of Science (Proceedings)* 64: S268–S278.

318

## Bibliography

Cutting, James E. (1986), *Perception with an Eye for Motion*. Cambridge, MA: MIT Press.

Darden, Lindley (1974), *Reasoning in Scientific Change: The Field of Genetics at Its Beginnings*. Ph.D. Dissertation, University of Chicago, Chicago, IL.

Darden, Lindley (1976), "Reasoning in Scientific Change: Charles Darwin, Hugo de Vries, and the Discovery of Segregation," *Studies in the History and Philosophy of Science* 7: 127–169.

Darden, Lindley (1977), "William Bateson and the Promise of Mendelism," *Journal of the History of Biology* 10: 87–106.

Darden, Lindley (1978), "Discoveries and the Emergence of New Fields in Science," in Peter D. Asquith and Ian Hacking (eds.), *PSA 1978*, v. 1. East Lansing, MI: Philosophy of Science Association, pp. 149–160.

Darden, Lindley (1980), "Theory Construction in Genetics," in Thomas Nickles (ed.), *Scientific Discovery, Case Studies*. Dordrecht: Reidel, pp. 151–170.

Darden, Lindley (1982a), "Artificial Intelligence and Philosophy of Science: Reasoning by Analogy in Theory Construction," in Thomas Nickles and Peter Asquith (eds.), *PSA 1982*, v. 2. East Lansing, MI: Philosophy of Science Association, pp. 147–165.

Darden, Lindley (1982b), "Aspects of Theory Construction in Biology," in *Proceedings of the Sixth International Congress for Logic, Methodology and Philosophy of Science*. Hanover: North Holland Publishing Co., pp. 463–477.

Darden, Lindley (1983), "Reasoning by Analogy in Scientific Theory Construction," in R. S. Michalski (ed.), *Proceedings of the 1983 International Machine Learning Workshop*, Urbana, IL: Department of Computer Science, University of Illinois, pp. 32–40.

Darden, Lindley (1986), "Reasoning in Theory Construction: Analogies, Interfield Connections, and Levels of Organization," in P. Weingartner and G. Dorn (eds.), *Foundations of Biology*. Vienna, Austria: Holder-Picher-Tempsky, pp. 99–107.

Darden, Lindley (1987), "Viewing the History of Science As Compiled Hindsight," *AI Magazine* 8 (2): 33–41.

Darden, Lindley (1990), "Diagnosing and Fixing Faults in Theories," in J. Shrager and P. Langley (eds.), *Computational Models of Scientific Discovery and Theory Formation*. San Mateo, CA: Morgan Kaufmann, pp. 319–346.

Darden, Lindley (1991), *Theory Change in Science: Strategies from Mendelian Genetics*. New York: Oxford University Press.

Darden, Lindley (1992), "Strategies for Anomaly Resolution," in Ronald Giere (ed.), *Cognitive Models of Science*, Minnesota Studies in the Philosophy of Science, v. 15. Minneapolis, MN: University of Minnesota Press, pp. 251–273.

Darden, Lindley (1995), "Exemplars, Abstractions, and Anomalies: Representations and Theory Change in Mendelian and Molecular Genetics," in James G. Lennox and Gereon Wolters (eds.), *Concepts, Theories, and Rationality in the Biological Sciences*. Pittsburgh, PA: University of Pittsburgh Press, pp. 137–158.

Darden, Lindley (1996), "Generalizations in Biology: Essay Review of K. Schaffner's *Discovery and Explanation in Biology and Medicine*," *Studies in History and Philosophy of Science* 27: 409–419.

Darden, Lindley (1998), "Anomaly-Driven Theory Redesign: Computational Philosophy of Science Experiments," in Terrell W. Bynum and James Moor (eds.), *The Digital Phoenix: How Computers are Changing Philosophy*. Oxford: Blackwell, pp. 62–78.

# Bibliography

Darden, Lindley (2001), "Discovering Mechanisms: A Computational Philosophy of Science Perspective," in Klaus P. Jantke and Ayumi Shinohara (eds.), *Discovery Science* (Proceedings of the 4th International Conference, DS2001). New York: Springer-Verlag, pp. 3–15.

Darden, Lindley (2002), "Strategies for Discovering Mechanisms: Schema Instantiation, Modular Subassembly, Forward/Backward Chaining," *Philosophy of Science (Supplement)* 69: S354–S365.

Darden, Lindley and Joseph A. Cain (1989), "Selection Type Theories," *Philosophy of Science* 56: 106–129.

Darden, Lindley and Michael Cook (1994), "Reasoning Strategies in Molecular Biology: Abstractions, Scans and Anomalies," in David Hull, Micky Forbes, and Richard M. Burian (eds.), *PSA 1994*, v. 2. East Lansing, MI: Philosophy of Science Association, pp. 179–191.

Darden, Lindley and Carl F. Craver (2002), "Strategies in the Interfield Discovery of the Mechanism of Protein Synthesis," *Studies in History and Philosophy of Biological and Biomedical Sciences* 33: 1–28.

Darden, Lindley and Nancy Maull (1977), "Interfield Theories," *Philosophy of Science* 44: 43–64.

Darden, Lindley, Dale Moberg, Satish Nagarajan, and John Josephson (1991), "Anomaly Driven Redesign of a Scientific Theory: The TRANSGENE.2 Experiments," *Technical Report 91-LD-TRANSGENE*. Laboratory for Artificial Intelligence Research, The Ohio State University, Columbus, OH.

Darden, Lindley and Roy Rada (1988a), "Hypothesis Formation Using Part-Whole Interrelations," in David Helman (ed.), *Analogical Reasoning*. Dordrecht: Reidel, pp. 341–375.

Darden, Lindley and Roy Rada (1988b), "Hypothesis Formation Via Interrelations," in Armand Prieditis (ed.), *Analogica*. Los Altos, CA: Morgan Kaufmann, pp. 109–127.

Darwin, Charles ([1859] 1966), *On the Origin of Species, A Facsimile of the First Edition*. Cambridge, MA: Harvard University Press.

Darwin, Charles (1868), *The Variation of Animals and Plants under Domestication*. 2 vols. New York: Orange Judd & Co.

Davis, Bernard D. (1980), "Frontiers of the Biological Sciences," *Science* 209: 78–89.

Davis, Randall and Walter C. Hamscher (1988), "Model-Based Reasoning: Troubleshooting," in H. E. Shrobe (ed.), *Exploring Artificial Intelligence*. Los Altos, CA: Morgan Kaufmann, pp. 297–346.

Didion, David (2003), "Relevant Bounds on Hierarchical Levels in the Description of Mechanisms," *History and Philosophy of the Life Sciences* 25: 5–25.

Dienert, F. (1900), "Sur la Fermentation du Galactose et sur l'Accoutamance des levures à ce Sucre," *Annales de l'Institute Pasteur* 14: 138–189.

Dietterich, Thomas G., B. London, K. Clarkson, and G. Dromey (1982), "Learning and Inductive Inference," in Paul R. Cohen and E. Feigenbaum (eds.), *The Handbook of Artificial Intelligence*, v. 3. Los Altos, CA: Morgan Kaufmann, pp. 323–511.

Dobzhansky, Theodosius (1937), *Genetics and the Origin of Species*. New York: Columbia University Press.

Dunbar, Kevin (1995), "How Scientists Really Reason: Scientific Reasoning in Real-World Laboratories," in R. J. Sternberg and J. E. Davidson (eds.), *The Nature of Insight*. Cambridge, MA: MIT Press, pp. 365–395.

Dunn, L. C. (1965), *A Short History of Genetics*. New York: McGraw-Hill.

Edelman, Gerald and V. Mountcastle (1978), *The Mindful Brain: Cortical Organization and the Group Selective Theory of Higher Brain Function*. Cambridge, MA: MIT Press.

Elliott, Kevin (2004), "Error as Means to Discovery," *Philosophy of Science* 71: 174–197.

Engert, E. and T. Bonhoeffer (1999), "Dendritic Spine Changes Associated with Hippocampal Long-Term Synaptic Plasticity," *Nature* 399: 66–70.

Fischer, E. (1894), "Einfluss der Konfiguration auf die Wirkung der Enzyme," *Berichte der deutschen chemische Gesellschaft* 27: 2985–2993.

Fogle, T. (2000), "The Dissolution of Protein Coding Genes in Molecular Biology," in Peter Beurton, Raphael Falk, and Hans-Jörg Rheinberger (eds.), *The Concept of the Gene in Development and Evolution*. New York: Cambridge University Press, pp. 3–25.

Foster, Patricia L. (1997), "Nonadaptive Mutations Occur on the F' Episome During Adaptive Mutation Conditions in *Escherichia coli*," *Journal of Bacteriology* 179: 1550–1554.

Foster, Patricia L. (2000), "Adaptive Mutation: Implications for Evolution," *BioEssays* 22: 1067–1074.

Foster, Patricia L. and John Cairns (1992), "Mechanisms of Directed Mutation," *Genetics* 131: 783–789.

Foster, Patricia L. and Jeffrey M. Trimarchi (1994), "Adaptive Reversion of a Frameshift Mutation in *Escherichia coli* by Simple Base Deletions in Homopolymeric Runs," *Science* 265: 407–409.

Franklin, Allan (1989), "The Epistemology of Experiment," in David Gooding, Trevor Pinch, and Simon Schaffer (eds.), *The Uses of Experiment: Studies in the Natural Sciences*. New York: Cambridge University Press, pp. 437–460.

Frey, U. and R. G. Morris (1998), "Synaptic Tagging: Implications for Late Maintenance of Hippocampal Long-Term Potentiation," *Trends in Neuroscience* 21: 181–188.

Friedberg, Errol C. (1997), *Correcting the Blueprint of Life: An Historical Account of the Discovery of DNA Repair Mechanisms*. Plainview, NY: Cold Spring Harbor Laboratory Press.

Fruton, Joseph S. (1972), *Molecules and Life: Historical Essays on the Interplay of Chemistry and Biology*. New York: Wiley-Interscience.

Galitski, T. and John R. Roth (1996), "A Search for a General Phenomenon of Adaptive Mutability," *Genetics* 143: 645–659.

Gallo, Robert (1991), *Virus Hunting, Aids, Cancer, and the Human Retrovirus: A Story of Scientific Discovery*. New York: Harper Collins Publishers, Basic Books.

Galton, Francis (1871), "Experiments in Pangenesis," *Proceedings of the Royal Society (Biology)* 19: 393–404.

Gamow, George (1954), "Possible Relation between Deoxyribonucleic Acid and Protein Structures," *Nature* 173: 318.

Gaudilliere, Jean-Paul (1993), "Molecular Biology in the French Tradition? Redefining Local Traditions and Disciplinary Patterns," *Journal of the History of Biology* 26: 473–498.

Gaudilliere, Jean-Paul (1996), "Molecular Biologists, Biochemists, and Messenger RNA: The Birth of a Scientific Network," *Journal of the History of Biology* 29: 417–445.

Genesereth, Michael (1980), "Metaphors and Models," *Proceedings of the First Annual National Conference on Artificial Intelligence.* Menlo Park, CA: American Association for Artificial Intelligence, pp. 208–211.

Gentner, Dedre (1983), "Structure Mapping – A Theoretical Framework for Analogy," *Cognitive Science* 7: 155–170.

Gilbert, G. K. (1896), "The Origin of Hypotheses, Illustrated by the Discussion of a Topographic Problem," *Science,* N. S., 3: 1–13.

Gilbert, Scott F. and Jessica A. Bolker (2001), "Homologies of Process and Modular Elements of Embryonic Construction," *Journal of Experimental Zoology* 291: 1–12.

Gilbert, W. and B. Müller-Hill (1966), "Isolation of the Lac Repressor," *Proceedings of the National Academy of Sciences* 56: 1891–1898.

Gilbert, W. and B. Müller-Hill (1967), "The Lac Operator Is DNA," *Proceedings of the National Academy of Sciences* 58: 2415–2421.

Glennan, Stuart S. (1992), *Mechanisms, Models, and Causation.* Ph.D. Dissertation, University of Chicago, Chicago, IL.

Glennan, Stuart S. (1996), "Mechanisms and the Nature of Causation," *Erkenntnis* 44: 49–71.

Glennan, Stuart S. (2002), "Rethinking Mechanistic Explanation," *Philosophy of Science (Supplement)* 69: S342–S353.

Glennan, Stuart S. (2005), "Modeling Mechanisms," in Carl F. Craver and Lindley Darden (eds.), Special Issue: "Mechanisms in Biology," *Studies in History and Philosophy of Biological and Biomedical Sciences* 36: 443–464.

Glymour, Clark (1980), *Theory and Evidence.* Princeton, NJ: Princeton University Press.

Goel, Ashok and B. Chandrasekaran (1989), "Functional Representation of Designs and Redesign Problem Solving," in *Proceedings of the Eleventh International Joint Conference on Artificial Intelligence,* Detroit, MI, pp. 1388–1394.

Golub, E. (1981), *The Cellular Basis of Immune Response,* 2nd ed. Sunderland, MA: Sinauer Associates.

Gould, Stephen J. and Elisabeth S. Vrba (1982), "Exaptation – A Missing Term in the Science of Form," *Paleobiology* 8: 4–15.

Greiner, Russell (1985), *Learning by Understanding Analogies.* Ph.D. Dissertation, Department of Computer Science, Stanford University, Stanford, CA.

Gros, Francois (1979), "The Messenger," in Andre Lwoff and Agnes Ullmann (eds.), *Origins of Molecular Biology: A Tribute to Jacque Monod.* New York: Academic Press, pp. 117–124.

Gros, Francois, Howard Hiatt, Walter Gilbert, Chuck G. Kurland, R. W. Risebrough, and James D. Watson (1961), "Unstable Ribonucleic Acid Revealed by Pulse Labeling of *E. coli,*" *Nature* 190: 581–585.

Hacking, Ian (1988), "On the Stability of the Laboratory Sciences," *Journal of Philosophy* 85: 507–514.

Hacking, Ian (1992), "The Self-Vindication of the Laboratory Sciences," in A. Pickering (ed.), *Science as Practice and Culture.* Chicago, IL: University of Chicago Press, pp. 29–64.

Bibliography

Hall, Barry G. (1990), "Spontaneous Point Mutations That Occur More Often When Advantageous Than When Neutral," *Genetics* 126: 5–16.

Hall, Barry G. (1997), "On the Specificity of Adaptive Mutations," *Genetics* 145: 39–44.

Hall, Zach W. (ed.) (1992), *An Introduction to Molecular Neurobiology*. Sunderland, MA: Sinauer Associates.

Hanson, Norwood Russell (1958), *Patterns of Discovery*. Cambridge: Cambridge University Press.

Hanson, Norwood Russell ([1961] 1970), "Is There a Logic of Scientific Discovery?" in H. Feigl and G. Maxwell (eds.), *Current Issues in the Philosophy of Science*. New York: Holt, Rinehart and Winston. Reprinted in B. Brody (ed.), *Readings in the Philosophy of Science*. Englewood Cliffs, NJ: Prentice-Hall, pp. 620–633.

Hanson, Norwood Russell (1963), *The Concept of the Positron: A Philosophical Analysis*. Cambridge: Cambridge University Press.

Harman, Gilbert (1986), *Change in View: Principles of Reasoning*. Cambridge, MA: MIT Press.

Harré, R. (1960), *An Introduction to the Logic of the Sciences*. London: Macmillan.

Harré, R. (1970), *The Principles of Scientific Thinking*. Chicago, IL: University of Chicago Press.

Harré, Rom and E. H. Madden (1975), *Causal Powers: A Theory of Natural Necessity*. Totowa, NJ: Rowman and Littlefield.

Harris, Reuben S., Simonne Longerich, and Susan M. Rosenberg (1994), "Recombination in Adaptive Mutation," *Science* 264: 258–260.

Hebb, D. O. (1949), *The Organization of Behavior*. New York: Wiley.

Hempel, Carl G. (1965), *Aspects of Scientific Explanation*. New York: The Free Press, Macmillan.

Hempel, C. G. (1966), *Philosophy of Natural Science*. Englewood Cliffs, NJ: Prentice-Hall.

Hendrickson, Heather, E. Susan Slechta, Ulfar Bergthorsson, Dan I. Andersson, and John R. Roth (2002), "Amplification Mutagenesis: Evidence that 'Directed' Adaptive Mutation and General Hypermutability Result from Growth with a Selected Gene Amplification," *Proceedings of the National Academy of Sciences* 99: 2164–2169.

Hesse, Mary (1966), *Models and Analogies in Science*. Notre Dame, IN: University of Notre Dame Press.

Hoagland, Mahlon B. (1955), "An Enzymic Mechanism for Amino Acid Activation in Animal Tissues," *Biochimica et Biophysica Acta* 16: 288–289.

Hoagland, Mahlon B. (1990), *Toward the Habit of Truth*. New York: Norton.

Hoagland, Mahlon B. (1996), "Biochemistry or Molecular Biology? The Discovery of 'Soluble RNA'," *Trends in Biological Sciences Letters (TIBS)* 21: 77–80.

Hoagland, Mahlon B., Paul Zamecnik, and Mary L. Stephenson (1959), "A Hypothesis Concerning the Roles of Particulate and Soluble Ribonucleic Acids in Protein Synthesis," in R. E. Zirkle (ed.), *A Symposium on Molecular Biology*. Chicago, IL: Chicago University Press, pp. 105–114.

Holland, John H. (1975), *Adaptation in Natural and Artificial Systems*. Ann Arbor, MI: University of Michigan Press.

Holyoak, Keith J. and Paul Thagard (1989), "Analogical Mapping by Constraint Satisfaction," *Cognitive Science* 13: 295–355.

Holyoak, Keith J. and Paul Thagard (1995), *Mental Leaps: Analogy in Creative Thought.* Cambridge, MA: MIT Press.

Hughes, Arthur (1959), *A History of Cytology.* New York: Abelard-Schuman.

Hull, David (1974), *Philosophy of Biological Science.* Englewood Cliffs, NJ: Prentice-Hall.

Hull, David (1980), "Individuality and Selection," *Annual Review of Ecology and Systematics* 11: 311–332.

Hull, David ([1981] 1984), "Units of Evolution: A Metaphysical Essay," in U. L. Jensen and R. Harré (eds.), *The Philosophy of Evolution.* Brighton: Harvester Press, pp. 23–44. Reprinted in Robert N. Brandon and Richard M. Burian (eds.), *Genes, Populations, and Organisms: Controversies over the Units of Selection.* Cambridge, MA: MIT Press, pp. 142–160.

Jablonka, Eva and Marion J. Lamb (1995), *Epigenetic Inheritance and Evolution: The Lamarckian Dimension.* New York: Oxford University Press.

Jacob, Francois (1988), *The Statue Within: An Autobiography.* New York: Basic Books.

Jacob, Francois and Jacques Monod (1961), "Genetic Regulatory Mechanisms in the Synthesis of Proteins," *Journal of Molecular Biology* 3: 318–356.

Janssens, F. A. (1909), "La theorie de la chiasmatypie," *La Cellule* 25: 389–411.

Jerne, Niels K. (1955), "The Natural-Selection Theory of Antibody Formation," *Proceedings of the National Academy of Sciences* 41: 849–857.

Jerne, Niels K. (1966), "The Natural Selection Theory of Antibody Formation: Ten Years Later," in John Cairns, Gunther S. Stent, and James D. Watson (eds.), *Phage and the Origins of Molecular Biology.* Cold Spring Harbor, NY: Cold Spring Harbor Laboratory of Quantitative Biology, pp. 301–312.

Johannsen, Wilhelm (1909), *Elemente der Exakten Erblichkeitslehre.* Jena: G. Fischer.

Josephson, John R., B. Chandrasekaran, J. Smith, and M. Tanner (1987), "A Mechanism for Forming Composite Explanatory Hypotheses," *IEEE Transactions on Systems, Man, and Cybernetics,* SMC- 17: 445–454.

Josephson, John R. and Susan G. Josephson (eds.) (1994), *Abductive Inference: Computation, Philosophy, Technology.* New York: Cambridge University Press.

Judson, Horace F. (1996), *The Eighth Day of Creation: The Makers of the Revolution in Biology.* Expanded Edition. Cold Spring Harbor, NY: Cold Spring Harbor Laboratory Press.

Karp, Peter (1989), *Hypothesis Formation and Qualitative Reasoning in Molecular Biology.* Ph.D. Dissertation, Stanford University, Stanford, CA. (Available as a technical report from the Computer Science Department: STAN-CS-89–1263.)

Karp, Peter (1990), "Hypothesis Formation as Design," in J. Shrager and P. Langley (eds.), *Computational Models of Scientific Discovery and Theory Formation.* San Mateo, CA: Morgan Kaufmann, pp. 275–317.

Kauffman, Stuart A. (1971), "Articulation of Parts Explanation in Biology and the Rational Search for Them," in Roger C. Buck and Robert S. Cohen (eds.), *PSA 1970.* Dordrecht: Reidel, pp. 257–272.

Kay, Lily E. (2000), *Who Wrote the Book of Life? A History of the Genetic Code.* Stanford, CA: Stanford University Press.

Kellogg, Vernon (1908), *Darwinism Today.* New York: Henry Holt.

Kemeny, J. and P. Oppenheim (1956), "On Reduction," *Philosophical Studies* 7: 6–17.

# Bibliography

Kettler, Brian and Lindley Darden (1993), "Protein Sequencing Experiment Planning Using Analogy," in L. Hunter, D. Searls, and J. Shavlik (eds.), *ISMB-93, Proceedings of the First International Conference on Intelligent Systems for Molecular Biology.* Menlo Park, CA: AAAI Press, pp. 216–224.

Keuneke, Anne (1989), *Machine Understanding of Devices: Causal Explanation of Diagnostic Conclusions.* Ph.D. Dissertation, Department of Computer and Information Science, The Ohio State University, Columbus, OH.

Keyes, Martha (1999a), "The Prion Challenge to the "Central Dogma" of Molecular Biology, 1965–1991, Part I: Prelude to Prions," *Studies in the History and Philosophy of Biological and Biomedical Sciences* 30: 1–19.

Keyes, Martha (1999b), "The Prion Challenge to the "Central Dogma" of Molecular Biology, 1965–1991, Part II: The Problem with Prions," *Studies in the History and Philosophy of Biological and Biomedical Sciences* 30: 181–218.

Kirkham, W. B. (1919), "The Fate of Homozygous Yellow Mice," *Journal of Experimental Zoology* 28: 125–135.

Kitcher, Philip (1981), "Explanatory Unification," *Philosophy of Science* 48: 507–531.

Kitcher, Philip (1984), "1953 and All That: A Tale of Two Sciences," *The Philosophical Review* 93: 335–373.

Kitcher, Philip (1987), "Ghostly Whispers: Mayr, Ghiselin, and the 'Philosophers' on the Ontological Status of Species," *Biology and Philosophy* 2: 184–192.

Kitcher, Philip (1989), "Explanatory Unification and the Causal Structure of the World," in Philip Kitcher and Wesley Salmon (eds.), *Scientific Explanation.* Minnesota Studies in the Philosophy of Science, v. 13. Minneapolis, MN: University of Minnesota Press, pp. 410–505.

Kitcher, Philip (1993), *The Advancement of Science: Science without Legend, Objectivity without Illusions.* New York: Oxford University Press.

Kitcher, Philip (1999), "The Hegemony of Molecular Biology," *Biology & Philosophy* 14: 195–210.

Kleiner, Scott A. (1993), *The Logic of Discovery: A Theory of the Rationality of Scientific Research.* Dordrecht: Kluwer.

Kohler, Robert E. (1982), *From Medical Chemistry to Biochemistry: The Making of a Biomedical Discipline.* New York: Cambridge University Press.

Korf, R. E. (1985), "An Analysis of Abstraction in Problem Solving," in J. J. Pottmyer (ed.), *Proceedings of the 24th Annual Technical Symposium.* Gaithersburg, MD: Washington, DC Chapter of the ACM, June 20, 1985, pp. 7–9.

Koshland, D. E. (1973), "Protein Shape and Biological Control," *Scientific American* 229: 52–64.

Kottler, Malcolm (1979), "Hugo de Vries and the Rediscovery of Mendel's Laws," *Annals of Science* 36: 517–538.

Kuhn, Thomas (1962), *The Structure of Scientific Revolutions.* Chicago, IL: University of Chicago Press.

Kuhn, Thomas (1970), *The Structure of Scientific Revolutions.* 2nd ed. Chicago. IL: University of Chicago Press.

Kuhn, Thomas (1974), "Second Thoughts on Paradigms," in Frederick Suppe (ed.), *The Structure of Scientific Theories.* Urbana, IL: University of Illinois Press, pp. 459–482.

Kuno, M. (1995), *The Synapse: Function, Plasticity, and Neurotrophism.* Oxford: Oxford University Press.

Lakatos, Imre (1970), "Falsification and the Methodology of Scientific Research Programmes," in Imre Lakatos and Alan Musgrave (eds.), *Criticism and the Growth of Knowledge*. Cambridge: Cambridge University Press, pp. 91–195.

Lakatos, Imre (1976), *Proofs and Refutations: The Logic of Mathematical Discovery*. J. Worrall and E. Zahar (eds.). Cambridge: Cambridge University Press.

Landman, Otto E. (1991), "The Inheritance of Acquired Characteristics," *Annual Review of Genetics* 25: 1–20.

Langley, Pat, Herbert Simon, Gary L. Bradshaw, and Jan M. Zytkow (1987), *Scientific Discovery: Computational Explorations of the Creative Process*. Cambridge, MA: MIT Press.

Laudan, Larry (1977), *Progress and Its Problems*. Berkeley: University of California Press.

Layton, Jill C. and Patricia L. Foster (2003), "Error-prone DNA Polymerase IV Is Controlled by the Stress-response Sigma Factor, RpoS, in *Escherichia coli*," *Molecular Microbiology* 50: 549–561.

Leatherdale, W. H. (1974), *The Role of Analogy, Model and Metaphor in Science*. New York: American Elsevier.

Lederberg, Joshua ([1959]1961), "Genes and Antibodies," *Science* 129: 1649–1653. Reprinted with a postscript in *Stanford Medical Bulletin* 19: 53–61.

Lederberg, Joshua (1965), "Signs of Life: Criterion-System of Exobiology," *Nature* 207: 9–13.

Lederberg, Joshua (1989), "Replica Plating and Indirect Selection of Bacterial Mutants: Isolation of Preadaptive Mutants in Bacteria by Sib Selection," *Genetics* 121: 395–399.

Lederberg, Joshua (1995), "Notes on Systematic Hypothesis Generation and Application to Disciplined Brainstorming," in *Working Notes: Symposium: Systematic Methods of Scientific Discovery*. AAAI Spring Symposium Series. American Association for Artificial Intelligence, Stanford, CA: Stanford University, pp. 97–98.

Lenat, Douglas (1976), *AM: An Artificial Intelligence Approach to Discovery in Mathematics as Heuristic Search*. Ph.D. Dissertation, Stanford University, Stanford, CA.

Leplin, Jarrett (1975), "The Concept of an *Ad Hoc* Hypothesis," *Studies in the History and Philosophy of Science* 5: 309–345.

Lewontin, Richard C. (1970), "The Units of Selection," *Annual Review of Ecology and Systematics* 1: 1–18.

Lloyd, Elizabeth (1984), "A Semantic Approach to the Structure of Population Genetics," *Philosophy of Science* 51: 242–264.

Lloyd, Elizabeth (1987), "Confirmation of Ecological and Evolutionary Models," *Biology and Philosophy* 2: 277–293.

Machamer, Peter (1998), "Galileo's Machines, His Mathematics and His Experiments," in Peter Machamer (ed.), *Cambridge Companion to Galileo*. New York: Cambridge University Press, pp. 53–79.

Machamer, Peter (2000), "The Nature of Metaphor and Scientific Description," in Fernand Hallyn (ed.), *Metaphor and Analogy in the Sciences*. Dordrecht: Kluwer, pp. 35–52.

Machamer, Peter (2004), "Activities and Causation: The Metaphysics and Epistemology of Mechanisms," *International Studies in the Philosophy of Science* 18: 27–39.

Machamer, Peter, Lindley Darden, and Carl Craver (2000), "Thinking About Mechanisms," *Philosophy of Science* 67: 1–25.

Machamer, Peter and Andrea Woody (1994), "A Model of Intelligibility in Science: Using Galileo's Balance as a Model for Understanding the Motion of Bodies," *Science and Education* 3: 215–244.

Mackie, John Leslie (1974), *The Cement of the Universe: A Study of Causation*. Oxford: Oxford University Press.

Maletic-Savatic, M., R. Malinow, and K. Svoboda (1999), "Rapid Dendritic Morphogenesis in CA1 Hippocampal Dendrites Induced by Synaptic Activity," *Science* 283: 1923–1926.

Malinow, R. (1998), "Silencing the Controversy in LTP?" *Neuron* 21: 1226–1227.

Manier, Edward (1969), "The Experimental Method in Biology, T. H. Morgan and the Theory of the Gene," *Synthese* 20: 185–205.

Marcum, James A. (2002), "From Heresy to Dogma in Accounts of Opposition to Howard Temen's DNA Provirus Hypothesis," *History and Philosophy of the Life Sciences* 24: 165–192.

Maull Roth, Nancy (1974), *Progress in Modern Biology: An Alternative to Reduction*. Ph.D. Dissertation, University of Chicago, Chicago, IL.

Maull, Nancy (1977), "Unifying Science Without Reduction," *Studies in the History and Philosophy of Science* 8: 143–162.

Mayo, Deborah G. (1996), *Error and the Growth of Experimental Knowledge*. Chicago, IL: University of Chicago Press.

Mayr, Ernst (1982), *The Growth of Biological Thought*. Cambridge, MA: Harvard University Press.

Mayr, Ernst and William Provine (eds.) (1980), *The Evolutionary Synthesis*. Cambridge, MA: Harvard University Press.

McHugh, T. J., K. I. Blum, J. Z. Tsien, S. Tonegawa, and M. A. Wilson (1996), "Impaired Hippocampal Representation of Space in CA1-Specific NMDARI Knockout Mice," *Cell* 87: 1339–1349.

Meheus, Joke and Thomas Nickles (eds.) (1999), *Scientific Discovery and Creativity: Case Studies and Computational Approaches*. Special Issue of *Foundations of Science* 4 (4).

Mendel, Gregor ([1865] 1966), "Experiments on Plant Hybrids." Translated from German and reprinted in Curt Stern and Eva Sherwood (eds.), *The Origin of Genetics, A Mendel Source Book*. San Francisco, CA: W. H. Freeman, pp. 1–48.

Mitchell, Tom M. (1982), "Generalization as Search," *Artificial Intelligence* 18: 203–226.

Mittler, John E. and Richard E. Lenski (1990a), "New Data on Excisions of Mu from *E. coli* MCS2 Cast Doubt on Directed Mutation Hypothesis," *Nature* 344: 173–175.

Mittler, John E. and Richard E. Lenski (1990b), "Mittler and Lenski Reply," *Nature* 345: 213.

Moberg, Dale and John Josephson (1990), "Appendix A: An Implementation Note," in J. Shrager and P. Langley (eds.), *Computational Models of Scientific Discovery and Theory Formation*. San Mateo, CA: Morgan Kaufmann, pp. 347–353.

Monaghan, Floyd and A. Corcos (1984), "On the Origins of the Mendelian Laws," *The Journal of Heredity* 75: 67–69.

Monod, J., J-P. Changeux, and F. Jacob (1963), "Allosteric Proteins and Cellular Control Systems," *Journal of Molecular Biology* 6: 306–329.

Monod, J., J. Wyman, and J-P. Changeux (1965), "On the Nature of Allosteric Transitions: A Plausible Model," *Journal of Molecular Biology* 12: 88–118.

Morange, Michel (1998), *A History of Molecular Biology*. Translated by Matthew Cobb. Cambridge, MA: Harvard University Press.

Morgan, Thomas Hunt (1909), "What are 'Factors' in Mendelian Explanations?" *American Breeder's Association* Report 5: 365–368.

Morgan, Thomas Hunt (1910a), "Chromosomes and Heredity," *American Naturalist* 44: 449–496.

Morgan, Thomas Hunt (1910b), "Sex-Limited Inheritance in *Drosophila*," *Science* 32: 120–122.

Morgan, Thomas Hunt (1911a), "An Attempt to Analyze the Constitution of the Chromosomes on the Basis of Sex-Limited Inheritance in *Drosophila*," *Journal of Experimental Zoology* 11: 365–413.

Morgan, Thomas Hunt (1911b), "Random Segregation versus Coupling in Mendelian Inheritance," *Science* 34: 384.

Morgan, Thomas Hunt (1917), "The Theory of the Gene," *American Naturalist* 51: 513–544.

Morgan, Thomas Hunt (1919), *The Physical Basis of Heredity*. Philadelphia, PA: J. B. Lippincott Co.

Morgan, Thomas Hunt (1926), *The Theory of the Gene*. New Haven, CT: Yale University Press.

Morgan, Thomas Hunt and Clara J. Lynch (1912), "The Linkage of Two Factors in *Drosophila* that Are Not Sex-Linked," *Biological Bulletin* 23: 174–182.

Morgan, Thomas Hunt, A. H. Sturtevant, H. J. Muller, and Calvin B. Bridges (1915), *The Mechanism of Mendelian Heredity*. New York: Henry Holt and Company.

Morowitz, Harold (1985), "Models for Biomedical Research: A New Perspective," Report of the Committee on Models for Biomedical Research, Washington, DC: National Academy Press.

Morowitz, Harold and Temple Smith (1987), "Report of the Matrix of Biological Knowledge Workshop, July 13–August 14, 1987," Santa Fe, NM: Santa Fe Institute.

Morris, R. G. M., P. Garrud, J. N. P. Rawlins, and J. O'Keefe (1982), "Place Navigation Impaired in Rats with Hippocampal Lesions," *Nature* 297: 681–683.

Nagel, Ernest (1961), *The Structure of Science*. New York: Harcourt, Brace and World.

Nersessian, Nancy J. (1993), "In the Theoretician's Laboratory: Thought Experimenting as Mental Modeling," in D. Hull, M. Forbes, and K. Okruhlik (eds.), *PSA 1992*, v. 2. East Lansing, MI: Philosophy of Science Association, pp. 291–301.

Newton-Smith, W. H. (1981), *The Rationality of Science*. Boston, MA: Routledge & Kegan Paul.

Nickles, Thomas (ed.) (1980a), *Scientific Discovery, Logic and Rationality*. Dordrecht: Reidel.

Nickles, Thomas (ed.) (1980b), *Scientific Discovery: Case Studies*. Dordrecht: Reidel.

Nickles, Thomas (1980c), "Introductory Essay: Scientific Discovery and the Future of Philosophy of Science," in Thomas Nickles (ed.), *Scientific Discovery, Logic and Rationality*. Dordrecht: Reidel, pp. 1–59.

Nickles, Thomas (1981), "What Is a Problem That We May Solve It?" *Synthese* 47: 85–118.

Nickles, Thomas (1987), "Methodology, Heuristics, and Rationality," in J. C. Pitt and M. Pera (eds.), *Rational Changes in Science*. Dordrecht: Reidel, pp. 103–132.

Nirenberg, M. W. and J. H. Matthaei (1961), "The Dependence of Cell-Free Protein Synthesis in *E. coli* upon Naturally Occurring or Synthetic Polyribonucleotides," *Proceedings of the National Academy of Sciences* 47:1588–1602.

O'Keefe, J. and J. Dostrovsky (1971), "The Hippocampus as a Spatial Map: Preliminary Evidence from Unit Activity in the Freely Moving Rat," *Brain Research* 34: 171–175.

Olby, Robert (1970), "Francis Crick, DNA, and the Central Dogma," in Gerald Holton (ed.), *The Twentieth Century Sciences*. New York: W. W. Norton, pp. 227–280.

Olby, Robert (1994), *The Path to the Double Helix: The Discovery of DNA*. Revised Edition. Mineola, NY: Dover.

Olton, D. S. and R. J. Samuelson (1976), "Remembrances of Places Passed: Spatial Memory in Rats," *Journal of Experimental Psychology: Animal Behavior Processes* 2: 97–116.

Oppenheim, Paul and Hilary Putnam (1958), "Unity of Science as a Working Hypothesis," in H. Feigl, M. Scriven, and G. Maxwell (eds.), *Concepts, Theories, and the Mind-Body Problem*, Minnesota Studies in the Philosophy of Science, v. 2. Minneapolis, MN: University of Minnesota Press, pp. 3–36.

Parascandola, Mark (1995), "Philosophy in the Laboratory: The Debate over Evidence for E. J. Steele's Lamarckian Hypothesis," *Studies in History and Philosophy of Science* 26: 469–492.

Pardee, Arthur B., Francois Jacob, and Jacques Monod (1959), "The Genetic Control and Cytoplasmic Expression of 'Inducibility' in the Synthesis of $\beta$−galatosidase," *Journal of Molecular Biology* 1: 165–178.

Pauling, Linus (1939), *The Nature of the Chemical Bond*. Ithaca, NY: Cornell University Press.

Pauling, Linus (1940), "A Theory of the Structure and Process of Formation of Antibodies," *Journal of the American Chemical Society* 62: 2643–2657.

Pauling, Linus and Robert B. Corey (1950), "Two Hydrogen-Bonded Spiral Configurations of the Polypeptide Chain," *Journal of the American Chemical Society* 72: 5349.

Piattelli-Palmarini, M. (1986), "The Rise of Selection Theories: A Case Study and Some Lessons from Immunology," in W. Demopoulos and A. Marras (eds.), *Language Learning and Concept Acquisition: Foundational Issues*. Norwood, NJ: Ablex Publishing Co., pp. 117–130.

Popper, Karl R. (1965), *The Logic of Scientific Discovery*. New York: Harper Torchbooks.

Popper, Karl (1968), *Conjectures and Refutations: The Growth of Scientific Knowledge*. New York: Harper Torchbooks.

Provine, William (1971), *The Origin of Theoretical Population Genetics*. Chicago, IL: University of Chicago Press.

Psillos, Stathis (2004), "A Glimpse of the *Secret Connexion*: Harmonizing Mechanisms with Counterfactuals," *Perspectives on Science* 12: 288–391.

Punnett, R. C. (ed.) (1928), *Scientific Papers of William Bateson*. Cambridge: Cambridge University Press, 2 vols.

*Bibliography*

Rescher, Nicholas (1996), *Process Metaphysics: An Introduction to Process Philosophy.* Albany, NY: State University of New York Press.

Rheinberger, Hans-Jörg (1997), *Experimental Systems: Towards a History of Epistemic Things. Synthesizing Proteins in the Test Tube.* Stanford, CA: Stanford University Press.

Rich, Alexander and James D. Watson (1954a), "Physical Studies on Ribonucleic Acid," *Nature* 173: 995–996.

Rich, Alexander and James D. Watson (1954b), "Some Relations Between DNA and RNA," *Proceedings of the National Academy of Sciences* 40: 759–764.

Roll-Hansen, N. (1978), "*Drosophila* Genetics: A Reductionist Research Program," *Journal of the History of Biology* 11: 159–210.

Rosenberg, Susan M. (1994), "In Pursuit of a Molecular Mechanism for Adaptive Mutation," *Genome* 37: 893–899.

Rosenberg, Susan M. and P. J. Hastings (2003), "Modulating Mutation Rates in the Wild," *Science* 300: 1382–1384.

Rosenberg, Susan M., Simonne Longerich, Pauline Gee, and Reuben S. Harris (1994), "Adaptive Mutation by Deletions in Small Mononucleotide Repeats," *Science* 265: 405–407.

Roth, John R., Eric Kofoid, Frederick P. Roth, Otto G. Berg, Jon Seger, and Dan I. Andersson (2003a), "Regulating General Mutation Rates: Examination of the Hypermutable State Model for Cairnsian Adaptive Mutation," *Genetics* 163: 1483–1496.

Roth, John R., Eric Kofoid, Frederick P. Roth, Otto G. Berg, Jon Seger, and Dan I. Andersson (2003b), "Letter to the Editor: Adaptive Mutation Requires No Mutagenesis – Only Growth Under Selection: A Response," *Genetics* 165: 2319–2321.

Rottenberg, A., M. Mayford, R. D. Hawkins, E. R. Kandel, and R. U. Muller (1996), "Mice Expressing Activated CaMKII Lack Low Frequency LTP and Do Not Form Stable Place Cells in the CA1 Region of the Hippocampus," *Cell* 87: 1351–1361.

Roush, W. (1997), "New Knockout Mice Point to Molecular Basis of Memory," *Science* 275: 32–33.

Rudge, David W. (1996), *A Philosophical Analysis of the Role of Natural Selection Experiments in Evolutionary Biology.* Ph.D. Dissertation, Department of History and Philosophy of Science, University of Pittsburgh, Pittsburgh, PA.

Rudge, David W. (1999), "Taking the Peppered Moth with a Grain of Salt," *Biology & Philosophy* 14: 9–37.

Ryan, F. J. (1955), "Spontaneous Mutation in Non-dividing Bacteria," *Genetics* 40: 726–738.

Salmon, Wesley (1984), *Scientific Explanation and the Causal Structure of the World.* Princeton, NJ: Princeton University Press.

Salmon, Wesley (1997), "Causality and Explanation: A Reply to Two Critiques," *Philosophy of Science* 64: 461–477.

Salmon, Wesley (1998), *Causality and Explanation.* New York: Oxford University Press.

Sarkar, Sahotra (1991), "Lamarck *Contre* Darwin, Reduction *Versus* Statistics: Conceptual Issues in the Controversy over Directed Mutagenesis in Bacteria," in Alfred I. Tauber (ed.), *Organism and the Origins of Self.* Dordrecht: Kluwer, pp. 235–271.

Sarkar, Sahotra (1996), "Biological Information: A Skeptical Look at Some Central Dogmas of Molecular Biology," in S. Sarkar (ed.), *The Philosophy and History of Molecular Biology: New Perspectives.* Dordrecht: Kluwer, pp. 187–231.

330

Schaffner, Kenneth (1974a), "Logic of Discovery and Justification in Regulatory Genetics," *Studies in the History and Philosophy of Science* 4: 349–385.

Schaffner, Kenneth (1974b), "The Peripherality of Reductionism in the Development of Molecular Genetics," *Journal of the History of Biology* 7: 111–139.

Schaffner, Kenneth (1974c), "The Unity of Science and Theory Construction in Molecular Biology," in R. J. Seeger and R. S. Cohen (eds.), *Philosophical Foundations of Science: Proceedings of Section L, AAAS 1969*. Boston Studies in the Philosophy of Science, v. 11. Dordrecht: D. Reidel Publishing, pp. 497–533.

Schaffner, Kenneth (1980), "Theory Structure in the Biomedical Sciences," *The Journal of Medicine and Philosophy* 5: 57–97.

Schaffner, Kenneth (1986a), "Computerized Implementation of Biomedical Theory Structures: An Artificial Intelligence Approach," in Arthur Fine and Peter Machamer (eds.), *PSA 1986*, v. 2. East Lansing, MI: Philosophy of Science Association, pp. 17–32.

Schaffner, Kenneth (1986b), "Exemplar Reasoning about Biological Models and Diseases: A Relation Between the Philosophy of Medicine and Philosophy of Science," *The Journal of Medicine and Philosophy* 11: 63–80.

Schaffner, Kenneth (1993), *Discovery and Explanation in Biology and Medicine*. Chicago, IL: University of Chicago Press.

Schank, Roger C. (1986), *Explanation Patterns: Understanding Mechanically and Creatively*. Hillsdale, NJ: Lawrence Erlbaum.

Scoville, W. B. and B. Millner (1957), "Loss of Recent Memory after Bilateral Hippocampal Lesions," *Journal of Neurology, Neurosurgery, and Psychiatry*. 20: 11–20.

Sembugamoorthy, V. and B. Chandrasekaran (1986), "Functional Representation of Devices and Compilation of Diagnostic Problem-solving Systems," in J. Kolodner and C. Reisbeck (eds.), *Experience, Memory, and Reasoning*. Hillsdale, NJ: Lawrence Erlbaum Associates, pp. 47–73.

Shapere, Dudley (1973), Unpublished MS. Presented at IUHPS-LMPS Conference on Relations Between History and Philosophy of Science. Jyväskylä, Finland.

Shapere, Dudley (1974a), "Scientific Theories and Their Domain" in Frederick Suppe (ed.), *The Structure of Scientific Theories*. Urbana, IL: University of Illinois Press, pp. 518–565.

Shapere, Dudley (1974b), "On the Relations Between Compositional and Evolutionary Theories," in F. J. Ayala and T. Dobzhansky (eds.), *Studies in the Philosophy of Biology*. Berkeley, CA: University of California Press, pp. 187–201.

Shapere, Dudley (1977), "Scientific Theories and Their Domains," in Frederick Suppe (ed.), *Structure of Scientific Theories*, 2nd ed. Urbana, IL: University of Illinois Press, pp. 518–562.

Shapere, Dudley (1980), "The Character of Scientific Change," in Thomas Nickles (ed.), *Scientific Discovery, Logic and Rationality*. Dordrecht: Reidel, pp. 61–101.

Shapere, Dudley and Gerald Edelman (1974), "A Note on the Concept of Selection," in F. J. Ayala and T. Dobzhansky (eds.), *Studies in the Philosophy of Biology*. London: Macmillan, pp. 202–204.

Shapiro, James A. (1995), "Adaptive Mutation: Who's Really in the Garden?" *Science* 268: 373–374.

Shepherd, Gordon M. (1994), *Neurobiology*. 3rd ed. New York: Oxford University Press.

331

Sherry, D. and S. Healy (1998), "Neural Mechanisms of Spatial Representation," in S. Healy (ed.), *Spatial Representation in Animals*. Oxford: Oxford University Press, 133–157.

Shi, S., Y. Hayashi, R. S. Petralia, S. H. Zaman, R. J. Wenthold, K. Svoboda, and R. Malinow (1999), "Rapid Spine Delivery and Redistribution of AMPA Receptors after Synaptic NMDA Receptor Activation," *Science* 284: 1811–1816.

Skipper, Robert A., Jr. (1999), "Selection and the Extent of Explanatory Unification," *Philosophy of Science* 66 (Proceedings): S196–S209.

Skipper, Robert A., Jr. (2000), *The R.A. Fisher-Sewall Wright Controversy in Philosophical Focus: Theory Evaluation in Population Genetics*. Ph.D. Dissertation, Department of Philosophy, University of Maryland, College Park, MD.

Skipper, Robert A., Jr. (2001), "The Causal Crux of Selection," *Behavioral and Brain Sciences* 24: 556.

Skipper, Robert A., Jr. (2002), "The Persistence of the R. A. Fisher-Sewall Wright Controversy," *Biology and Philosophy* 17: 341–367.

Skipper, Robert A., Jr. and Roberta L. Millstein (2005), "Thinking about Evolutionary Mechanisms: Natural Selection," in Carl F. Craver and Lindley Darden (eds.), Special Issue: "Mechanisms in Biology," *Studies in History and Philosophy of Biological and Biomedical Sciences* 36: 327–347.

Sober, Elliott (1984), *The Nature of Selection*. Cambridge, MA: MIT Press.

Stahl, F. W. (1988), "A Unicorn in the Garden," *Nature* 335: 112–113.

Steele, E. J. (1981), *Somatic Selection and Adaptive Evolution: On the Inheritance of Acquired Characters*. 2nd ed. Chicago, IL: University of Chicago Press.

Steele, Edward J., Robyn A. Lindley, and Robert B. Blanden (1998), *Lamarck's Signature*. Reading, MA: Perseus Books.

Stent, Gunther (1964), "The Operon: On Its Third Anniversary," *Science* 144: 816–820.

Stent, Gunther (1969), *The Coming of the Golden Age: A View of the End of Progress*. Garden City, NY: American Museum of Natural History Press.

Stern, C. and E. Sherwood (eds.) (1966), *The Origin of Genetics, A Mendel Source Book*. San Francisco, CA: W. H. Freeman.

Stevens, C. F. (1996), "Spatial Learning and Memory: The Beginning of a Dream," *Cell* 87: 1147–1148.

Sticklen, J. (1987), *MDX2: An Integrated Medical Diagnostic System*. Ph.D. Dissertation, Department of Computer and Information Science, The Ohio State University, Columbus, OH.

Strickberger, Monroe (1985), *Genetics*. 3rd ed. New York: Macmillan.

Studer, Kenneth E. and Daryl E. Chubin (1980), *The Cancer Mission: Social Contexts of Biomedical Research*. Sage Library of Social Research, v. 103. Beverly Hills, CA: Sage Publications.

Sturtevant, A. H. (1913), "The Linear Arrangement of Six Sex-linked Factors in *Drosophila*, as Shown by Their Mode of Association," *Journal of Experimental Zoology* 14: 43–59.

Suppe, Frederick (1974), *The Structure of Scientific Theories*. Urbana, IL: University of Illinois Press.

Suppe, Frederick (1979), "Theory Structure," in Peter Asquith and H. Kyburg (eds.), *Current Research in Philosophy of Science*. East Lansing, MI: Philosophy of Science Association, pp. 317–338.

Sutton, Walter ([1903] 1959), "The Chromosomes in Heredity," *Biological Bulletin* 4: 231–251. Reprinted in J. A. Peters (ed.), *Classic Papers in Genetics*. Englewood Cliffs, NJ: Prentice-Hall, pp 27–41.

Tabery, James G. (2004), "Synthesizing Activities and Interactions in the Concept of a Mechanism," *Philosophy of Science* 71: 1–15.

Talmadge, D. W. (1957), "Allergy and Immunology," *Annual Review of Medicine* 8: 239–256.

Talmadge, D. W. (1986), "The Acceptance and Rejection of Immunological Concepts," *Annual Review of Immunology* 8: 239–256.

Temin, Howard M. (1963), "The Effects of Actinomycin D on Growth of Rous Sarcoma Virus in Vitro," *Virology* 20: 577–582

Temin, Howard M. (1964a), "The Participation of DNA in Rous Sarcoma Virus Production," *Virology* 23: 486–494.

Temin, Howard M. (1964b), "Homology between RNA from Rous Sarcoma Virus and DNA from Rous Sarcoma Virus-infected Cells," *Proceedings of the National Academy of Sciences* 52: 323–329.

Temin, Howard M. (1964c), "Nature of the Provirus in Rous Sarcoma," *National Cancer Institute Monograph* 17: 557–570.

Temin, Howard M. (1971), "Guest Editorial. The Protovirus Hypothesis: Speculations on the Significance of RNA-directed DNA Synthesis for Normal Development and Carcinogenesis." *Journal of the National Cancer Institute* 46: III-VIII.

Temin, Howard M. (1977), "The DNA Provirus Hypothesis: The Establishment and Implications of RNA-directed DNA Synthesis," in *Nobel Lectures in Molecular Biology, 1933–1975.* New York: Elsevier, pp. 509–529. Presented December 12, 1975.

Temin, Howard M. and Satoshi Mizutani (1970), "RNA-dependent DNA Polymerase in Virions of Rous Sarcoma Virus," *Nature* 226: 1211–1213.

Thagard, Paul (1988), *Computational Philosophy of Science*. Cambridge, MA: MIT Press.

Thagard, Paul (1992), *Conceptual Revolutions*. Princeton, NJ: Princeton University Press.

Thagard, Paul (2003), "Pathways to Biomedical Discovery," *Philosophy of Science* 70: 235–254.

Thaler, David S. (1994), "The Evolution of Genetic Intelligence," *Science* 264: 224–225.

Thieffry, Denis and Richard M. Burian (1996), "Jean Brachet's Alternative Scheme for Protein Synthesis," *Trends in the Biochemical Sciences* 21 (3): 114–117.

Thompson, Paul (1983), "The Structure of Evolutionary Theory: A Semantic Approach," *Studies in the History and Philosophy of Science* 14: 215–229.

Tolman, E. (1948), "Cognitive Maps in Rats and Men," *Psychological Review* 55: 189–208.

Tolman, E. and C. Honzick (1930), "Introduction and Removal of Reward and Maze Performance in Rats," *University of California Publications in Psychology* 4: 257–275.

Toulmin, Stephen (1972), *Human Understanding*. v. 1. Princeton, NJ: Princeton University Press.

Tsien, J. Z., D. E. Chen, D. Gerber, C. Tom, E. Mercer, D. Anderson, M. Mayford, and E. R. Kandel (1996a), "Subregion- and Cell Type-Restricted Gene Knockout in Mouse Brain," *Cell* 87: 1317–1326.

Tsien, J. Z., P. T. Huerta, and S. Tonegawa (1996b), "The Essential Role of Hippocampal CA1 NMDA Receptor-Dependent Synaptic Plasticity in Spatial Memory," *Cell* 87: 1327–1338.

Vorzimmer, Peter (1970), *Charles Darwin, The Years of Controversy*. Philadelphia, PA: Temple University Press.

Vrba, Elisabeth S. and Stephen J. Gould (1986), "The Hierarchical Expansion of Sorting and Selection: Sorting and Selection Cannot Be Equated," *Paleobiology* 12: 217–228.

Vries, Hugo de ([1889] 1910), *Intracellular Pangenesis*. Translated by C. S. Gager. Chicago, IL: Open Court.

Vries, Hugo de ([1900] 1966), "The Law of Segregation of Hybrids." Translated from German and reprinted in C. Stern and E. Sherwood (eds.), *The Origin of Genetics, A Mendel Source Book*. San Francisco, CA: W. H. Freeman, pp. 107–117.

Vries, Hugo de ([1903–04; 1909–1910] 1969), *The Mutation Theory*. 2 vols. Translated by J. B. Farmer and A. D. Darbishire. New York: Kraus Reprint Company.

Waters, C. Kenneth (1990), "Why the Anti-reductionist Consensus Won't Survive the Case of Classical Mendelian Genetics," in Arthur Fine, Micky Forbes, and Linda Wessels (eds.), *PSA 1990*, v. 1. East Lansing, MI: Philosophy of Science Association, pp. 125–139.

Waters, C. Kenneth (1994), "Genes Made Molecular," *Philosophy of Science* 61: 163–185.

Watson, James D. ([1962] 1977), "The Involvement of RNA in the Synthesis of Proteins," in *Nobel Lectures in Molecular Biology 1933–1975*. New York: Elsevier, pp. 179–203.

Watson, James D. (1965) *Molecular Biology of the Gene*. New York: W. A. Benjamin.

Watson, James D. (1968), *The Double Helix*. New York: New American Library.

Watson, James D. (1970), *Molecular Biology of the Gene*. 2nd ed. New York: W. A. Benjamin.

Watson, James D. (1977), *Molecular Biology of the Gene*. 3rd ed. New York: W. A. Benjamin.

Watson, James D. (2000), *A Passion for DNA: Genes, Genomes, and Society*. Cold Spring Harbor, NY: Cold Spring Harbor Laboratory Press.

Watson, James D. and Francis Crick (1953a), "A Structure for Deoxyribose Nucleic Acid," *Nature* 171: 737–738.

Watson, James D. and Francis Crick (1953b), "Genetical Implications of the Structure of Deoxyribonucleic Acid," *Nature* 171: 964–967.

Watson, James D., Nancy H. Hopkins, Jeffrey W. Roberts, Joan Argetsinger Steitz, and Alan M. Weiner (1988), *Molecular Biology of the Gene*. 4th ed. Menlo Park, CA: Benjamin/Cummings.

Weber, Marcel (2005), *Philosophy of Experimental Biology*. New York: Cambridge University Press.

Weismann, A. (1892), *The Germ-Plasm, A Theory of Heredity*. Translated by W. N. Parker and H. Rönfeldt. New York: Charles Scribner's Sons.

Westfall, Richard S. (1977), *The Construction of Modern Science: Mechanisms and Mechanics*. New York: Cambridge University Press.

White, Abraham, Philip Handler, Emil L. Smith, and DeWitt Stetten, Jr. (1954), *Principles of Biochemistry*. New York: McGraw-Hill.

Williams, Mary (1970), "Deducing the Consequences of Evolution: A Mathematical Model," *Journal of Theoretical Biology* 29: 343–385.

Wilson, Edmund B. (1900), *The Cell in Development and Inheritance*. 2nd ed. New York: Macmillan.

Wilson, M. A. and B. McNaughton (1993), "Dynamics of the Hippocampal Ensemble Code for Space," *Science* 261: 1055–1058.

Wimsatt, William (1972), "Complexity and Organization," in Kenneth F. Schaffner and Robert S. Cohen (eds.), *PSA 1972, Proceedings of the Philosophy of Science Association*. Dordrecht: Reidel, pp. 67–86.

Wimsatt, William (1976), "Reductive Explanation: A Functional Account," in Robert S. Cohen (ed.), *PSA 1974*. Dordrecht: Reidel, pp. 671–710. Reprinted in Elliott Sober (ed.) (1984), *Conceptual Issues in Evolutionary Biology: An Anthology*. 1st ed. Cambridge, MA: MIT Press, pp. 477–508.

Wimsatt, William C. (1981), "Robustness, Reliability, and Overdetermination," in M. Brewer and B. Collins (eds.), *Scientific Inquiry and the Social Sciences*. San Francisco, CA: Jossey-Bass, pp. 124–163.

Wimsatt, William (1987), "False Models as Means to Truer Theories," in Matthew Nitecki and Antoni Hoffman (eds.), *Natural Models in Biology*. New York: Oxford University Press, pp. 23–55.

Woodward, James (2002), "What Is a Mechanism? A Counterfactual Account," *Philosophy of Science (Supplement)* 69: S366–S377.

Woodward, James (2003), *Making Things Happen: A Theory of Causal Explanation*. New York: Oxford University Press.

Zamecnik, Paul C. (1953), "Incorporation of Radioactivity from DL-Leucine-1-C[14] into Proteins of Rat Liver Homogenate," *Federation Proceedings* 12: 295.

Zamecnik, Paul C. (1958), "The Microsome," *Scientific American* 198 (March): 118–124.

Zamecnik, Paul C. (1962a), "History and Speculation on Protein Synthesis," *Proceedings of the Symposia on Mathematical Problems in the Biological Sciences* 14: 47–53.

Zamecnik, Paul C. (1962b), "Unsettled Questions in the Field of Protein Synthesis," *Biochemical Journal* 85: 257–264.

Zamecnik, Paul C. (1969), "An Historical Account of Protein Synthesis, with Current Overtones – A Personalized View," *Cold Spring Harbor Symposia on Quantitative Biology* 34: 1–16.

Zamecnik, Paul C. (1976), "Protein Synthesis – Early Waves and Recent Ripples," in A. Kornberg, B. L. Horecker, L. Cornudella, and J. Oro (eds.), *Reflections in Biochemistry*. New York: Pergamon Press, pp. 303–308.

Zamecnik, Paul C. (1979), "Historical Aspects of Protein Synthesis," *Annals of the New York Academy of Sciences* 325: 269–301.

Zamecnik, Paul C. (1984), "The Machinery of Protein Synthesis," *Trends in Biochemical Sciences (TIBS)* 9: 464–466.

Zola-Morgan, S. and L. Squire (1993), "Neuroanatomy of Memory," *Annual Review of Neuroscience* 16: 547–563.

# Index

abduction, 289; *see also* retroduction

Abir-Am, Pnina, 229

Abrahamsen, Adele, xxii, 3, 8, 279, 283, 292, 297, 298, 308, 314

abstraction, xiv, xxiv, 6, 28, 30, 38, 62, 68, 86, 94, 108, 112, 121, 152, 182–196, 198, 200, 201, 203, 215, 216, 225, 229, 230, 234, 235, 238, 244, 265, 281, 282, 284, 297, 298, 309

vs. generalization, 28, 297

action potential, 18, 25, 36, 42, 51, 52, 55, 58, 276

activities, xi, xii, 3, 4, 13–20, 22, 24–36, 41–48, 50, 51–54, 58, 59, 63, 67, 68, 70, 72, 74–79, 83–92, 108–110, 112, 123, 130, 131, 133, 139–143, 149, 153, 158, 210, 219, 239, 273, 274–279, 281–285, 287, 288, 290–293, 295, 296, 299, 305, 306, 311, 312, 327, 333

activity-enabling properties, 89–91, 109, 110, 275, 276, 288, 293

signature, 89, 92, 275, 288

Ada, G. L., 195, 204, 313

adaptation problem, xiii, 6, 182, 185, 196, 197, 201–203

adaptive mutation, xiv, xix, 6, 7, 249, 252, 255–262, 264, 303, 305; *see also* directed mutation; *see* Table 11.1

Alberts, Bruce, 102, 118, 313

Allchin, Douglas, xxii, 3, 8, 248, 302, 304, 308, 313

Allemang, Dean, 207, 219, 225, 313

Allen, Garland, 160, 167, 313

allosteric regulation, 5, 91, 132, 133, 139, 141–143, 145; *see* Figure 5.2

analogy, xiii, 2, 5, 6, 38, 130, 151–154, 156–166, 168, 182, 184, 185, 188, 200, 203–205, 208, 210, 214, 216, 220, 225, 226, 239, 246, 271, 281, 284–286, 292–294, 310, 319; *see also* strategies for producing new ideas; strategies for mechanism discovery; strategies for theory construction; *see* Figure 6.1

Anderson, D., 64, 333

Andersson, Dan I., 259, 260, 264, 266, 267, 313, 323, 330

anomaly, xii, xiv, xv, xix, xxii, xxiv, 2, 3, 6, 7, 26, 30, 34, 61, 62, 79, 80, 81, 85, 88, 149, 158, 162, 207–225, 229, 232–237, 240, 242–246, 248, 249, 251–265, 271, 272, 298, 301–306; *see also* strategies for anomaly resolution; anomaly driven theory redesign

model, 7, 214, 215, 224, 234, 236, 241, 243, 244, 245, 262, 297, 302, 303

monster, xiv, 214, 218, 221, 222, 224, 234

special-case, xiv, 7, 70, 229, 243, 245, 257, 258, 262, 302, 303

anomaly driven theory redesign, 80, 234; *see also* strategies for anomaly resolution

Anscombe, G. E. M., 18, 37, 313

Archimedes, 27

Arnold, A. J., 190, 204, 313

artificial intelligence, 185, 207, 225, 305, 306; *see also* computational philosophy of science; heuristics

Ayala, F. J., 169, 204, 206, 313

Bailer-Jones, D., 13

Baker, Jason, xxii, 5, 8, 248, 271, 313

337

*Index*

Baltas, A., 13
Baltimore, David, 236, 238, 239, 244, 245, 314
Barinaga, M., 54, 62, 314
Barrionnevo, G., 40
Barker, Matthew, xxiii
Barnett, Leslie, 94, 121, 318
Bateson, B., 158, 167, 314
Bateson, William, xiii, 5, 114, 121, 135, 146,
    153, 155–161, 163, 164, 167, 168,
    175–177, 180, 231, 235, 245, 314, 317,
    319, 329
Beatty, John, 170, 184, 204, 248, 258, 264,
    297, 303, 308, 314
Bechtel, William, xxii, 2, 3, 5, 8, 15, 16, 37,
    48, 49, 51, 53, 62, 66, 84, 93, 109, 121,
    170, 238, 245, 273, 279, 283, 290–292,
    296–298, 304, 308, 314
Belozersky, Andrei N., 33, 37, 79, 93, 315
Berg, Paul, 244, 246, 267, 315, 330
biochemistry, xv, 1, 5, 66–68, 70, 72, 73, 78,
    84, 95, 97, 101, 115, 122, 124, 127–129,
    131–133, 141–145, 178, 272, 323
Bjedov, Ivana, 257, 264, 315
black box; *see also* gray box; glass box;
    in mechanism sketch, xii, 5, 30, 34, 48, 61,
    86, 87, 99, 105, 106, 112, 115, 118–120,
    156, 282, 285, 290, 291, 299, 301,
    306
Blackwell, R., 151, 167, 315
Blanden, Robert B., 267, 312, 332
blending inheritance, 174
Blum, K. I., 63, 327
Bogen, James, xxii, 4, 8, 13, 42, 50, 62, 109,
    121, 261, 264, 271, 276, 278, 283, 290,
    308, 315
Bolker, Jessica, 287, 310, 322
Bourgeois, S., 141, 146, 315
Boveri, Theodor, 114, 121, 136, 137, 146, 160,
    161, 167, 315
Brachet, Jean, 8, 37, 72, 93, 96, 316, 333
Bradie, Michael, 182
Bradshaw, Gary L., 227, 326
Brandon, Robert N., 2, 8, 15, 37, 66, 93, 184,
    188, 204, 315
Brenner, Sydney, 82, 93, 94, 121, 304, 308,
    315, 318
Brewer, William F., 303, 308, 315
Bridewell, Will, 305, 308, 315
Bridges, Bryn A., 256, 257, 264, 265, 315
Bridges, Calvin B., 115, 123, 138, 143, 146,
    147, 160, 162, 167, 168, 247, 316, 328

Brisson, Dustin, 259, 264, 316
Brush, Stephen, xxii, 161, 166, 167, 316
Buchanan, Bruce, xxiii, 2, 8, 316
Burian, Richard M., xxii, 2, 8, 13, 15, 37, 66,
    67, 72, 93, 96, 201, 204, 303, 308, 316,
    333
Burnet, F. M., 182, 195, 196, 204, 316
Bylander, Tom, 222, 225, 316

Cain, Joseph A., xiii, xxi, xxiii, 5, 6, 86, 94,
    182, 188, 204, 216, 226, 230, 246, 254,
    263, 265, 274, 278, 284, 309, 316,
    320
Cairns, John, 249–259, 261–263, 265, 266,
    287, 309, 316, 317, 321
Carlson, Elof A., 160, 167, 317
Carmadi, G., 13
Carothers, Eleanor E., 137, 146, 317
Cartwright, Nancy, 16, 37, 276, 283, 309,
    317
Casey, David, 182
Caspersson, T. O., 72
Castle, William E., 223, 225, 317
causality, xi, xxii, 2, 4, 13, 15, 16, 18, 19, 37,
    39, 49, 55, 106, 133, 134, 139, 141, 143,
    145, 165, 171–173, 179, 180, 187–194,
    199, 210, 219, 223, 224, 238, 283, 286,
    298, 308, 313, 315
Cech, T. R., 70, 93, 317
central dogma of molecular biology, xiv, xv,
    xvii, 7, 28–30, 83, 87, 112, 232–234, 236,
    238, 240, 242, 244, 245, 248, 249,
    251–255, 259, 264, 281, 303; *see* Figure
    1.3, Figure 1.4, Figure 3.1, Figure 3.3,
    Figure 3.4, Figure 3.5, Figure 10.5,
    Figure 10.6, Figure 10.7
Chadarevian, Soraya de, 67, 74, 93, 94, 102,
    121, 317
Chandrasekaran, B., xxiii, 214, 216, 219, 220,
    222, 225–227, 316, 317, 322, 324, 331
Changeux, J.-P., 139, 141, 142, 147, 197, 198,
    204, 317, 328
Chapuis, N., 50, 62, 317
Charniak, Eugene, 211, 225, 317
chemical neurotransmission; *see* Figure 1.1,
    Figure 1.2
    mechanism of, 14, 25, 51
Chen, D. E., 64, 333
Chinn, Clark A., 303, 308, 315
chromosome theory, 5, 106, 114, 115, 119,
    127, 132, 136–138, 143–145, 156–158,

338

genetics; *see also* molecular biology
 Mendelian, xii–xiv, 3, 4, 7, 98–100, 102,
  103, 105, 107, 109, 113, 115, 118, 120,
  154, 172, 176, 179, 180, 218, 221, 230,
  232–236, 245
 population, xiii, 172, 177, 179, 180
Gentner, Dedre, 185, 205, 286, 310, 322
Gerber, D., 64, 333
Gilbert, G. K., 151, 168, 322
Gilbert, Scott, 286, 287, 310, 322
Gilbert, Walter, 95, 141, 147, 310, 322
glass box
 in mechanism schema, 282, 299, 301, 306
Glennan, Stuart S., xxiii, 2, 9, 13, 16, 38, 66,
  95, 109, 122, 273, 277, 278, 282, 289,
  290, 296, 310, 322
Glymour, Clark, 210, 226, 322
Goel, Ashok, 222, 226, 322
Goldberg, Nathaniel, 98
Golub, E., 202, 205, 322
Gould, Stephen J., 183, 190, 192, 194, 205,
  206, 322, 334
gray box
 with functional role, 258, 282, 285, 290,
  291, 299, 301, 306
Greiner, Russell, 185, 205, 322
Grene, Marjorie, 8, 37, 93, 182
Gros, Francois, 67, 82, 95, 304, 310, 322
Gustav, N., 195, 204, 313

Hacking, Ian, 55, 62, 63, 322
Hagan, Joel, 182
Hall, Barry G., 256, 259, 266, 323
Hall, Nancy S., xxii, 13, 40, 65, 98
Hall, Zach W., 38, 323
Hamscher, Walter C., 22, 208, 226, 320
Hanson, Norwood R., 2, 9, 151, 168, 183, 205,
  282, 310, 323
Harman, Gilbert, 263, 266, 305, 310, 323
Harré, Rom, 2, 9, 152, 165, 168, 286, 292,
  310, 323
Harris, Reuben S., 257, 266, 267, 323, 330
Hastings, P. J., 257, 267, 330
Hawkins, R. D., 63, 330
Hayashi, Y., 63, 332
Healy, S., 50, 63, 332
Hebb, D. O., 43, 63, 323
Hempel, Carl G., 99, 122, 149, 150, 168, 323
Hendrickson, Heather, 260, 266, 323
Henson, Pamela, xxii, 182
Herbert, Sandra, 65

heredity
 mechanisms of, xii, 5, 112, 119, 280; *see*
  Figure 4.1
Herman, David, 219, 225, 317
Hesse, Mary, 150, 152, 163, 168, 210, 216,
  226, 323
heuristics, 2, 8, 149, 165, 166, 207, 210, 263,
  294, 297
Hiatt, Howard, 95, 310, 322
Hick, Darren, xxiii
hierarchy, xii, xviii, 2, 4, 14, 25, 26, 28, 37, 45,
  58, 59, 62, 103, 121, 136, 152, 165, 172,
  186, 190–192, 201, 222, 223, 225, 280,
  298, 309, 318; *see also* levels; *see*
  Figure 2.4
Hoagland, Mahlon B., 67, 69, 73, 75, 77, 78,
  87, 95, 288, 323
Holland, John H., 185, 205, 323
Holmes, L., 13
Holyoak, Keith J., 3, 9, 185, 205, 216, 226,
  285, 286, 310, 323, 324
Honzick, C., 49, 64, 333
Huerta, P. T., 64, 334
Hughes, Arthur, 101, 122, 135, 147, 324
Hull, David, 6, 98, 104–106, 120, 122,
  187–190, 192, 204, 205, 248, 324
Hume, David, 34

immunology, xv, 1, 3, 6, 182, 195, 201, 272,
  274, 285
independent assortment, xii, 98, 100, 107, 108,
  113, 114, 118, 120, 131, 137, 159, 161,
  177, 235, 236, 245; *see* Figure 4.1
information flow, 76, 82, 86, 87, 232, 248,
  253
inheritance of acquired characters, 183, 254
instructive type theories, 202, 263
interactions, 16, 17, 19, 41, 51, 70, 109, 123,
  127, 130, 131, 143, 157, 186, 190, 194,
  198, 201, 277, 286, 312, 333
interfield connections, xiii, 2, 5, 103, 112,
  132–134, 143–145, 151, 153, 156, 157,
  160, 162–167, 178, 271; *see also*
  interfield theories
interfield theories, xii, 2, 5, 106, 114, 115, 127,
  132–135, 137, 138, 143–145, 153, 157,
  170, 178, 179, 271; *see also* multifield
  theories; *see* Table 5.1
interrelations, 3, 4, 127, 131, 150, 156; *see*
  *also* interfield connections
Iseda, Tetsuji, 13, 40

Maull, Nancy, xii, xxi, xxiii, 5, 66, 94, 100, 106, 115, 121, 127, 128, 145, 147, 153, 157, 160, 165, 168, 178, 180, 216, 226, 236, 246, 320, 327
Mayford, M., 63, 64, 330, 333
Mayo, Deborah G., 302, 311, 327
Mayr, Ernst, 177, 180, 205, 325, 327
McDermott, Drew, 211, 225, 317
McGuire, J. E., 13
McHugh, T. J., 59, 60, 63, 327
McNaughton, B., 58, 64, 335
mechanism, xi–xiii, xv, xvii, xxi, xxii, 1–7, 13–22, 24–36, 40–61, 65–93, 98–120, 161, 171–173, 175, 179, 180, 187, 195, 196, 202, 238, 248–264, 271–306, 308; *see also* activities; entity: working; productive continuity; explanation; *see* Figure 2.1
  characterization of, 4, 67, 106, 271–273, 280, 306
  features of, xv, 3, 4, 7, 20, 21, 272, 280, 301; *see* Table 12.1
  vs. machine, 280, 281
mechanism schema, xii, xv, xvii, 1, 2, 4, 6, 7, 14, 28–33, 36, 41, 47, 48, 52, 56, 58, 66–68, 70, 72, 76, 79, 81–84, 86–88, 93, 100, 101, 107–109, 111, 112, 151, 153, 184, 230–232, 248, 272, 278, 281–290, 292, 297–303
mechanism sketch, 7, 13, 30, 32, 34, 40–44, 47, 48, 54, 60, 61, 70, 109, 112, 116, 117, 282, 285, 286, 290–293, 295, 299, 301, 306
Meheus, Joke, 66, 96, 327
Mendel, Gregor, xviii, 99, 101, 107, 108, 112–115, 119–123, 131, 135, 136, 153–156, 159, 161, 162, 167–169, 172, 174–176, 180, 181, 230, 233, 235, 245–247, 314, 317, 327, 332, 334; *see* Figure 6.2, Figure 9.2
Mendel's first law: *see* segregation
Mendel's second law: *see* independent assortment
Mercer, E., 64, 333
Meselson, Matthew, 93, 308, 315
metaphor: *see* analogy
microreduction: *see* reduction
Miller, S., 249, 265, 309, 317
Millner, Brenda, 57, 63, 331
Millstein, Roberta L., 6, 10, 278, 279, 312, 332
Mitchell, Tom M., 215, 227, 327

Mittal, Sanjay, 222, 225, 316
Mittler, John E., 256, 266, 327
Mizutani, Satoshi, 239, 247, 333
Moberg, Dale, 207, 222–224, 227, 246, 320, 327
model, 2, 6, 112, 282, 296
model organism, 101, 116, 131, 175, 273
molecular biology, xi, xii, xiv, xv, 1, 3, 4, 7, 13–15, 17, 19, 25, 26, 28, 33, 35–37, 66, 67, 70, 73, 77, 83, 89, 98–104, 107–109, 112, 113, 115–120, 143, 163, 195, 198, 202, 229, 232–234, 236, 240–242, 245, 248, 249, 251–253, 272, 273, 281, 285, 289; *see also* central dogma of molecular biology; Watson, James; Crick, Francis
molecular genetics: *see* molecular biology
Monaghan, Floyd, 114, 122, 327
Monod, Jacques, 80, 81, 91, 95, 96, 117, 122, 139, 141, 142, 145, 147, 304, 310, 322, 324, 328, 329
Monod–Wyman–Changeux theory, 139
Monster-barring, see anomaly, monster
Morange, Michel, 33, 38, 67, 80, 82, 96, 101, 102, 115, 117, 122, 328
Morgan, Gregory, xxii, 98, 248, 271
Morgan, Thomas H., xiii, xviii, 5, 101, 104, 113–115, 123, 131, 137, 138, 147, 153, 155, 156, 160–162, 167, 168, 177, 231, 235, 236, 246, 247, 328; *see* Figure 10.2, Figure 10.3
Morowitz, Harold, 208, 227, 298, 311, 328
Morris, R. G., 44, 52, 57, 62, 63, 321, 328
Mountcastle, V., 197–200, 203, 204, 321
Muller, H. J., 115, 123, 147, 160, 162, 167, 168, 247, 317, 328
Muller, R. U., 63, 330
Müller-Hill, B., 141, 147, 322
multifield theories, xiii, 2, 170, 178; *see also* interfield theories; interconnections

Nagarajan, Satish, 246, 320
Nagel, Ernest, 36, 38, 99, 103, 123, 127, 147, 328
natural selection, xiii, xv, 5–7, 171, 173, 174, 182–184, 186–192, 194, 195, 198, 201, 248, 249, 252, 254, 255, 264, 278, 279, 284, 285; *see also* evolutionary synthesis; Darwin, Charles
NeoDarwinism, 249, 259, 264
NeoLamarckism, 248, 254

strategies for experimentation, 4, 278, 291, 295, 296, 300; *see* Figure 2.5, Figure 2.6
strategies for revision, 272, 301; *see* strategies for anomaly resolution
strategies for theory assessment: *see* strategies for evaluation
strategies for theory change, xiv, 6, 215, 225; *see* Figure 9.1 (cf. Figure 12.1)
Straumanis, Joan, xxii, 98
Strickberger, Monroe, 231, 247, 332
Studer, Kenneth E., 229, 237, 238, 247, 332
Sturtevant, A. H., 115, 123, 147, 160, 162, 168, 169, 247, 328, 332
Suppe, Frederick, 144, 148, 169, 182, 183, 206, 228, 246, 331, 332
Suskin, Alana, 229
Sutton, Walter, 114, 119, 123, 136, 137, 148, 160, 161, 169, 333
Svoboda, K., 63, 327, 332

Tabery, James G., 98, 106, 109, 123, 277, 312, 333
Talmadge, D. W., 195, 206, 333
Tanner, M., 216, 226, 324
Temin, Howard M., 7, 10, 236, 237, 238, 239, 244, 247, 297, 333
Thagard, Paul, 3, 9, 10, 185, 205, 207, 208, 216, 226, 228, 285, 286, 310, 323, 324, 333
Thaler, David, 13, 248, 254, 255, 267, 333
theories
    change of, xviii, 4, 7, 93, 207–209, 212–214, 229, 233, 234, 236, 244
    representation of, xiv, 1, 207, 218, 229, 235, 299
    types of, 182–184, 203, 263, 271; *see also* selection type theories; instructive type theories
Thieffry, Denis, 67, 72, 96, 333
Thinus-Blanc, C., 62, 317
Thompson, Paul, 184, 206, 333
tobacco mosaic virus (TMV), 74, 232
Tolman, E., 49, 50, 64, 333
Tom, C., 64, 333
Tonegawa, S., 63, 64, 327, 334
Tooze, J., 239
Toulmin, Stephen, xxiii, 128–130, 148, 207, 228, 333
transcription: *see* protein synthesis
translation: *see* protein synthesis
Trimarchi, J. M., 256, 266, 321

Tsien, J. Z., 59, 60, 63, 64, 327, 333, 334

undulatory hypothesis, 157
unification, 10, 63, 107, 108, 122, 145, 146, 166, 184, 246, 310, 312, 332
unity of science, xiii, 144, 145, 164, 178
Urban, Nathan, 13, 40

Vollmer, Sara, 229
Vorzimmer, Peter, 173, 180, 334
Vrba, Elisabeth S., 183, 190, 192, 194, 205, 206, 322, 334
Vries, Hugo de, 113, 123, 135, 154, 155, 167, 169, 173–175, 177, 180, 181, 187, 204, 206, 231, 247, 319, 325, 334

Waters, C. Kenneth, 98, 107, 108, 111, 120, 124, 334
Watson, James D., xvii, 25, 28, 29, 31, 32, 39, 67, 69, 70, 72–77, 81–83, 86, 87, 90, 94–96, 101, 107, 112, 115, 116, 118, 119, 121, 124, 145, 232, 236, 238, 240, 247, 252, 253, 267, 281, 285, 288, 287, 293, 310–313, 318, 322, 330, 334
Watts-Tobin, R. J., 94, 121, 318
Weaver, Warren, 101
Weber, Marcel, 276, 292, 312, 334
Weismann, August, 136, 148, 334
Wenthold, R. J., 63, 332
Westfall, Richard S., 312, 334
White, Abraham, 124, 334
Williams, Mary, 184, 206, 335
Wilson, Edmund B., 101, 114, 124, 135, 148, 161, 335
Wilson, M. A., 58, 63, 64, 327, 335
Wimsatt, William C., xxiii, 2, 10, 15, 16, 39, 66, 96, 106, 108–110, 124, 210, 228, 296, 312, 335
Woodward, James, xxii, 42, 50, 62, 109, 121, 261, 264, 278, 283, 290, 308, 312, 315, 335
Woody, Andrea, 3, 10, 35, 38, 327
Wyman, J., 141, 142, 147, 328

Zaman, S. H., 63, 332
Zamecnik, Paul C., xvii, 31, 32, 39, 65, 67, 69–75, 77, 81, 83, 86, 87, 89, 95–97, 288, 295, 323, 335
Zola-Morgan, S., 57, 64, 335
Zytkow, Jan M., 227, 336